T0332111

The Botany of Mangroves

Second Edition

Mangroves are distinctive tropical plant communities that occupy the intertidal zone between sea and land. They are of major ecological importance, have economic value as a source of food and raw materials, and serve as a buffer from flooding and climate change-induced sea level rise. Mangroves are under threat from pollution, clearance, and overexploitation, and increasing concern has driven demand for an improved understanding of mangrove species.

This book provides an introduction to mangroves, including their taxonomy, habitat-specific features, reproduction, and socioeconomic value. Fully updated to reflect the last two decades of research, this new edition of a key text includes newly documented taxa, new understandings of vivipary and the evolution of mangrove species, and a rich set of color illustrations. It will appeal to researchers and students across a range of disciplines, including botany, ecology, and zoology.

P. B. Tomlinson is Professor Emeritus at Harvard Forest, Harvard University, and Crum Professor of Tropical Botany at the National Tropical Botanical Garden, Hawaii. He is a leading scholar on the botany of tropical plants, and has published widely on plant anatomy and morphology across a diverse range of communities and species, including palms, aborescent monocotyledons, seagrasses, gymnosperms, and mangroves.

Frontispiece: *Rhizophora* x *harrisonii*, Tivives, Costa Rica. ("*If I have seen further than you. . .it is by standing on the shoulders of Giants.*" Isaac Newton, 1675.)

The Botany of Mangroves

Second Edition

P. B. TOMLINSON

CAMBRIDGE
UNIVERSITY PRESS

University Printing House, Cambridge CB2 8BS, United Kingdom

Cambridge University Press is part of the University of Cambridge.

It furthers the University's mission by disseminating knowledge in the pursuit
of education, learning, and research at the highest international levels of excellence.

www.cambridge.org
Information on this title: www.cambridge.org/9781107080676

© Cambridge University Press 1986, 2016

First published 1986
First paperback edition 1994
Reprinted 2004
Second Edition 2016

Printed in the United Kingdom by TJ International Ltd. Padstow Cornwall

A catalog record for this publication is available from the British Library

Library of Congress Cataloging-in-Publication Data
Names: Tomlinson, P. B. (Philip Barry), 1932– author.
Title: The botany of mangroves / P.B. Tomlinson.
Description: Second edition. | New York : Cambridge University Press, 2016. |
Includes bibliographical references and index.
Identifiers: LCCN 2016015459 | ISBN 9781107080676 (Hardback : alk. paper)
Subjects: LCSH: Mangrove plants. | Mangrove ecology.
Classification: LCC QK938.M27 T66 2016 | DDC 583/.763–dc23 LC
record available at https://lccn.loc.gov/2016015459

ISBN 978-1-107-08067-6 Hardback

Contents

The plate section appears between pages 178 and 179.

Preface to Second Edition of *The Botany of Mangroves*

Mangroves form the "rain forests of the sea" (Warne, 2011) and impact the lives of millions of people in the tropics and in over 120 nation states, as a source of food, raw material, and even recreation, as they have done for millennia. This book is therefore presented as an updated introduction to the study and identification of plants of the mangal, or such of it that still remains! In an era in which humanity has become aware of the effects of global warming on climate change, it is salutory to realize that one effect – rising sea levels – places mangroves at the front line of such change. Knowledge about them is, more than ever, absolutely vital.

Originally written in 1986, with a paperback reprint including an appendix in 1994, *The Botany of Mangroves* has achieved steady sales for 30 years as a primer for the study of plants in tropical intertidal forests. This success reflects the biological uniqueness of this distinctive tropical community. *Mangal* has become the accepted name for the community; *mangroves* for the constituent plants. Their economic value has always been understood, but increasingly their sustainability has been under threat because of frequent overexploitation and conversion. Clearly, however, the original book has become out-of-date in some respects so that the invitation from Cambridge University Press to produce this second edition was received with enthusiasm. The specialized nature of the mangal is reflected in its exclusion by Peter Ashton from his monumental survey of South-East Asian rain forests (Ashton, 2014) largely because of space limits and the subject being too far outside his personal research experience. He kindly defers to *The Botany of Mangroves* as an alternative source.

However, in the course of the last three decades, exploitation of mangroves has proceeded at such a pace that there is now global concern for their very existence. Destruction and conversion have displaced many of those millions whose livelihood is determined by the existence of mangal, either as an ever-present resource, or even as living space (Warne, 2011). This concern has been expressed in the development of organizations devoted to their study. Expanded interest in establishing detailed knowledge is provided by the *World Atlas of Mangroves* (Spalding et al., 2010). This book is the result of an international collaborative effort, involving many contributors, which produced a socioeconomic overview on a pantropical scale. The book charts, in great detail, the distribution of mangrove communities and a country-by-country listing of mangrove species. The book is based on many individual reports and ground-truthing of mangrove locations, but particularly emphasizes the extent of mangal destruction by agronomic and other practices. This book demonstrates the concern of ecologists and botanists for the future sustainability of these tidal forests. It mentions the existence of an International Society for the Study of Mangrove Ecosystems, founded in 1990 in Japan, but which, in 2009, included 38 institutions, in 90 countries, with a membership of over 1000 individuals. Spalding et al. (2010) thus reflects the global concern for the continued development of an understanding of the mangal ecosystem and its plant constituents, their association with resident and visitor wild life, and the effects of human and natural disturbances.

The need for an updated edition of *The Botany of Mangroves* as a description of the plants of the mangal is justified in this global context. As a primer, it retains much of the morphological descriptions, which intrinsically do not change, but also needs to describe recent advances in many relevant disciplines. Recent applications of molecular systematic methods are now included, but change little the appreciation that the mangrove habitat has been occupied by a diversity of unrelated taxa. A major addition made possible by similar methods allows a discussion of the way in which genetic interchange is possible between populations that are in varying degrees discontinuous, especially those long separated by plate tectonic events. The existence of hybridization among certain taxa, long suspected on morphological grounds, has also been verified by appropriate molecular methods. Other advances include increased knowledge of biological features, especially of vivipary, some of this coming from my own research. However, not all included knowledge is necessarily new because I add a historical introduction of the early development of mangrove naming and description made possible by a translation of the *Herbarium Amboinense* of Rumphius (Beekman, 2011) demonstrating that old knowledge also can be new-found. This work of the seventeenth century originally drew the attention of western eyes to the strange intertidal forests of the tropics.

The recent available literature on mangroves reflects the continuing activity of research workers in many disciplines. An online search relevant to the subject of common mangrove species can produce several hundred citations and hence a realization that this edition cannot survey it all. I have retained much of the early literature as a historical record, but have been very selective of the massive new literature. Selection has therefore been limited to review articles or those seen by me as adding new but basic information to my descriptions. This means I inevitably must apologise to those numerous authors whose work I have not cited. Rollet (1981) produced a bibliography on mangrove research for the period 1600–1975 that included over 6000 citations. This number has been easily doubled in the last 30 years. The members of the International Association mentioned earlier have been very busy and productive!

In conclusion, there is no shortage of water on Earth; the problem is that most of it is salty. We should therefore have as complete an understanding as possible of those few trees that have mastered an existence that allows them to flourish at the intersect of land and sea. This book introduces them.

Acknowledgments

To the First Edition. An introduction to mangrove ecosystems followed almost naturally from my interest in field botany and a position at Fairchild Tropical Garden (1960–1971), which allowed reasonable freedom of research objectives with one of the largest areas of mangroves at hand in South Florida. Dr. A. M. Gill on the staff of Fairchild Garden (1967–1970) did much to stimulate my interest through his own ecologically oriented research. Financial support from 1965 to the present has come from the Maria Moors Cabot Foundation for Botanical Research and subsequently from the Atkins Garden Funds, both of Harvard University. In addition, I have received travel support from the National Geographic Society (Tomlinson, 1982a) and the National Science Foundation, Office of International Programs (United States–Australia Scientific Collaborative Program). These awards facilitated the limited knowledge I had gained in the Caribbean area to be broadened by visits to the richer mangal of the Indo–Pacific region. Several institutions and many individuals have contributed facilities and field assistance. They include the Australian Institute of Marine Science, Townsville, Queensland (AIMS), and the Division of Botany, Department of Forests, Lae, Papua New Guinea (PNG). Individuals who have made fieldwork possible include J. S. Bunt, B. Clough, N. C. Duke (AIMS), J. S. Womersley, E. E. Henty, M. Galore, R. J. Johns (PNG), and Dr. J. Davey (University of Brisbane). Professor Engkik Soepadmo of the University of Malaysia, Kuala Lumpur; Dr. Paul Chai and the staff of the Forestry Department, Sarawak; and Mr. J. M. Maxwell of the Singapore Botanic Garden all were helpful on a visit to Malaysian mangroves. Jorge Jiménez was my host when I visited Costa Rican mangroves. Richard Primack of Boston University was a stimulating field companion and a guide to the theoretical and applied aspects of floral biology. The libraries and collections of the Harvard University Herbaria served as an indispensable source of systematic, geographic, and historical information. Peter Stevens, Curator of the Harvard University Herbaria, was particularly helpful in a critical review of the systematic descriptions. Adrian Juncosa made helpful comments on early drafts and supplied some material for illustration. John Sperry provided discussions of mangrove physiology. The morphological illustrations in Section B are the work of Priscilla Fawcett, Botanical Illustrator at Fairchild Tropical Garden, supplemented in Massachusetts by the drafting skills of Susan White and Elizabeth Bullock. Ms. D. R. Smith worked without complaint to maintain some coherence throughout the many draft versions of the book.

In addition to the figures prepared especially for this book, a number of sources have been used for illustration. Permission to reproduce these figures has been given by the following authors and/or journal editors: *American Fern Journal* (Figs. B.60, B.61); *American Journal of Botany* and A. M. Juncosa (Fig. 8.5); *Annals of Botany*, London (Figs. 4.4, B.51, B.52); *Biotropica* (Figs. 5.10, 5.11, B.66, B.67, B.71, B.72, B.74, B.76); *Biological Journal of the Linnean Society* and A. G. Marshall (Fig. 7.2); *Botanical Journal of the Linnean Society* (Fig. 4.8); *Bulletin of the Fairchild Tropical Garden* and I. Olmsted (Figs. 1.10, 1.11, 1.12); *Contributions Herbarium Australiense* (Figs. B.66, B.70); *Journal of the Arnold Arboretum* (Fig. B.21); *Vegetatio* and V. Semeniuk (Fig. 1.7); J. Cramer and Helen Correll (Figs. B.6, B.15, B.29, for figures that first appeared in D. S. Correll and H. B. Correll, Flora of the Bahama Archipelago); and Figures B.8, B.19, B.24, B.25, B.26, B.28, B.37, B.38 first appeared in P. B. Tomlinson's *Biology of Trees Native to Tropical Florida*, copyright by the author. In addition, previously unpublished diagrams and photographs were supplied by J. S. Bunt and N. C. Duke (Figs. 1.8, 3.5); A. M. Gill (Figs. 5.4, B.41, B.62, B.75); and A. M. Juncosa (Figs. 1.3, 1.4, 4.10, 5.5, 8.6, B.22, B.23, B.55, B.57, B.58, B.80). All other figures are either original or redrawn.

To the Second Edition. Some continuity with the old version is seen in the further support of individuals who appeared in it. These include Norman Duke [NCD], Malcolm Gill [AMG], Adrian Juncosa [AMJ], Ingrid Olmsted [IO], and Jeff Vincent [JRV], who have supplied illustrations in the form of color transparencies that appeared initially as black-and-white images, but for the most part now appear here in color. To this company can now be added individuals who have supplied old and new color images; they include Aaron Ellison (with Elizabeth Farnsworth) [AE], Jack Fisher [JBF], Doug Goldman (DG), Jay Horn [JWH], David Lee [DL], Tracy Magellan [TMM], and Tokushiro Takaso [TT]. Their included initials in the captions acknowledge this help. All other photographic illustrations are my own, often now reproduced in color. Line drawings by the late Priscilla Fawcett [PF] are reproduced as in the first edition, with a few added from other sources. For continued help with the mammoth new literature, I benefitted enormously from the professional assistance of Judy Warnement and her staff at the Harvard University Herbarium Library [HUH]. Guidance through the nomenclatural maze was provided by K. Ghandi, at the same location. Help in resolving numerous computer problems was provided by Brett Huggett, Madelaine Bartlett, Benjamin Burnham, and John Stevenson, Sr. The support of the staff of the Montgomery Botanical Center, Coral Gables, Florida, especially Patrick Griffith and Tracy Magellan, has been continually sustaining.

I, a culpable mistake-maker, apologise for any deficiencies.

Figures in Section A (General Account) are numbered according to chapter (e.g., Fig. 4.6 is the sixth illustration in Chapter 4); figures in Section B (Detailed Description by Family) are numbered consecutively with the prefix B (i.e., B.1–B.69). Tables in Section A are also numbered according to chapter (e.g., Table 2.2 is the second table in Chapter 2).

Part I

General Account

1   Historical Prelude

Mangrove vegetation was encountered by early travelers from temperate regions as plants that would prevent easy landfall. In particular, they would have been visible in the estuaries of major tropical rivers that would have provided easy access to interior regions. This familiarity generated a number of still current myths, notably the impenetrability of the mangal, as voiced by Dampier (*Voyage au nouvelle monde*, 1723): "Where this sort of tree grows, it is impossible to march by reason of these stakes, which grow so mixed one among the other that I have, when forced to go through them, gone half a mile and never stepped foot on the ground, stepping from root to root" (Fig. 1.4). This route is demonstrated in the Frontispiece. They can, however, be explored by shallow draught boat along their numerous tributary creeks. Penetrability also depends on the size and maturity of the forest. In stands with tall trees up to a canopy height of 30 m, the common absence of any undergrowth allows easy access by walking or wading under, rather than over, the flying-buttress roots of *Rhizophora* (Plate 13A). With the settlement of Europeans in the tropics, the study and description of mangroves made information about them necessary and, with it, their naming.

Outstanding in this respect is the monumental work of the Georg Everhard Rumpf (1627–1702), of German origin, known to posterity by the Latinized form *Rumphius*, and in the format of the six-volume *Herbarium Amboinense* ("The Amboinese Hebarium"), a description of the plants of Amboina, a colony of the Dutch East India Company, where he was a merchant. Corner (1966), in his *The Natural History of Palms*, has provided a succinct summary of the extraordinary history of this work and the unlikely circumstances under which it ever appeared. Corner's book is an appropriate source because Rumphius himself emphasized palms as the group first described in Volume 1.

...the leading Dutch merchant Georg Everard Rumpf (1627–1702) was writing his Herbarium Amboinense. He started in 1662, but never saw its publication; he never saw, indeed, the final manuscript. Tragedy upon tragedy beset this unfortunate man who became known to botany as Rumphius. In 1670 he grew blind. On 17 February 1674, an earthquake destroyed the larger part of the town of Amboina. His wife and younger daughter were killed; and a clerk, in his account of the devastation, has recorded these words: 'Very sad it was to perceive that man sitting near these his bodies, and to hear his lament, both on this accident and his blindness'. In 1687 a fire ravaged the town and Rumphius' library, manuscripts and illustrations for the Herbarium Amboinense were destroyed. He commenced to repair the loss by means of assistants who were put at his disposal by the authorities, to copy out what he could remember. The first six books were dispatched in 1690 to Holland via Batavia in Java, but the ship was sunk by the French and its

3

cargo lost. Fortune smiled because the governor general of Batavia, Camphius, had caused a copy to be made and this arrived in Holland in 1696. Rumphius completed his task in 1701, and the last books reached their destination. On 19 May 1702, the Governor of Amboina wrote that 'nothing more was to be expected of that old gentleman, having lived his years'. His grave was destroyed by a party of English soldiers who hoped to find gold buried under the stone. A second monument, erected in 1824 to honour the pioneer naturalist, 'remained until it was hit by a bomb in the last world war and was smashed'. The Herbarium Amboinense was eventually published by the Dutch botanist Johannes Burmann (1706–1779) in the years 1741–1755 and an index was supplied in 1769.

His monument in Amboina was the site for old pilgrimages, but now new ones, as a more recent monument has been erected.

Herbarium Amboinense is profusely illustrated, with a text jointly in Latin and medieval Dutch. Until recently it existed in rare book collections, but has been translated into English – over 250 years later! – by the Dutch scholar E. M. Beekman (2011), making this work, which includes a wealth of economic and ethnobotanical information, more accessible. The book was not written from the limited approach of a medieval herbal, but included extensive descriptions of the structure and biological attributes and habitat locations of plants that were largely unfamiliar to botanists of that era (Fig. 1.1). Rumphius had a particular fascination with mangroves, as remarkable

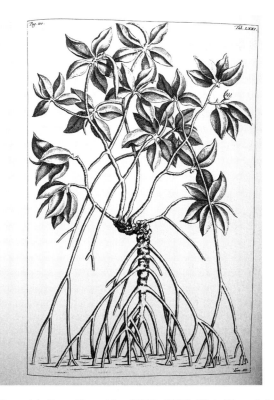

Figure 1.1. From Rumphius (1741–1755), Plate 71, vol. 2, Book 4, with the habit of *Mangium candelaria* in the original.

Table 1.1 Rumphius' Terminology using Mangium: Five "Genera"

1.	*Mangium legitimum*, "proper Mangium" with four further "species"
	1a. *Mangium celsum*, "tall or lofty mangium"
	1b. *Mangium minus*, "small mangium"
	1c. *Mangium digitatum*, "fingered mangium"
	1d. *Mangium candelarium*, "candle mangium"
2.	*Mangium caseolare*, "cheese mangium"
3.	*Mangium album*, "white mangium"
4.	*Mangium fru[c]ticans*, "shrub-like mangium"
5.	*Mangium ferreum*, "hard or iron mangium"

trees growing in salt water. He named them (Table 1.1), using *Mangium* as a generic name with an ecological rather than a systematic connotation. The work provides insights into the general level of botanical understanding of that period and, together with its predecessor *Hortus Malabaricus* of Rheede (1678–1703), drew to the attention of European botanists the great wealth and diversity of tropical floras, providing a source of information and direction for the later exploration of the tropics worldwide in the eighteenth and nineteenth centuries.

Identifying Rumphius' Plant Names

Although *Herbarium Amboinense* predated the starting point of modern botanical nomenclature (Linnaeus, 1753), it still had some influence on subsequent naming despite difficulties in providing equivalent modern names of precise application. This exercise was attempted in great detail by Merrill (1917), with extended discussion (Table 1.2). The interpreted nomenclature is an essential tool in Beekman's translation, especially in its marginal notes. Scrutinizing the illustrations, without commenting on their quality, one discovers how they portray biological features of appreciable accuracy even from the point of view of modern investigation. The translated version is a work of complex scholarship, especially as it is intended to be acceptable to an English-speaking reader of the period in which the original was written (E. M. Beekman, personal communication). The layout for the most part cannot be seen as systematic in the modern sense but relates to groups of plants with similar uses. It begins with the palms, as easily recognized, economically important, and perhaps most iconic of tropical plants. The mangroves are discussed as an ecological group, readily categorized by their exclusively intertidal location.

Merrill (1917), in trying to recognize the species described in *Herbarium Amboinense*, discusses many of the uncertainties of identification and he was critical of their representation, describing many of them as "crude," but if one steps back from the niceties of modern nomenclature (which itself has its own necessary uncertainties) and accepts them without the unfair advantage of over 300 years of

Table 1.2 Rumphian Names, in English Translation (Beekman, 2011) and their Presumed Modern Equivalents. Sources: Merrill (1917), Ding Hou (1958)

Avicenniaceae
Mangium album, "the white mangi tree" [= *Avicennia officinalis* L. (Merrill, p. 456)]

Lythraceae
Mangium ferreum mas, "the iron mangi tree" [= *Pemphis acidula*, Forst]
Mangium ferreum porcellanicum, "the iron mangi tree" [= *Pemphis acidula*, Forst]

Myrsinaceae (Primulaceae)
Mangium corniculatum, "the horned mangi tree" [= *Aegiceras corniculatum* (L.), Blanco (Merrill, p. 413)]
Mangium fru[c]ticans, "the horned mangi tree," second species of this genus but not illustrated [= *Aegiceras floridum*, Roem and Schultes]
Mangium fru[c]ticans II parvifolium; *Mangium floridum*; *Mangium parvifolium II* [all probably = *Aegiceras floridum*, Roem and Schultes]

Rhizophoraceae
Mangium candelarium or *arcuatum*, "candle-bearing mangi tree" [= *R. apiculata*, Blume (Merrill, p. 387)]
Mangium digitatum, "the fingered mangi tree" [= *Bruguiera sexangula* (Lour.), Poir (Merrill, p. 389)]
Mangium celsum, "the small mangi tree" [= composite species of *Bruguiera gymnorrhiza* (L.), Lamk (espec. "fruits" = seedlings) and *B. cylindrica* (L.), Blume (flowers)]
Mangium celsum, "the clove-shaped mangi tree" [= *Bruguiera cylindrica* (L.), Blume (Merrill, p. 388–389)]
Mangium celsum, "the fingered mangi tree" [= *Bruguiera sexangula* (Lour.), Poir (Merrill, p. 389) but *Bruguiera gymnorrhiza* (L.), Lamk (Ding Hou, 1958)]
Mangium celsum, "the tall mangi tree or lalary wood" [= *Bruguiera gymnorrhiza* (L.) Lamk]
Mangium minus [= ?*Bruguiera gymnorrhiza* (L.) Lamk (Ding Hou, 1958)]
Mangium caryophylloides I [= *Bruguiera cylindrica*) (L.), Blume]
Mangium caryophylloides II parvifolium [= *Ceriops tagal* (Perr.), C. B. Rob (Merrill, p. 386)]
Mangium caryophylloides III latifolium [= *Ceriops tagal* (Perr.), C. B. Rob (Merrill, p. 386)]

Sonneratiaceae
Mangium caseolare album, "the white cheese-bearing[*] tree" [= *Sonneratia alba* Sm.] (Merrill, p. 383)]
Mangium caseolare rubrum, "the red cheese-bearing[*] tree" [= *Sonneratia caseolaris* (L.), Engler] (Merrill, p. 383)]

---

[*] "cheese-bearing" because Rumphius was reminded by the shape of the fruits of small cheeses, perhaps natural for Dutchmen.

post-Rumphian research, there is great pleasure in using this book if one has some familiarity with the plants described in it, and may even be drawn to marvel at the accuracy of their representation. The example to which attention is drawn is the very accurate rendition of the seedling of *Rhizophora* (Fig. 1.2) in relation to a recent study of this topic (Tomlinson and Cox, 2000), as discussed in the later description of this genus.

Mangroves in Rumphius

In *Herbarium Amboinense*, the section on mangroves treats them as an ecological group of plants, unfamiliar to European botanists because they were trees that grow within the tidal influence of the sea. Rumphius admired mangroves because of their location in salt

Figure 1.2. From Rumphius (1741–1755), Plate 72, vol. 2, Book 4, with an illustration of *Rhizophora apiculata*, showing the characteristic basal hook.

water and the presence of roots arising from branches, features unknown outside the tropics. He also recognized them as a distinct ecological group, in contrast to most of the other plants he described that were categorized largely according to their uses. He referred to the mangrove plants as "mangi-mangi." This can be interpreted as either the vegetation *in toto* or the individual species. The vernacular of his time would have been Malay, in which language a plural is produced by repeating a noun, thus mangi mangi would seem to mean mangroves collectively. In modern parlance, the distinction is made between mangal as the community and mangroves as the plants in order to avoid any confusion, although in context, readers usually have no difficulty. Rumphius Latinized the singular as *Mangium*, which he used largely to refer to the plants as an ecological group. Beekman (2011) has faithfully retained this terminology in his translation; for example, *Mangium candelarium*, "The candle-bearing mangi tree," which turns out to be *Rhizophora*; the "candles" are the pendulous hypocotyls of the viviparous seedlings (Plate 14B). The result is an association of some dozen species that are currently distributed among several unrelated families (Table 3.1). *Nypa*, the mangrove palm, is described elsewhere, and appropriately so, in the introductory palm chapters. Thus, my own treatment of an ecological assemblage has followed the same nonsystematic context. However, Rumphius also added a number of other plants under the same category that had no particular association with coastal vegetation (e.g., *Eucalyptus deglupta* Blume, "the many-colored mangi tree" – the rainbow eucalypt of cultivation). The word "mangium" in modern use is found only in *Acacia mangium* Willd. (originally *Mangium montanum*, "the mountain mangi tree," Herb. Amb. T. 81), an inland Asian tree and a modern source of fuel wood. He says these additions could have been described elsewhere, but "preferred not to." Nor are the mangrove associates or members of "back-mangrove" communities described, as is done in this book (Table 3.2). Other than these, mangium has a purely ecological connotation. The names Rumphius used and their modern equivalents, in so far as they have been identified, are listed in Table 1.1, 1.2. In the translation of Rumphius, the terms genus and species are used, but this should not be interpreted in a modern connotation. His usage was summarized by him but translated in Beekman (2011) as follows:

Rumphius and Mangrove Biology

One can understand some of the frustration expressed by Merrill (1917) that may have led to his rather severe criticism, resulting from his inability to name Rumphian species with precision. This was largely due to the absence of discernable diagnostic features, although for the most part, genera can be identified. In addition, some of the illustrations were from a mixture of taxa or were even copied from earlier works; for example, the *Hortus Malabaricus* of Rheede. Hence, the comment by Merrill in his notes on *Mangium* that the text of their chapters was acceptable but the illustrations were "very crude." Rumphius, after all, was blind and did not see the plates in their final form. However, this does not mean that we should dismiss them as worthless. If one approaches them from the perspective of what useful information can be gleaned from them, they can be seen to represent the growth and development of taxa without too much concern for precise nomenclature.

Architecture. The drawing including the habit of *Rhizophora* (Herb. Amb. Tab. 71, our Fig. 1.1) is very accurate and shows the obconical base of a rooted trunk progressively widened upward and associated with the development of aerial roots of increasing diameter upward; that is, in reverse order of their age. Biomechanically, the canopy of the tree is thus supported by a series of aerial roots of increasing size acting as flying buttresses (Plate 13A). Clearly, the illustration by Rumphius was drawn from nature. Rumphius also shows aerial roots arising from distal branches, which does occur, but exceptionally and largely as a response either to direct damage of the crown or where the canopy above a crown is opened, as by storm damage. Such roots in tall trees, even if they can reach the ground, may be of little architectural significance, although they may be very important in the development of a shrub-like habit in open or marginal communities, notably so in

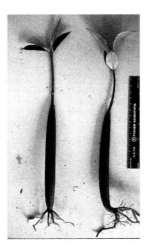

Figure 1.3. Seedlings of *Bruguiera gymnorrhiza*: left, planted vertically, axis remains straight; right, planted horizontally, erected by basal hook.

the Florida Everglades (Plate 1). This form of rooting is also the reason plants of limited stature can survive in a vegetative state, as described by Guppy (1906) for *Rhizophora* x *selala* in Fiji. Once again, Rumphius presents what is easily observed.

Seedling dispersal and establishment. We agree with Merrill that the comments of Rumphius on seedling development are accurate, although we must point out that both of them referred to these structures as "fruits" (the "candles" of Rumphius), although they are actually the extended hypocotyls of viviparous seedlings. Particularly accurate is the Rumphian illustration of an established seedling of *Rhizophora* with its characteristic basal "hook" (Fig. 1.2). The development of this basal hook is easily demonstrated by field observation and experiment (Fig. 1.3), and we have much to say about this in relation to the description of the viviparous condition in Chapter 10. In his description of seedlings on the tree, Rumphius implies that they "remain upright" (strictly, pendulous; Plate 14B), and where they hang low enough "to touch the ground," they send out roots to become "new little trees." This is extremely unlikely, but more significantly and "they are tossed elsewhere by the flowing tide, where they will nevertheless root and establish themselves." This aspect of dispersal and establishment was to become a major focus of ecological research, but with several miss-steps along the way, as discussed in Chapter 7.

Conclusion

We may find fault with the publications of a blind botanist of the seventeenth century, but we might better be employed in extracting original information from his work in

Figure 1.4. Rhizophora forest, Biscayne Bay, Miami, FL. (From a color transparency by AMJ.)

a more positive frame of mind. Rumphius did not have the opportunity to see the plates published in his name and to continue the many and careful observations that he had made initially. Perhaps it is as well; he could easily have scooped us all and many hypotheses about mangrove biology need not have existed and an experimental approach to plant ecology might have begun earlier!

2 Ecology

Mangrove and Mangal

The mangrove community is not set apart from adjacent vegetation; it derives mineral nutrients from the fresh water flowing into it from the hinterland and supplies biomass to adjacent marine communities resulting from litterfall and tidal scour (Fig. 2.1). This defining characteristic is most easily recognized in the tidal fluctuation that determines much of mangrove biology (Figs. 2.2, 2.3). This importance is much emphasized in the scientific literature and that relating to economic valuation (Chapter 11).

The word "mangrove" has been used to refer either to the constituent plants of tropical intertidal forest communities or to the community itself. Usually there is no contextual confusion; otherwise, some qualification, such as "mangrove plants" or "mangrove community," is needed. MacNae (1968) proposed "mangal" as a term for the community, leaving "mangrove" for the constituent plant species, and this usage is widely adopted. "Mangal associate," "mangrove associate," and "back mangal" may also be used as a logical extension of this distinction between species and community. Mangroves in the more limited sense may thus be defined as tropical trees restricted to intertidal and adjacent communities, or subject to the indirect influence of tides. This influence can be unexpected, as with storm surges; *Rhizophora* propagules can be carried several miles inland during hurricanes in the Florida Everglades.

Mangrove plants are distinctive, according to Shi et al. (2005), because they exhibit "the most remarkable morphological specializations" of vivipary, salt secretion, and aerial roots, although all these characters are not necessarily found together in every species. Molecular evidence provided by these and many other authors show multiple origins of these characters; that is, a remarkable convergence of seemingly unrelated characters.

Another definition of mangal is a community that contains mangrove plants. This is less nonsensical in its circularity than it at first appears, because strict mangroves are characterized by their high fidelity to the ecotone influenced by tides (Figs. 2.1, 2.2, 2.3). Tidal influence can be interpreted narrowly, simply to mean the shoreline inundated by the extremes of tides, or it can more widely refer to riverbank communities where tides cause some fluctuation but no salinity. Mangroves can penetrate inland extensively along riverbanks (Plate 1A).

This book is about mangroves and not mangal. This restriction is appropriate because there is need for a uniform systematic, morphological, geographic, and biological

Figure 2.1. Hinchinbrook Island, Queensland, at the turn of the tide. Mangrove detritus carried by the ebbing tide is visible as scum on the surface of the sea; terrestrial organic nutrients en route to marine environments.

Figure 2.2. The mangrove habitat. *Sonneratia* and *Avicennia* at low tide; Semetan, Sarawak.

Figure 2.3. Same mangrove habitat shown in Figure 2.2, but at high tide.

treatment of mangroves distinct from those numerous general treatments of mangal that emphasize community aspects, although frequently with attention to details of the biology of individual species. The literature on the subject of ecology is extensive and cannot be reviewed here, but some elementary discussion seems necessary as a background to other topics. The community is so distinctive that it is often considered outside the scope of generalized discussion of tropical forests (Ashton, 2014).

Mangal is a discrete community but with a close physiographic relationship to other strand communities. Schimper (1891), in his description of coastal communities in the Indo-Malayan region, distinguished mangrove from "*Nypa* formation," "*Barringtonia* formation," and "*Pes-caprae* formation" (from *Ipomoea pes-caprae*). The transition between these is usually abrupt, but elements of all communities are sometimes intermixed and some species from the non-mangal are treated here as "mangal associates." A more gradual transition between mangal and inland freshwater swamp forest may again intermingle species from contrasted habitats. Corner (1978) describes this interrelationship well for South Johore and Singapore. Schimper's emphasis on mangal as only one of a number of discrete coastal communities is appropriate; he considered that many biological processes that are particularly well expressed in mangroves, such as dispersal, salt tolerance, xerophytism, and gas exchange, also occur more generally. These generalized statements can obviously be extended to features of floral biology and plant–animal interactions.

From the mangrove forest, one can move easily to contrasted marine and shoreline communities. Toward the sea on shallow sloping shores, there may be a submerged seagrass meadow with scattered mangrove seedlings in the shallow parts, sometimes

Figure 2.4. Salt desert in the back mangal. Low *Ceriops tagal* in the background; halophytes (e.g., *Batis* and *Sesuvium*) in the middle ground, fairly abruptly transitional to the area from which plants are totally excluded by persistent high salinity. This corresponds to the salt flat in Figure 2.6. (From a color transparency by AMJ.)

indicative of mangrove encroachment. Where seagrasses are found among mangroves in an appropriate substrate (Plate 13B), they are likely to be shaded out eventually; therefore, the penetration is not very far. Along the coast, one may move to beach communities with *Ipomoea*, *Suriana*, or *Tournefortia* in front, but trees like *Barringtonia*, *Thespesia*, and *Tilipariti elatum* behind. *Casuarina* forest may grow behind mangal but only when the soil is sandy and well aerated. Rocky coasts or coasts with exposed coral are unsuited to mangroves except where plants can root in silt-filled depressions. As a consequence of changes in topography along a coastline, mangal is not necessarily continuous over long distances, and this can have important consequences for genetic exchange within mangrove species.

On drier coasts (with low levels of rainfall), salt accumulation may produce an inland sterile salt flat or "salt desert," with a fringe of strict halophytes (Fig. 2.4). The back mangal is very complex, as the possible adjacent terrestrial communities are numerous and diverse and the transition may be abrupt or gradual. Consequently, the number of mangal-associate species is large. I have referred to or illustrated those that I have found most commonly mentioned.

Mangal subtypes. Within the strict mangal, a complex subdivision is possible, even with the relatively few species available. Such a synopsis can be found in Chapman (1976). The *Nypa* formation of Schimper (1891) is most distinctive and easily recognized. Competitive exclusion of other mangroves seems related to the rhizomatous habit of *Nypa*. Much of the description of other mangrove communities is concerned with the relative abundance or dominance of given species or even contrasted vegetation types distinguished by differing habits but made up of one species. Otherwise, lack of uniformity is, in fact, a measure of the plasticity of mangroves and their ability to colonize an enormous range of habitats. Instability and change are the most consistent characteristics of mangal.

Mangal concentration. Although mangroves may grow throughout the tropics in suitable areas, with the exception of the Central Pacific, particular regions are noted for the broad extent of mangal. Typically, these are the estuaries of large rivers that run over a shallow continental shelf. Examples are the mouths of the Ganges and Brahmaputra rivers (the Sundarbans) mainly in Bangladesh, the Fly and Purari rivers in Papua New Guinea, and the Mekong Delta in Vietnam. The two largest tropical rivers, the Amazon and the Congo, do not develop such extensive estuarine mangal for physiographic reasons, although mangroves are certainly present (Spalding et al., 2010). The Florida Everglades is a drainage basin that gradually changes from fresh water to an extensive mangal at its seaward margin, forming the largest expanse of mangroves outside the tropics (Plate 1).

Mangal characteristics. A distinctive character of mangal is its physiognomic diversity. The vegetation itself appears uniform, but the community has a distinct aspect and atmosphere. Mangal can be studied from the air. Plate 17A represents a rather upscale use of the helicopter for a preliminary survey, but direct access must be by boat and on foot. Travel into the mangrove is best done using shallow-draft boats in drainage creeks (Fig. 2.5), taking care not to become stranded at low tide. Ideally, initial access is provided by inflatable boat easily transported to a site, but limited by the width of channels. Commercial use requires bigger boats, but still shallow draft (Plate 16B). Stepping out of them is hardly "landing." The substrate is often a firm to soft mud into which the traveler may sink, so that walking is difficult if not impossible. Progress is hindered chiefly by the looping aerial roots of *Rhizophora*, usually the most abundant component especially in a seaward direction (e.g., Plate 1F, G). The traveler has the choice of walking with care on the slippery crowns of these arching stepping stones or working almost hand-over-hand around them. Rarely in a mature stand of tall trees does one get the comfort of a real forest (e.g., Plate 13A).

The substrate can also be either muddy or richly organic with peat made up largely of accumulated underground portions of mangrove root systems. In muddy peaty substrates, a disturbance produces a strong smell of hydrogen sulfide (H_2S), indicating the completely anaerobic property of water-logged soil (Nickerson and Thibodeau, 1985). The oxidation-reduction potential from aerobic (+700 mV) to extreme anaerobic (−300 mV) provides a range of values that influences the soil chemistry (Clough et al., 1983). As the value drops, oxygen is first reduced to water; then at the lowest levels, carbon dioxide is reduced to methane. The ease with which ions function as electron acceptors determines the intermediate progression: nitrate (NO_3^-) – nitrogen (N_2), manganous (Mn^{4+}) – manganic (Mn^{2+}), ferrous (Fe^{3+}) – ferric (Fe^{2+}), sulfate (SO_4^{2-}) – sulfide (S^{2-}). The progressive accumulation of H_2S was measured by

Figure 2.5. Mangal is best explored in a small, shallow-draft boat with an outboard motor; this permits access via narrow stretches of open water (cf. Plate 11B).

Kryger and Lee (1995) in a reclaimed area. Once the supply of precipitating free metals was converted to sulphides, hydrogen sulphide was released as a gas.

A peculiar effect of this process is that the rhizosphere, that is, the boundary layer surrounding the living, absorbing portion of the mangrove root system, can actually become oxidized by oxygen released from within the root at the end of its long passage from air via the aerial roots, as demonstrated by Thibodeau and Nickerson (1986), who sealed aerial roots to prevented air entry and thus disrupted the chemistry.

Canopy height depends on climate, topography, and the extent of human disturbance. Mature undisturbed forest develops a high, dense canopy; the trees have tall boles. Aerial roots may be limited in their development, and passage may be relatively easy. There is little stratification and only an understorey of, at most, suppressed seedlings, except strikingly below canopy gaps. The canopy itself is monotonous because of the uniform leaf shape, size, and texture (Plate 8). In disturbed or impoverished sites, plants are stunted and scrubby and form an open or closed community. To force a way through continuous mangrove scrub is particularly difficult; fortunately, there are few spiny plants, although dead snags of stems and roots can be quite sharp. Crabs and mudskippers may be abundant and conspicuous, even on the trees. There may be communities of the tailor ant (*Oecophylla*, Plate 3H) to avoid, as well as nests of bees and wasps (Plate 3E). Where the mangal is dense but with a tall canopy, the interior can be dark, but this is often a welcome coolness relative to the bright heat of open water. The staff of the Inshore Productivity Group at the Australian Institute of Marine Science wear white dungarees for field clothes because this increases their mutual visibility (Fig. B.66).

Probably the best account of mangal is by MacNae (1968), even though he deals solely with Indo-Pacific mangal, but written by somebody with extensive first-hand familiarity with mangal and its plants and animals. Not only is it a valuable scientific treatment but also communicates the author's enthusiasm for the subject that has obviously sustained him in the field. The paper is written by a person who saw, in detail, the vegetation he describes. Early cosmopolitan treatments are those of Schimper (1891), Haberlandt (1910), Walsh (1974), and Chapman (1976). Regional treatments, often extensively illustrated, are exemplified by Davis (1940) for South Florida, Semeniuk et al. (1978) for western Australia, Karsten (1890, 1891) and Schimper (1891) for Indo-Malaya, Walter and Steiner (1936) for East Africa, Percival and Womersley (1975) for Papua New Guinea, and Bunt and his associates (Bunt et al., 1982) for Queensland. Chapters with varying degrees of completeness by individual contributors in Chapman (1977a) deal, respectively, with the mangal of the eastern United States (Reimold, 1977), middle and South America (West, 1977), Africa and the Indian subcontinent (Blasco, 1977; Chapman, 1977b; Zahran, 1977), Indo-Malesia (Chapman, 1977c), the Pacific Islands (Hosokawa et al., 1977), and Australasia (Saenger et al., 1977). A general account of the ecology of mangroves written by van Steenis is included in Ding Hou's account of the Rhizophoraceae for Flora Malesiana (1958); Lugo and Snedaker (1974) provide a more modern ecological summary. A number of international symposia have been organized, some of them leading to extensive published proceedings (Walsh et al., 1975). More recent treatments may be exemplified by Duke (2006) and those cited in Spalding et al. (2010), whose bibliography is too extensive to be repeated here.

Mangrove associates. Some accounts deal with cryptogams (algae, bryophytes, lichens). Fungi are rarely discussed, even though there may be some significant pathogens of mangroves in the fungal flora; for example, a gall disease *Cylindrocarpon didyum* (Hartog) Wallenw. (Olexa and Freeman, 1978); in the Gambia, a related lethal species was estimated to have lost firewood valued at US \$40 million (Teas, 1982). The study of saprophytic fungi is important because these organisms (together with bacteria) convert lingo-cellulose into energy sources for other organisms in the food web (Ulken, 1983). Epiphytic vegetation, usually only the vascular epiphytes, is not often described, largely because it is less accessible and a more casual association, but occasionally it can be spectacular (Plate 2D).

Fauna

Treatments of the animals found in mangal are usually limited to a list of species without much discussion of the interaction between animals and mangroves. MacNae (1968) is exceptional in this respect. He discusses the environment of the mangal in terms of animal microhabitats. He distinguishes the tree canopy, rot holes in trunks and branches, the surface of the soil, the soil subsurface, and permanent and semi-permanent pools, to which should be added water courses themselves. This gives some idea of habitat diversity. Faunistic lists may be long because both terrestrial and marine faunas should be included. Saenger et al. (1977), for example, list the following numbers of animals for the Australasian mangal and tidal salt marshes: mollusca, 95 species; crustacea, 65 determined species; worms, 97 species; birds, 242 species. Almost all the species in the census are listed as "visitors" (seen in mangal or salt marshes) or "associated" (utilizing the community but not restricted to it). One species of noddy (*Anous*) is recorded as nesting in mangroves. The significant bird fauna is the one that is "exclusive" to the plant community; only 14 species of birds are listed in this category. Although the most conspicuous animals in the mangal are indeed non-resident, there is a distinct fauna of vertebrates, as described by Luther and Greenberg (2009), that are permanent residents and live exclusively among mangroves. Most are quite rare, but clearly habitat destruction is fateful for them.

Spalding et al. (2010) lay emphasis on some of the larger animals, notably those considered dangerous (crocodiles and tigers) or those that may come into the mangal to browse (hippopotami – Plate 3B, deer, and especially goats). They also present records for bird species, again mentioning that few are permanent residents and many are migratory. Direct plant–animal interaction can best be perceived in nectarivorous birds (Kondo et al., 1987). Other birds may feed in the mangrove habitat, mainly on insects and other invertebrate faunas. An example of direct dependence pointed out by Saenger et al. (1977) is the mistletoe bird (*Dicaeum hirundinaceum*), which feeds on the fruits of a mistletoe (*Amyema* spp., Loranthaceae). This bird is common in mangal in northern and northeastern Australia, only where *Amyema* species are parasitic in mangroves; it occurs elsewhere in terrestrial communities. The mistletoe bird presumably is a major cause of infection of mangroves by vascular parasites. In contrast, marine animals are associated with mangal largely because they share the same substrate requirement. Some larger animals may be characteristic: mudskippers, crabs, oysters, and snails. MacNae (1968) discusses these animals in terms of their adaptation to the mangal environment. Oysters on *Rhizophora* roots (Plate 2B, C) provide the catechism that they "grow on trees," but snails also grow on trunks and roots! (Plate 2A). Mudskippers, which are gobioid fish, are characteristic of mangal and related areas. Several genera and species are involved; some can even climb trees. Crabs are said to be big consumers of fallen leaves, but choosing older and partly decomposed ones (Warne, 2011).

Crocodiles (Plates 3D, 20F) are probably the most dangerous large animals, even though they are protected as endangered because their populations have been reduced enormously by hunters. There are a few poisonous snakes. There are few large terrestrial animals, but tigers characterize the mangal of the Ganges Delta, which for them is almost a refugium (Plate 3G). The "tiger fences" (Plate 3A) in the Sundabans of Bangladesh are not very reassuring as tigers can swim. Aquatic mammals can sometimes be seen, as with hippopotami (Plate 3B) and seals (Plate 3C). Other mammals are deer, raccoons, and bats, but these should all be described as visitors. Bats play an important role in the reproduction of *Sonneratia* in some parts of its range.

Indirectly, a number of marine animals may influence mangal vegetation. Crabs destroy seedlings and prevent regeneration. *Acrostichum* and other plants are characteristic associates on the mounds made by the lobster *Thalassina anomala*, which is common throughout Southeast Asia. *Acrostichum* generally is treated as a weedy genus, favoring disturbed sites (Plate 20).

There is some controversy about the significance of faunal predators on the aerial roots of *Rhizophora* (Ribi, 1982). These roots branch opportunistically; that is, they remain unbranched until they take root distally, or until some environmental stress induces regenerative or proliferative branching by a process that is not understood developmentally. The stress may be climatic (a simple drying out of the apical meristem) but seems more often to be the result of attacks of predatory animals. Common examples in the Florida–Caribbean region, where the phenomenon has been studied, are the isopod *Sphaeroma terebrans* and the beetle *Poecilips rhizophorae*. Where these attacks are extensive (e.g., *Sphaeroma*, Plate 3F), the mangroves have been considered to be threatened (Rehm and Humm, 1973). Other authors (Simberloff et al., 1978) have suggested that the attacks are beneficial, as the roots branch if attacked and an extensive branched root system is necessary for the survival of the tree. The ultimate question is: When does enough presumed beneficial effect become too much and a detrimental effect? The example specifies the general question of the effect of the total epifauna on the aerial roots of mangroves. Other authors have revealed the extent of tree mortality from unknown causes (Jiménez et al., 1985).

In general, there is very little precise information about plant–animal interactions in mangroves; the most significant exception relating to pollination is discussed in Chapter 9. Even here, the lack of close correlation is indicated by the observation that pollinators are generalized for mangroves. The honey-eaters that visit the flowers of *Bruguiera gymnorrhiza* in Queensland (Tomlinson et al., 1979) are replaced by sunbirds in East Africa (Davey, 1975). One peculiar association suggested by Primack and Tomlinson (1980) is that honey-eaters may be attracted to the shoots of *Rhizophora stylosa* by the sugary secretion that originates in the stipules. In addition, birds may eat insects on the leaves and branches; this is presumably beneficial if the insects are phytophagous. One further reason for limited plant–animal interaction in mangal is that mangroves are almost exclusively dispersed by water; fruits and seeds are rarely attractive to animals. Little attempt has been made to list insects in mangroves, but the numbers would be proportionately large. The insect fauna of the mangal has not been studied extensively, although insects are probably the main predators of mangrove leaves, as described later. Mosquitoes and midges are the most notable and offensive insects of the mangal and may make the community totally inhospitable to humans, as in South Florida in the summer months. Where the mosquitoes are hosts to diseases like malaria and dengue fever, the mangal is truly fever-ridden. Again, MacNae is exceptional in his discussion of the insect fauna. He describes the activities of tailor ants (*Oecophylla* species Plate 3H), which may be beneficial to a tree because they eat (but also tend) scale insects. He discusses the behavior of fireflies, a characteristic feature of Asian mangroves and now a tourist attraction. The ubiquity of ants is reflected in a "mangrove ant" (*Camponotus anderseni*).

Herbivory

Despite the evident high salt concentration of mangrove leaves, there can be extensive herbivory. This has been the source of suggestions for planting mangroves so they can be used as browse for domestic animals (camels, goats, water buffalo).

Insect herbivory of mangroves can be considerable, as shown in the survey by Murphy (1990) of the limited mangroves of Singapore. The results obtained by Johnstone (1981) from a study in the Port Moresby region of Papua New Guinea show that as much as 20 percent of leaf tissue could be consumed by herbivores, but with no correlation between the amount eaten and such factors as specific diversity and density, seasonality, nitrogen pollution, and the calorific value of leaf tissue. His observation that the amount was also independent of the chloride concentration of leaves suggests that the leaf salinity of mangroves does not discourage predators. The results are of interest because it is usually assumed that mangrove biomass finds its way into food chains largely via detritus feeders; alternatives to that particular food chain must exist. Johnstone (1981), however, reports that the ant *Oecophylla* does not affect the amount of leaf tissue eaten by herbivores, even though its activities have suggested a protective role.

There seem to be no studies of the leaf consumption of larger herbivores, although MacNae (1968) mentions that leaf-eating monkeys occupy mangal. In Malaysia, proboscis monkeys are a tourist attraction. One specialized example is *Aratus pisonii*, a Brazilian tree-climbing crab that has been observed to consume mangrove wood pulp. This species makes a virtue of necessity, as it seems to be confined by its terrestrial predators to the mangal canopy during its adult life (Lacerda, 1981).

Analytical Methods

Quantified analyses of mangal are numerous. They rely largely on profile diagrams based on transects usually at right angles to the shore, so that the most extreme transition is portrayed. In profile diagrams, the species are usually so few and also distinct in architecture and morphology (e.g., aerial root systems) that an accurate visual impression of zonation can be conveyed. Often transects carry further information that is important in ecological interpretation. Figure 2.6 is a stylized example. Tidal ranges are almost always indicated. Particularly helpful is the substrate composition. Gradients in salinity and pH are less helpful unless there is some indication of ranges and frequencies, that is, if they are based on repeated measurements. Sometimes the distribution of marine animals (shellfish and crabs) is included, or the distribution of mosquito eggs and their larvae can be shown. The range of plant species within the zonation can also be added but is not represented in Figure 2.6. Figure 2.7 is a schematic profile from the wet coast of Queensland, transitional inland to wet forest and here showing generic distribution, but such images cannot necessarily be uniformly representative.

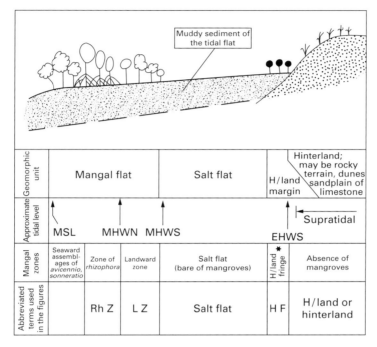

Figure 2.6. Schematic profile of tidal flat in northwestern Australia, illustrating some geomorphological and mangrove terms. Asterisk indicates that the zone may be absent from drier regions. MSL, mean sea level; MHWN, mean high water at neap tide; MHWS, mean high water at spring tide; EHWS, extreme high water at spring tide. (From Semeniuk, 1983.)

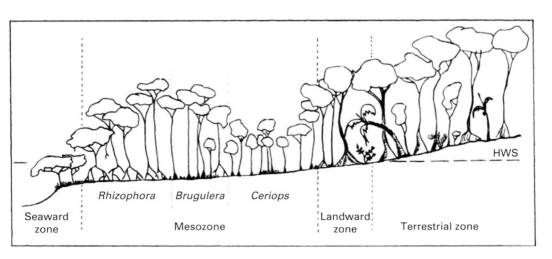

Figure 2.7. Schematic and generalized profile of a tidal flat in northeastern Australia (Queensland) to contrast zonation in a region of high rainfall with the region shown in Figure 2.6. The seaward zone is usually *Avicennia*. The salt flat in Figure 2.6 corresponds to the *Ceriops* zone in Figure 2.7. It is assumed that rainfall is the controlling factor, but site (like distance up the river), latitude, tidal range, and substrate affect the zonation. Profiles similar to Figure 2.6 therefore can occur in eastern Australia in drier areas. (Courtesy of N. C. Duke.)

Profile diagrams are particularly helpful for comparative purposes. They have demonstrated that there is no constant sequence for mangroves of contrasted regions, even though individual species tend to prefer a more seaward or a more landward site. However, the example that shows *Rhizophora* as the most seaward species in one locality may be contrasted with adjacent localities where *Avicennia* or *Sonneratia* occupies the seaward limit of the mangal. Profile diagrams are sometimes used as evidence for succession, as the zonation in space is often accepted as a zonation in time. Perhaps there is no more confusing or controversial topic in mangrove ecology than succession (Lugo, 1980). It would be inappropriate to attempt to discuss the issue, least of all in this book, but the essay by Snedaker (1982) is relevant.

The transect method is sometimes complemented by a plan view, essentially introducing a more three-dimensional picture, but much less accessible. The one used by Watson (1928), which shows zones following shoreline contours, has been reproduced several times, but the fact that it is purely imaginary and used to illustrate the concept of inundation classes is not usually stated. Watson (1928) mentions the suitability of mangroves for aerial survey because of their level topography and the clarity of discrete types. The whole west coast of peninsular Malaysia was surveyed in 1927–1928 from a height of 2000 ft.; if still extant, these photographs would be of enormous ecological value. The studies by Semeniuk (1980, 1983) of mangroves in northwest Australia used a diversity of approaches, ranging from ground to aerial surveys. These studies show the intricate relation of topography, substrate, and tidal influence.

Dynamic approaches to the study of mangal underlie most analyses. They attempt to locate the complex of characters that influence distribution and estimate the relative successional status of different species. The frequently described zonation of mangroves (Fig. 2.7) fosters this approach and provides the stimulus for such studies. Interpretative analysis often leads to the production of flow charts that connect zones within the mangal, with mangal-associated vegetation, and often with certain aspects of the ecotone, such as salinity, substrate composition, and topography.

Zonation and Succession

The existence of zones, often monospecific, is evident in mangal even to the most superficial observer. Profile diagrams produce the impression that zonation is a regular series of vegetational bands parallel to the coastline. However, any regular zonation is modified by local topography, which determines tidal and freshwater runoff, and by sediment composition and stability (Semeniuk, 1980, 1983). Lugo and Snedaker (1974) have illustrated how community structure may be elaborated in the absence of much floristic diversity. They recognized six different mangal types in South Florida determined by topography, even though only four mangrove species are involved (cf. Plate 1A, C, E). In view of the meandering and often intersecting runoff creeks that cut irregularly across the zonation, it is not surprising that the vegetation becomes more readily interpretable as a mosaic. Where mangroves run along riverbanks or front steep shores, there is insufficient space for zonation to develop. Where conditions are such that one species dominates a large shallow-water area, as in a *Nypa* swamp (Plate 17D), there may be only a truncated zonation at the margin of the swamp. Generalized descriptions of the ecological distribution of mangroves indicate that most common species can form pure stands in at least some locations (van Steenis, 1958). Species of the genus *Rhizophora* that grow together show pronounced zonation in West Africa (Savory, 1953), Queensland, and New Guinea, but the zonation is easily disrupted.

Further instability in the vegetation pattern is introduced by variation in the substrate, which ranges from coarse coral fragments to peat. Along high-energy shores, or in estuaries subject to irregular runoff, sediment redistribution is highly variable and the environment very unstable.

Because of the diversity of mangrove environments, it is difficult to present a consistent view of the relative position of each species in terms of ecological requirements. A precise description of mangal in one region can easily be contradicted by an equally precise analysis in an adjacent region. A common assumption is that the zonation of species in space represents their succession in time. Chapman (1976) has been the chief advocate of the zonation-equals-succession school of thought, but he includes numerous cautionary disqualifiers. The idea of mangroves representing a successional stage in the development of some climax terrestrial community is strongly embedded in the ecological literature and comes from the analogy of a freshwater "hydrosere" leading from open lake to terrestrial system (Johnstone, 1983). This concept is an ecological straw man put there for historical as much as hypothetical reasons, as it derives from the Clementsian school of plant ecology. Johnstone, for example, uses it to make the point that the mangal is a community with its own climax and denies alternative "steady-state" notions of mangal, apparently because for him a climax must be found. Curiously, in the same volume in which his paper appears there is an analysis by Woodroffe (1983), which documents a steady-state *Rhizophora*-dominated community on Grand Cayman established on a Pleistocene substrate and persisting and expanding by the formation of mangrove peat, which can be roughly dated by radiocarbon analysis. This community has persisted while sea levels have risen at least 2 m in about 2000 years, building up deposits of peat exceeding 4 m in some places. In Biscayne Bay, Florida, mangrove peat is found beyond the seaward limit of the existing *Rhizophora* forest, showing that the zone has either shifted landward or shrunk. Davis (1940) has suggested that the mangroves themselves determine geomorphological processes because of an assumed ability to trap and retain sediments. Other authorities (Thom, 1967, 1975) maintain that geomorphological processes are the primary cause of any vegetational migration.

The chief deficiency in these discussions seems to be lack of a direct demonstration of real successional processes in time (rather than imaginary ones in space). One reason for this is the absence of datable growth rings in mangrove woods (Plate 10) so that a static analysis cannot be transcribed into a dynamic one by comparing the relative ages of trees in the community. This deficiency is stultifying to the whole of tropical ecology (Ashton, 2014).

Multifactorial Analysis

The complexity of the mangal ecosystem is dramatically portrayed by Lugo (1980), who summarizes interrelationships by means of a diagram that incorporates marine and terrestrial ecosystems; interlinked topics include substrate, climatic, and topographic variables, recurrent cycles, and infrequent (cataclysmic) events in an elaborate reticulum. The diagram is not reproduced here, but its complexity can be appreciated by the fact that there are 35 interconnected factors and the pathways between them seem almost limitless. Bunt and Williams (1981) define 29 "association groups" in their analysis of 35 species that occur in 1391 sites in northeastern Australia. They emphasize that any unidimensional approach to species associations is misleading.

The study by Semeniuk (1983) of the ecological distribution of mangroves in northwestern Australia is particularly instructive. Although it represents a rather simple (i.e., floristically depauperate) system, it raises most of the questions concerning zonation. In this location, the habitat is somewhat marginal for mangroves because of relatively low levels of rainfall and runoff. In general, there is a seaward assemblage of *Avicennia* and *Sonneratia* followed by a *Rhizophora* zone transitional to the zone of greatest diversity, which itself grades into an extensive salt flat, always devoid of mangroves. At the landward fringe of the salt flat there may or may not be a second reduced mangrove zone finally transitional to terrestrial communities beyond the influence of tides. Semeniuk's analysis of physiography and hydrology provides evidence that ground-water salinities are an important influence on mangrove distribution. This statement finds support in a later study by Ball (1988). The simplest evidence is that where ground-water seepage is abundant, there is a landward fringe, otherwise it is absent. The explanation of distribution by correlation is strong here, but what, we might ask, is the limiting factor? Is it rainfall, which is the source of ground water? Is it topography, which directs ground-water flow, or edaphic conditions, which modify it? Is it the dispersibility of mangrove propagules, which determines initially whether a species can find a given site (Rabinowitz, 1978a), or is it simply seedling size (Rabinowitz, 1978b)? Is it tidal influence, which determines the initial supply of sodium chloride? Is it the physiological adaptation of the species to salt concentrations over a given range of values (Ball, 1988)? Is it coastal geomorphology, which provides contrasting sediment types and redistributes them during storms (Thom, 1967, 1975)?

It is, of course, no one of these factors alone, but their effect in concert that establishes the vegetation pattern. The ecologist needs a sympathetic audience when it is appreciated that s/he is unable to specify the precise effects of 35 variables in determining vegetation zonation.

Brief mention may be made of some factors that can influence mangrove zonation; these might be contrasted on the basis of whether the influence is abiotic or biotic. Most of the factors can work in concert, as they are not mutually exclusive.

Abiotic Factors Influencing Zonation

Geomorphology. Detailed analyses of coastal geomorphology by Thom and co-workers in Mexico and Australia (Thom, 1967) established the direct dependence of mangrove

Table 2.1 Watson's Inundation Classes (Port Swettenham = Port Klang)

Inundation Class	Flooded by	Height above Datum Line (feet)		Times Flooded per Month	
		From	To	From	To
1	All high tides	0	8	56	62
2	Medium high tides	8	11	45	59
3	Normal high tides	11	13	20	45
4	Spring high tides	13	15	2	20
5	Abnormal (equinoctial tides)	15	—	—	2

distribution and coastal diversity from a dynamic point of view. Semeniuk (1980, 1983) documents the variation of this dependence in considerable detail in northwestern Australia. These results may be interpreted to show that mangal distribution follows and does not determine the dynamics of topography so that mangroves in the environments studied do not override abiotic land-building processes. They still leave unresolved any explanation of zonation in terms of differential biological adaptation to contrasted physiographic factors.

Inundation classes. Watson (1928), in his classic studies of mangal in the Malay Peninsula, defined zones as regional sets influenced by a combination of numbers and kinds of tidal inundation (inundation classes), according to a simple scheme (Table 2.1). Watson emphasizes that these classes refer to the conditions in the Klang area, but they have been applied and elaborated elsewhere (Chai, 1982), because over the long term, they can be precisely measured. It becomes possible to ascribe given species to a narrow range of inundation classes and this implies that a species may be more or less restricted to a given class. The restriction can be modified by other factors, notably type of substrate. However, the analysis in no way accounts for this restriction in a functional sense.

Physiological responses to gradients. Numerous authors have concentrated on physiological adaptations that would restrict the range of conditions under which a species might grow. The usual approach is to measure an element of the environment that can be presumed to have a direct physiological effect and establish correlations with the distribution of different species. The most usual parameter considered has been salinity, as it can be measured easily and often correlates closely with species distribution. Salinity refers to the numbers of grams of dissolved salts in 1000 g of seawater, with chloride concentration used as an index. Values are usually expressed in parts per thousand (0/00) and range from 33 to 38 0/00 in the open ocean. Another commonly used but not very precise parameter is the dilution (or concentration) relative to seawater. Mangroves can adjust to about 90 0/00 (Cintron et al., 1978) in the more salt-tolerant species like *Avicennia* (i.e., about 2.5 times the concentration of seawater), but higher tolerances have been claimed.

For many ecologically constraining factors, it is the extremes that are limiting, rather than the average; this is particularly clear with the salt concentration in the soil. This may determine zonation at the margin of a salt flat, for instance, but the range of salinities within the extended seaward zones subjected to frequent inundation (Classes 1 to 3 in Watson's scheme) may be so small as to be physiologically insignificant. Other limiting factors must be sought, such as propagule size and competition, which may be defined as biotic factors.

Biotic Factors Influencing Zonation

Propagule sorting. An early attempt to explain mangal zonation by Rabinowitz (1978b) suggested that propagules sort by size, with the heavier propagules occupying the seaward habitats, because they are less readily dispersed landward and also become established more easily in deeper water and more frequently inundated sites. Support for this hypothesis comes from reciprocal transplant experiments, which suggest that mangroves are interchangeable between zones, as they grow as seedlings equally well in the "wrong" zone; therefore, they must be constrained by some biotic factor to their normally restricted ecological range. The hypothesis is based on the assumption that because there is no precise explanation of physiological adaptation to the physical conditions of a given mangal habitat, a biological explanation must be provided. It also assumes that the establishment of a plant as a seedling is sufficient to account for its persistence as a mature tree. We will see later that mangroves can be grown successfully in fresh water for experimental purposes (Chapter 10; Plate 15); they appear indifferent to water quality and substrate.

Contrary evidence to the propagule-sorting hypothesis is easily provided when species with small propagules can be shown to occupy the seaward margin of the mangal. This occurs in many parts of Australasia where *Avicennia* and *Sonneratia*, which have relatively small propagules, are the seaward dominants (Chai, 1982). Furthermore, some species range widely throughout the zones, such as *Avicennia* in northwestern Australia (Semeniuk, 1983).

Competition. If two species have exactly the same responses to physical circumstances, can one exclude the other by some biological mechanism? This possibility is implied in the propagule-sorting hypothesis of Rabinowitz, because seedlings planted in the "wrong zone" do not survive competitively. Under these circumstances, discussion of whether mangal structure is the result of either a steady-state or a successional process (Lugo, 1980) becomes less philosophical and can be discussed in terms of interaction between species. In this sense, it seems appropriate to look at the ecological status of mangroves in relation to concepts of "gap phase" dynamics and succession in tropical communities.

Ecological Status of Mangroves

Numerous authors have compared mangroves with plants of other communities; for example, salt marsh, salt desert, swamp forest, and xerophytic vegetation. Their designation as plants subject to "physiological drought" by Schimper (1891) is based on an

Table 2.2 Comparison of Pioneer and Mature-phase Species and Communities with Mangroves and with Mangal

Species	Pioneer	Mature	Mangrove
Seed size	Small	Large	Usually large
Seed number	Numerous	Few	Often numerous
Dispersal agent	Often abiotic (e.g., wind)	Usually biotic	Always abiotic (water)
Dispersibility	Wide	Limited	Wide
Geographic range	Broad	Narrow	Broad
Seed production	Continuous	Discontinuous	Sometimes continuous
Seed dormancy and viability	Long	Short	Short (?)
Seedlings	Light-demanding, not dependent on seed reserves	Not light-demanding, dependent on seed reserves	Light-demanding(?), dependent on reserves
Reproductive maturity	Early	Late	Early
Life span	Short	Long	Probably long
Leaf size	Often large	Medium or small	Medium
Leaf palatability	High	Low	Low
Wood	Soft, light	Hard, heavy	Hard, heavy
Architecture	Model-conforming	Not model-conforming	Model-conforming
Crown shape	Uniform	Varied	Uniform
Competitiveness	For light	For many resources	For light and other resources
Pollinators	Not specific	Highly specific	Not specific
Flowering period	Prolonged or continuous	Short	Prolonged or continuous
Breeding mechanism	Inbreeding favored	Outbreeding favored	Inbreeding favored
Community	Pioneer	Mature	Mangal
Floristic composition	Poor	Rich	Poor
Stratification	Absent	Well developed	Absent
Age composition	Even-aged	Uneven-aged	Even-aged (?)
Large stems	Absent	Present	Usually absent
Undergrowth	Dense	Sparse	Absent
Climbers	Few	Many	Few
Epiphytes	Few	Many	Few

Sources: Budowski, 1965; Gomez-Pompa and Vazquez-Yanes, 1974; UNESCO, 1978; Ewel, 1980; Primack and Tomlinson, 1980; Whitmore, 1983.

incomplete knowledge of their water relations, a subject dealt with later (Chapter 10). It is better to view mangroves as forming unique communities so that mangal is not readily compared with other vegetation in any constructive way. Insufficient attention has been paid to the biology of individual species. A more synthetic approach to their study can be adopted, in which the status of individuals (mangroves) is emphasized at the expense of a consideration of the status of the community (mangal).

Mangroves and mangal can be compared to nonsaline pioneer and mature-phase plants in tropical forests (Table 2.2). Mangroves share an interesting mixture of the attributes of the species and community in these contrasted phases. They have clearly pronounced characteristics of pioneer species in their reproductive biology but of mature-phase species in some aspects of their community structure and vegetative growth. An alternative designation would be to say that they have the properties of r-selected species in finding their habitat but of K-selected species in maintaining it, where the concept of r- and K-selection introduced by MacArthur and Wilson (1967) refers to species that either maximize their intrinsic rate of population increase (r) or maintain populations at the maximum carrying capacity (K) of the environment. Mangroves have their cake and eat it!

The mixture of characteristics listed in Table 2.2 suggest that mangroves initially have problems similar to those of pioneer species in locating a habitat that is patchy, varied, and available for only limited periods. Once established – and establishment has led to the development of unique biological features, notably vivipary – the mangrove community that develops has unusual properties; it has little structure because there is no further successional development. No understorey develops, there is no stratification, and much competition is intraspecific, while species distribution is strongly influenced by edaphic factors such as the degree of salinity and frequency of inundation. The analysis included here (Table 2.2) can be compared with the account in Ashton (2014; p. 79) in which a more detailed survey for Malayan Dipterocarp forests contrasts pioneer with climax forest species in many characteristics that underlie the successional biological strategies that can account for the dynamics of Asian forests.

On this basis, we might categorize mangroves as pioneer species, primarily because of their reproductive capabilities, but they form a community without succession, as shown by their performance at maturity in the vegetative state. This, at least, is a hypothesis that should be capable of development. Furthermore, as a result of this characterization of the ecological status of mangroves and in view of the relative instability of their habitat, it is easy to see how they can be said to have "invasive" propensities, as is discussed by Fourqurean et al. (2009).

3 Floristics

Categorization

Mangal is a community that is defined both by its constituent species and its environment, hence the original use of the word "mangrove" to define both the plants and the vegetation. This leads to a definition of mangroves as "trees characteristically found in tidal swamps." Excluding herbs is not artificial; the only likely candidate for a herbaceous mangrove is *Crenea patentinervis* (Lythraceae). Just as mangal is difficult to delimit in its transition to terrestrial and other seashore communities, it becomes rather arbitrary to determine the floristic limits of the group of plants that should be included in an account of mangroves. If one establishes the limits of the community by the extent of tidal influence, salinity, or type of substrate, one inevitably includes elements that are more characteristic of adjacent freshwater swamp forest, strand and beach communities, estuarine riverbank forest, and salt flats. Here we define and set limits between three groups: major elements of mangal, minor elements of mangal, and mangal associates, using fairly rigid criteria to distinguish them. A controlling and unique feature of mangal is the diurnal fluctuation of the seawater level, as explained later in the discussion of vivipary.

The following analysis is a personal one, but recognizes well-established biogeographic principles. Other analyses confirm them, but may differ somewhat in taxonomic circumscription (Wang et al., 2003; Spalding et al., 2010).

Major Elements of Mangal ("Strict" or "True" Mangroves)

Major elements are recognized because they possess all or most of the following features:

1. Complete *fidelity* to the mangrove environment; that is, they occur only in mangal and do not extend into terrestrial communities.
2. Play a *major role* in the structure of the community and have the ability to form pure stands.
3. *Morphological specialization* that adapts them to their environment; the most obvious are aerial roots, associated with gas exchange, with vivipary of the embryo less easily explained.

4.   Some physiological mechanism for *salt exclusion* so that they can grow in seawater; they frequently visibly excrete salt.
5.   *Systematic isolation* from terrestrial relatives. Strict mangroves are separated from their relatives, at least at the generic level and often at the subfamily or family level. For minor mangroves, the isolation is mostly at the generic level.

Although criterion 1 is the most significant, the importance of the group of traits is stressed because many occur singly in other groups of plants. For example, many swamp forest plants develop aerial roots associated with gas exchange, and all halophytes have some degree of salt exclusion, although not necessarily the same as that in mangroves. Vivipary among seed plants, on the other hand, is well developed only in mangroves. It occurs occasionally in other flowering plants, notably in seagrasses (Elmquist and Cox, 1996). Criterion 5 might seem artificial, irrelevant, and derived only *a posteriori*, but comparative taxonomic study supports it as a correlative feature so that the more faithful a taxon is to the mangrove community, the more isolated it seems to be from its relatives. This may reflect the age of the taxon as a plant specialized to mangal and its consequent lengthy period for ecological isolation, which leads to genetic isolation and thence evolutionary divergence from terrestrial relatives.

Minor Elements of Mangal

These minor species are distinguished by their inability to form a conspicuous element of the vegetation. They may occupy peripheral habitats and only rarely form pure communities.

Table 3.1 lists the first two categories of mangroves with some of their attributes relating to the criteria I have erected. It is clear that these plants do not have equivalent ecological roles. *Nypa*, for example, forms rather distinctive pure communities in quiet estuaries (Plate 17D). At the same time, it is the only palm that has a viviparous fruit as an indication of its degree of specialization (Plate 18). *Kandelia* is an uncommon element in mangal but occasionally does form pure stands.

Mangrove Associates

The list of plants in Table 3.2 is arbitrary and could certainly be extended, especially if non-woody plants were to be included. It has been drawn from a number of sources with extensive lists of mangrove associates (Watson, 1928; Chai, 1982) and from my own experience (Tomlinson, 2001). If herbaceous or subwoody terrestrial plants were included, most members of the tropical beach community would be added, many of which are pantropical. One could then include such plants as *Achyranthes, Entada, Ipomoea, Pluchea, Remirea, Sesuvium,* and *Suriana*, species of which are dispersed in part by sea currents and are characteristic of open coastal communities in the tropics. Such a larger list could be drawn from sources that discuss seed dispersal by ocean

Table 3.1 True Mangroves

Family	Genus	Mangrove spp.	Aerial Roots	Vivipary
Major Components				
Arecaceae (Palmae)	*Nypa*	1	–	+
Monotypic subfamily within the family				
Avicenniaceae	*Avicennia*	8	++	+
Old monogeneric family, now subsumed in Acanthaceae, but clearly isolated				
Combretaceae	*Laguncularia*	1	+	–
	Lumnitzera	2	+	–
Tribe Lagunculariae (including Macropteranthes = non-mangrove)				
Rhizophoraceae	*Bruguiera*	6	++	++
	Ceriops	2	++	++
	Kandelia	2	–	++
	Rhizophora	8	++	++
Collectively form tribe Rhizophorae, a monotypic group, within the otherwise terrestrial family				
Sonneratiaceae	*Sonneratia*	5	++	–
Old family including only one other genus (Duabanga), now subsumed in Lythraceae as a distinct tribe				
Total: 9 genera, 35 species				
Minor components				
Euphorbiaceae	*Excoecaria*	1 (–2)	–	–
Genus incudes ~35 non-mangrove taxa				
Lythraceae	*Pemphis*	1	–	–
Genus distinct in the family				
Malvaceae	*Camptostemon*	2	+	–
Formerly in Bombacaceae, now an isolated genus in subfamily Bombacoideeae				
Meliaceae	*Xylocarpus*	2	++	–
Genus of 3 species, one non-mangrove, forms tribe Xylocarpaeae with Carapa, a non–mangrove				
Myrtaceae	*Osbornia*	1	–	–
An isolated genus in the family				
Pellicieraceae	*Pelliciera*	1	–	+
Monotypic genus and family of uncertain phylogenetic position				
Plumbaginaceae	*Aegialitis*	2	–	+
Isolated genus, at times segregated as family Aegialitidaceae				
Primulaceae	*Aegiceras*	2	–	+
Formerly an isolated genus in Myrsinaceae				
Pteridaceae	*Acrostichum*	2	–	–
A fern somewhat isolated in its family				
Rubiaceae	*Scyphiphora*	1	–	–
A genus isolated in the family				
Total: 11 genera, ~19 species				

currents (Guppy, 1917; Ridley, 1930). These elements, however, are not inhabitants of strict mangrove communities and may occur only in transitional or disturbed vegetation. Grasses, rushes, and sedges occur only when they penetrate the more open parts of mangroves from adjacent freshwater or saline marshes. Examples are *Monanthochloe*, *Paspalum distichum* L., *Phragmites karka* (Roxb.) Trin. ex Steud. (Poaceae), *Juncus*

Table 3.2 Mangrove Associates

Family	Genus	Mangal/Coastal spp.	Terrestrial spp.	
Acanthaceae	*Acanthus*	3 (IWP)	30	Back mangal
Anacardiaceae	*Gluta*	1 (IWP)	20	Back mangal
Apocynaceae	*Cerbera*	3 (IWP)	3	Back mangal
	Rhabdadenia	1 (AEP)	3	Back mangal
Arecaceae (Palmae)	*Calamus*	1 (IWP)	~350	Back mangal
	Oncosperma	1 (IWP)	4	Back mangal
	Phoenix	1 (IWP & AEP)	12	Coastal swamps
	Raphia	1 (AEP & IWP)	2	Coastal swamps
Asteraceae	*Pluchea*	2 (AEP)	40	Wetlands
	Tuberostylis	2 (AEP)	2	Back mangal
Batidaceae	*Batis*	1 (–2) (P)	0	Salt marsh
Bignoniaceae	*Amphitecna*	1 (AEP)		Back mangal, riverbanks
	Anemopaegna	1 (AEP)	30	Back mangal, riverbanks
	Dolichandrone	1 (IWP)	9	Back mangal, riverbanks
Celastraceae	*Cassine*	1 (IWP)	80	Wet coastal communities
Combretaceae	*Conocarpus*	1 (IEP)	2	Back mangal, wet coastal communities
	Terminalia	1 (IWP)	200	Beach, coastal communities
Ebenaceae	*Diospyros*	1 (IWP)	400	Back mangal to rain forests
Euphorbiaceae	*Glochidion*	1 (AEP)	300	Back mangal
	Hippomane	1 (AEP)		Beach communities
Fabaceae (=Leguminosae) (Caesalpinioideae)				
	Cynometra	2 (IWP)	70	Back mangal and inland
	Caesalpinia	2 (P)	40	Back mangal
(Papilionoideae)				
	Aganope	1 (IWP)	6	Back mangal
	Dalbergia	2 (AEP)	300	Back mangal
	Derris	1 (IWP)	50	Back mangal
	Inocarpus	1 (IWP)	3	Back mangal
	Intsia	1 (IWP)	8	Back mangal
	Mora	1 (AEP)	10	Back mangal
	Pongamia	1 (IWP)	2 (IWP)	Back mangal, coastal communities
Flacourtiaceae	*Scolopia*	1 (IWP)	37	Back mangal, riversides inland
Goodeniaceae	*Scaevola*	2 (1 IWP, 1 AEP)	90	Beach communities
Guttiferae (Calophyllaceae)				
	Calophyllum	1 (IWP)	250	Beach communities
Lecythidaceae	*Barringtonia*	2 (IWP)	40	Back mangrove, riverbanks, inland
Malvaceae (Malvoideae)				
	Palitiri (*Hibiscus*)	1 (P)	200	Beach and coastal communities

Table 3.2 (*cont.*)

Family	Genus	Mangal/Coastal spp.	Terrestrial spp.	
	Pavonia	1 (AEP)	200	Back mangal, coastal communities
	Thespesia	2 (P)	15	Beach, coastal communities
Malvaceae (Brownlowiodeae)				
	Brownlowia	2 (IWP)	30	Swamp forest
Melastomataceae				
	Ochthocharis	1 (IWP)	5	Back mangal
Meliaceae	*Amoora*	1 (AEP)	20	Back mangal
Myristicaceae	*Myristica*	1 (IWP)	120	Back mangal, coastal communities
Pandanaceae	*Pandanus*	2 (AEP & IWP)	300	Back mangal, coastal swamps
Primulaceae (Myrsinaceae)				
	Ardisia	1 (IWP)	250	Back mangal
	Myrsine	1 (IWP)	10	Back mangal, coastal swamps
Rubiaceae	*Rustia*	1 (AEP)	300	Back mangal, coastal swamps
Rutaceae	*Merope*	1 (IWP)		Back mangal
Sapindaceae	*Allophylus*	1 (IWP)	190	Diverse
Sapotaceae	*Pouteria*	1 (IWP)	30	Swamp forest

IWP, Indo-West Pacific; AEP, Atlantic–Eastern Pacific; IEP, Indo-East Pacific; P, Pacific.

roemerianus L. (Juncaceae), *Cyperus javanicus* Houtt. (Cyperaceae). Mangroves otherwise form forests with no understorey except their own seedlings; there is no understorey of herbs. The fern *Acrostichum* may become dominant in disturbed sites; it exists in the undisturbed mangrove by virtue of its ability to colonize elevated sites that are not inundated at high tide (Plate 20). In Malaysia, for example, it occurs on the tall mounds made by the burrowing lobster *Thalassina*. Ferns do not otherwise occur as terrestrial plants in mangroves, but may exist as epiphytes in adjacent communities. The existence of free-living gametophytes in ferns has to be considered when their distribution is discussed (e.g., Plate 20E).

A feature distinguishing mangrove associates from "true mangroves" is the much greater diversity of leaf form, size, and texture, partly shown in Figure 3.1; cf. Plate 8.

Systematic Position of Mangroves and their Associates

Because mangal associates and specialized groups play so inconspicuous a role in the basic structure of mangrove forests, it may be argued that any discussion or description

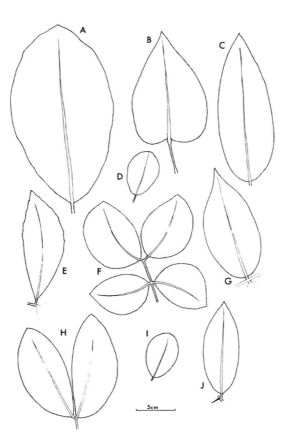

Figure 3.1. Leaf outlines of some species from the back mangal. Range of size and shape is greater than in the front mangal. A. *Heritiera littoralis*; B. *Brownlowia argentata*; C. *Inocarpus fagifer*; D. *Cassine viburnifolia*; E. *Allophylus cobbe* (one leaflet); F. *Intsia bijuga*; G. *Mora oleifera* (one leaflet); H. *Cynometra ramiflora*; I. *Glochidion littorale*; J. *Merope angulata*, with axillary spine.

of them is irrelevant. They do interact with mangroves at two important levels: they may share (or compete for) the same pollinators, and they may share the same predators and parasites so that, as alternative hosts, their influence may not be entirely negligible. In any evolutionary sense, some associates may be capable of further specialization as major mangrove components; they may give clues to the evolutionary pathway by which the highly specialized adaptive syndrome of mangroves has been achieved. In this respect, the distinctive method of germination of *Barringtonia* suggests a preadaptation to the mangrove habitat.

Gymnosperms play no role in these communities, although the cycad *Cycas rumphii* frequently occurs in communities where the topography rises abruptly behind mangroves; it will tolerate some salt spray but is typically a denizen of dry hillsides. Cycads also uniformly have hypogeal germination, not a feature of true mangroves and their closest relatives. Watson (1928) records occasional specimens of *Podocarpus polystachyus* and *Agathis alba* in sandy coastal sites. Conifers in general have little or no tolerance of salt.

Specialized Groups

Three additional synusiae or "guilds" characteristic of tropical forests but sparingly represented in mangal deserve mention: climbers, epiphytes, and parasites.

Climbers. Plants that are not self-supporting do not occur in the strict mangrove community. However, a number of scrambling or viny plants rooted in the back mangal can extend their aerial shoots into the seaward zone. Several leguminous vines fall into this category: *Aganope*, *Caesalpinia*, *Dalbergia*, and *Derris. Smythea lanceata* (Tul.) Summerh. (Rhamnaceae) has a wide distribution in the eastern tropics but is not very common. In the New World, there is the apocynaceous vine *Rhabdadenia* and the woody bignoniaceous lianes *Anaemopegma* and *Phryganocydia*. *Calamus erinaceus* (Arecaceae/Palmae) in Malaya is described as a "mangrove rattan," and may be extensive in the back mangal. This species and the cosmopolitan *Caesalpinia bonduc* are grapnel climbers and are the most spiny element of the mangrove community. Spines or prickles are (fortunately) rare, but do occur in the nonclimbing *Merope* and some palms and pandans. *Acanthus* is often scrambling and has the status of a mangrove "thistle" because of its spiny leaves and stems; it may prefer canopy gaps.

Other recorded climbing plants are the orchid *Vanda* and the climbing fern *Lygodium scandens*, both in Malaysia. We can account for the absence of climbers in mangal because their slender stems have wide vessels. Subject to extreme water tension (Chapter 8), the xylem is highly vulnerable to cavitation.

Epiphytes. Most vascular epiphytic plants are intolerant of salt; thus one encounters only a limited range of species, mainly in the back mangal, relatively high in the canopy, and in areas transitional to adjacent terrestrial communities where the epiphytes are more characteristic (e.g., Plate 2D). Some epiphytic orchids grow within a few feet of the high-water mark, however. The list that could be drawn up is potentially long; the following are simply representative of mangal in Sarawak, from Chai (1982) with some additions. Pridgeon (2014) comments: "There are no truly marine orchids, some species of *Brassavola*, *Dendrobium*, *Myrmechophila*, and other genera are epiphytic on mangroves"; some more are listed next. Benzing (1990) also comments on their existence in mangroves. Benzing and Davidson (1979) made a special study of the effects of salt on some epiphytic bromeliads that can occur in mangroves in South Florida; despite the statement that they can be "dense" on mangroves, it is suggested that *Rhizophora mangle* supports few or no epiphytes because of an axenic bark response, even though seedlings of *Tillandsia pauciflora* can be experimentally germinated on its bark if well watered.

Examples of Mangrove Epiphytes

Dicotyledons
    Asclepiadaceae
        Dischidia spp.
    Ericaceae
        Vaccinium piperifolium Sleum.
    Melastomataceae
        Medinilla crassifolia Reinw. ex Bl.
    Rubiaceae
        Hydnophytum formicarum Jack
        Myrmecodia tuberosa Jack
    Urticaceae
        Poikilospermum suaveolens (Bl.) Merr.
Monocotyledons
    Orchidaceae
        Species of *Agrostyphyllum, Brassavola, Dendrobium, Dipodium, Eria, Luisia, Myrmechophila, Podochilus, Taeniophyllum, Trichoglottis, Vanilla*
Ferns and fern allies
    Aspleniaceae
        Asplenium macrophyllum Sw.
        Asplenium nidus L.
    Davalliaceae
        Humata cf. repens Diels
    Lycopodiaceae
        Lycopodium carinatum Desv.
        Lycopodium phlegmaria L.
    Oleandraceae
        Nephrolepis acutifolia (Desv.) Chr.
    Polypodiaceae
        Crypsinus spp.
        Drymoglossum piloselloides (L.) Presl.
        Platycerium coronarium (Koenig) Desv.

In the New World, bromeliads (species of *Aechmea, Tillandsia,* and *Vriesia*) are occasional epiphytes that transgress from nearby terrestrial communities, indicating that they can withstand a limited amount of salt. A list of mangrove epiphytic ferns in the New World has not been drawn up. Less attention has been given to the epiphytic nonvascular cryptogams, but Nakanischi (1964) lists examples that grow on *Kandelia* in southern Japan.

Parasites. These are not a large component, but mention may be made of species in the Loranthaceae of *Amyema* [e.g., *A. mackayense* (Blakely) Dans.] and *Lysianthus* [e.g., *L. subfalcata* (Hook.) Barlow subsp. *maritima* Barlow] in eastern Australia and adjacent Papua New Guinea (Barlow, 1966). These may be described as mangrove "mistletoes" and seem indiscriminate in their choice of host.

Nomenclature

The names of plants provide the key to knowledge about them in the scientific literature; in turn, correct identification is necessary before correct names can be applied. Consequently, precise nomenclature, identification, and systematics are all basic to precise research endeavor. One justification for writing this book is to provide a base line of reliable data for the use of other biologists, and no little time has been spent trying to make this as accurate as possible, within the limits posed by constant name changes. Always, such activity must be seen as a work in progress.

Mangroves were certainly known to the ancient peoples (MacNae, 1968), but significant study began with European colonization in the sixteenth and seventeenth centuries. As is described in detail in the historical introduction (Chapter 1), the earliest published account of them is found in the *Hortusi indicus malabaricus* of H. van Rheede tot Drakenstein (Rheede, 1678-1703) but most influentially in the *Herbarium Amboinense* of Georg Everhard Rumpf (Rumphius, 1741–1755; Beekman, 2011). The names used by Rheede and Rumphius are of no scientific significance because they predate the publication of Linnaeus' *Species Plantarum* (1753), which is the starting point for names of higher plants according to the International Rules of Botanical Nomenclature. Rumphius included, in his comprehensive genus *Mangium*, elements that are now listed in genera as disparate as *Aegiceras*, *Avicennia*, *Bruguiera*, *Pemphis*, *Rhizophora*, and *Sonneratia*. In using Rumphian descriptions, Linnaeus himself initially retained a similar broad ecological view so that his *Rhizophora* included species that are now included in at least five modern genera. The only Linnaean name for a mangrove that is still valid is *Rhizophora mangle*, based by Linnaeus on the description by Patrick Browne in the *Civil and Natural History of Jamaica* (1765). Linnaeus' assumption that one species of *Avicennia* was pantropical led to some nomenclatural confusion between New and Old World species.

Although the number of plant species growing in mangal is small, the nomenclature is still sometimes confused for some of them, either for historical reasons or because of incomplete systematic knowledge despite continuous attempts to improve it. An example is the paper by Booberg (1933), which is an update of the species listed by Schimper (1891). This approach justifies including some synonymy in later descriptions. Most early systematic work was done entirely on the basis of herbarium specimens, thus numerous descriptions of the same species under different names came into being. Fortunately, through the activities of monographic specialists and intelligent fieldwork, considerable stability has been achieved. Nevertheless, there can be confusion where older, invalid names have been retained. For example, the nomenclature in the classic, much-cited study of Malaysian mangroves by Watson (1928) is about 50 percent incorrect by modern standards. A reader unaware of this would find a comparison of Watson's results with those in modern works very difficult. In some instances, Watson mentions some still to be described formally (e.g., *Acrostichum speciosum* and *Bruguiera hainesii*). The papers by Wyatt-Smith (1953a, b, 1954) were, in part, intended to correct some of Watson's errors and omissions and to update his nomenclature. None of these statements is a critique of the significant information in the earlier literature; our current assessment will equally be subject to the same limitations.

Systematics and Distribution

Stability in systematic botany is achieved only gradually and is constantly subject to new discovery based on further exploration. *Rhizophora* x *lamarckii*, described by Montreuzier in 1860, was thought to be endemic to New Caledonia until it was recognized as an apparent hybrid (*R. stylosa* x *apiculata*) and shown to have a wide distribution (Tomlinson and Womersley, 1976; Duke, 2006). The recent detailed exploration of Australia, especially Queensland, has indicated the likely existence of previously undescribed taxa in *Sonneratia* (Duke and Jacques, 1987) and *Avicennia* (Duke, 1991a; Duke, Benzie et al., 1998). The separation of *Kandelia* into two species has been confirmed by extensive evidence (Sheue, Liu et al., 2003). This is welcome news for systematic botanists intent on cataloguing biological diversity. Hybrids are described in detail later but some long-recognized, but only now named, formally include *Rhizophora* x *selala* in Fiji (Tomlinson et al., 1978) and *Sonneratia* x *gulngai* in Queensland (Duke, 1984). Ground truthing in New Caledonia has provided evidence of almost complete hybridization between co-occurring species in *Rhizophora* (Duke, 2010).

Taxonomic uncertainties still exist despite our apparent familiarity with these common tropical plants, largely because they have wide distributions and few field-workers have had the opportunity to explore species throughout their range. Often the characters that separate species are not very obvious; the physiognomic uniformity of mangrove plants is confusing to the beginner (Plate 8). Mistaken identifications in herbaria are common and are made even by specialists. Consequently, much of our systematic knowledge of mangrove plants is surprisingly modern, even though they have been studied for over 300 years.

Details of geographic distribution are given in Part II, but brief mention should be made of the relation between geography and systematics. Knowledge of the geographic range of a species is often incomplete, therefore checklists for given areas, as in Spalding et al., 2010, are constantly in need of revision (e.g., the legend to Fig. 4.4). In Queensland, for example, one can contrast the list made by Jones (1971) with the results of a thorough survey by Bunt et al. (1982) to appreciate that the list of mangrove species in this area had grown almost 40 percent in the intervening period. Latitudinal and longitudinal limits of species are often unclear, even though the information is of considerable phytogeographic interest. The range of many species of mangrove eastward into the Pacific from the Indo-Malayan region is not well established; the available early information is ably summarized by Fosberg (1975). The discontinuities in the ranges of some species are suspect (e.g., *Bruguiera hainesii*), and the disjunct ranges of different species within one genus (e.g., *Aegialitis* [Fig. 4.3] and *Camptostemon*) need precise verification. A survey of the distribution of widespread species from an examination of herbarium specimens may be all that is possible initially and provides good preliminary documentation; for example, *Rhizophora* (Figs. B.51, B.52), but it requires detailed confirmation by ground study. For example, we will not know the status of intraspecific taxa in the *Avicennia marina* complex until a thorough field survey throughout its total range has been carried out – from East Africa to New Zealand, from Indochina to New South Wales. This would provide a simple but important systematic base line. Thus, more than modesty recognizes the present account as nothing more than a progress report. Field verification of current systematic and phytogeographic information is a continuing need.

Identification

Field identification of mangroves is of considerable practical importance to foresters and fieldworkers, and there are numerous regional field keys (Watson, 1928; Wyatt-Smith, 1953a,b, 1954, 1960; Allen, 1956; Stearn, 1958; Jonker, 1959; Jones, 1971; Percival and Womersley, 1975; Semeniuk et al., 1978), with an update in Spalding et al. (2010). The key in Duke (2006) has accompanying illustrations. Because mangal has few species and many of them are illustrated in this account, the beginner may not need much help in learning the common mangroves. They can provide for students an introductory lesson in tropical botany because confidence can be built with so few taxa to work on.

The following keys are intended as a preliminary aid to the characters that identify mangrove genera, largely using readily visible vegetative characters. After a genus is located, there are detailed keys to species in the individual descriptions in Part II.

Key to Dicotyledonous Genera of Strict Mangroves (Major and Minor) Atlantic-Caribbean-Eastern Pacific (AEP) Region

1A. Leaves opposite; flowers small (1–1.5 cm); aggregated aerial roots usually present. 2
1B. Leaves alternate, flowers showy, large (l0–12 cm); terminal; aerial roots absent, trunk base expanded and fluted. . . . .*Pelliciera*
2A. Stipules present, overwrapping, leaving a conspicuous annular scar above petiole insertion; flowers small (1 cm), seedlings conspicuous; viviparous; aerial roots forming arching loops from trunk and branches. . . . .*Rhizophora*[*]
2B. Stipules absent; seedlings not conspicuously viviparous; aerial roots from underground cable roots. 3

* *Pantropical genera.*

3A. Leaf undersurface with a fine glaucous gray indumentum; leaf base grooved; petiole without glands; aerial roots as pointed pneumatophores . . . . *Avicennia*[*]

3B. Leaf undersurface glabrous; petiole with a pair of glands just below insertion of blade; pneumatophores, if present, blunt, not pointed . . . . *Laguncularia*

Indo-West Pacific (IWP) region

1A. Leaves compound, paripinnate, with two or three pairs of leaflets . . . .
. . . . *Xylocarpus*

1B. Leaves simple 2

2A. Leaves opposite 3

2B. Leaves alternate 7

3A. Stipules present 4

3B. Stipules absent 5

4A. Stipules overwrapping, falling, standing above the leaf insertion, leaving an annular nodal leaf scar . . . . Mangrove Rhizophoraceae (*Bruguiera*, *Ceriops*, *Kandelia*, *Rhizophora*)

4B. Stipules interpetiolar, not overwrapping, persistent . . . . *Scyphiphora*

5A. Leaves with a basal groove; stem lacking nodal glands 6

5B. Leaves without a basal groove; stem usually with a pair of nodal glands . . . . *Sonneratia*

6A. Leaves aromatic, with translucent punctate dots . . . . *Osbornia* (uncommon)

6B. Leaves not aromatic, without translucent dots; glaucous undumentum on lower surface of leaves . . . . *Avicennia* (abundant)

7A. Leaves with a broad encircling attachment, leaving an annular scar; plants shrubby, with basally swollen trunks . . . . *Aegialitis*

7B. Leaves with a narrow, non-encircling attachment; plants usually trees, but without a swollen basal trunk 8

8A. Stipules present, leaves 15–30 cm long; undersurface silvery white . . . . *Heritiera*

8B. Stipules absent, leaves less than 15 cm; not silvery white underneath 9

9A. Milky latex exuding from cut surfaces; leaf margin slightly notched . . . .
. . . . *Excoecaria*

9B. Milky latex absent, leaf margin entire 10

10A. Lower leaf surface with an indumentums of minute scales; leaf blade abruptly inserted on petiole . . . . *Camptostemon*

10B. Lower leaf surface without a scaly indumentum; leaf blade gradually inserted on petiole 11

11A. Leaves with conspicuous hairs on both surfaces, usually small (<3 cm); usually shrubby plants at seaward margin of mangal . . . . *Pemphis*

---

[*] *Pantropical genera.*

11B. Leaves glabrous, usually >3 cm; shrubs to tall trees not exclusive to seaward margin of mangal 12

12A. Leaves ovate, not fleshy, with extended glandular translucent dots; fruits elongated, one-seeded, cryptoviviparous. . . . . *Aegiceras*

12B. Leaves obovate, fleshy, without translucent dots; fruits not elongated, drupe-like; not cryptoviviparous. . . . . *Lumnitzera*

Many species of dicotyledons are likely to be encountered in the back mangal, so that any synoptic key would be incomplete and probably misleading. The following are the common distinctive genera and their most obvious diagnostic features:

Spiny or prickly plants include *Acanthus* (leaf and stem prickles); *Merope* (axillary thorns); *Caesalpinia*, which includes spiny vines with bipinnate leaves; and *Calamus*, which is a spiny climbing palm. Other climbing plants are *Derris* and *Aganope*.

Shrubby plants include *Batis* (with succulent leaves); *Ochthocharis* with prominently veined leaves; *Allophylus*, which has trifoliate leaves with the leaflet margins dentate; *Cassine* with small white flowers; and *Scaevola* with silvery hairy leaves.

Leguminous trees include *Intsia*, *Pongamia* = *Millettia* (with pinnate leaves), *Inocarpus* with simple leaves, and *Cynometra* with bijugate leaves.

Other trees with compound leaves are *Dolichandrone* and *Amoora*.

Trees with white latex include *Cerbera* (with frangipani-like leaves) and *Pouteria* (with round fruits); trees with black-spotted leaves and a clear sap turning black are likely to be *Gluta*, and should be avoided.

Trees with a scaly or stellate indumentum on the lower surface include *Brownlowia*, *Heritiera*, *Palitiri*, and *Thespesia*.

Bearing in mind that the back mangal has imprecise limits, other diagnostic features of mangrove associates that might be included in a more comprehensive (and hence unwieldy) key could be:

Vines with a milky sap (*Rhabdadenia*), with numerous prickles (*Caesalpinia*), with short woody tendrils and pea-like flowers (*Dalbergia*).

Shrubby or creeping plants either with succulent leaves (*Batis*), or opposite leaves and daisy-like flowers (*Tuberostylis*).

Trees with a milky sap (*Hippomane* – take care!), with pinnate leaves and four leaflets (*Mora*), with small cone-like fruits, alternate leaves, and petiolar glands (*Conocarpus*), with cordate leaves and irregular (*Palitiri*) or globose (*Thespesia*) capsular fruits.

Bark Characters

An identification character much employed by field biologists is the physiognomy of the bark in mature stems and roots (Duke, 2006). A sampling is provided in

Plate 4, showing the trunk surface appearance of some common mangroves. This can contrast even seemingly related taxa (Plate 4A versus I) and may distinguish aerial roots in *Rhizophora*. The appropriateness of the description of the bark of *R. apiculata* as "crocodile skin" can be checked against the original (Plate 20F versus Plate 4A).

4 Biogeography

The early literature on mangrove biogeography (Chapman, 1977a; Barth, 1982) has been largely superseded by the *World Atlas of Mangroves* (Spalding et al., 2010). This shows dot maps of the global distribution of the mangal, based on satellite imagery and ground truthing. The distribution of the species themselves is listed on a country-by-country basis, convenient politically but insufficient to supply precise locations that might be of value ecologically. This minor deficiency is compensated to some extent by more detailed published local studies that are cited, and especially in the treatments of regional areas, largely produced to provide identification and ecological analysis. Examples include those for Australia (with several regional publications), Malaysia, the Philippines, and Costa Rica. A broad picture is attempted in the present account, but recent research has considerably expanded an understanding of the paleobiogeography of mangroves, often based on molecular analysis of genetic diversity (Triest, 2008). This allows a discussion of the influence of plate tectonics that can explain much of past migration and present distribution of mangroves, as presented in the descriptions of individual taxa (Part II). The newly risen discipline of paleophytogeography, with the aid of genetic study, has greatly expanded ideas about past migration of mangroves.

An Overview

Taxonomic identity and nomenclature of species has been detailed in Chapter 3. Here we deal largely with the "true mangroves," as defined there, and which can be said to form the core of an understanding of mangrove biogeography. In contrast, Barth (1982) took a broad view and included in his lists many species here referred to as "mangal associates." Field study complemented by existing collections is still the most consistent source of information; the wide distribution of most taxa makes the task of accumulating precise field knowledge impossible by one individual, although detailed studies of small regions have been conducted (Duke, 2010, in New Caledonia; Jiménez and Soto, 1985, in Costa Rica).

Mangroves are essentially tropical, occupy two separate hemispheric regions, and are more abundant in the Old World than in the New World tropics.

A. The Indo-West Pacific (IWP) complex, variously referred to as the "Eastern" or "Old World" group (essentially the eastern hemisphere), including East Africa,

India, Southeast Asia, Australia, and the Western Pacific. The total number of true mangrove species in this area is 40.

B.   The Atlantic Eastern Pacific (AEP) complex, variously referred to as the "Western" or "New World" group (essentially the western hemisphere), including West Africa, Atlantic South America, the Caribbean, Florida, Central America, and Pacific North and South America, with one species recently extended into the western Pacific. The total number of true mangrove species in this area is only eight, although there is a local concentration of species that are incipient mangroves in western Colombia.

In both regions, the numbers are larger for both groups in Wang et al. (2003) because of a more inclusive distribution, but the principles remain the same.

Principles

In considering mangrove distribution, their unique biological feature should be emphasized. This general information provides the backdrop to later detailed elaboration.

1.   Mangroves can only migrate along hospitable coastlines.
2.   Dispersabilty of propagules is always by water and limits the extent of colonization, even though it is seemingly very efficient for most species; additionally, it can be conditioned by ocean currents.
3.   Latitudinal distribution is limited by cold temperatures, again conditioned by ocean currents.
4.   Major land barriers to distribution include the projection of southern landmasses into high latitides. Mangroves have not yet rounded the Cape of Good Hope and Cape Horn. As a result, for example, the floras of East and West Africa have no common species.
5.   The Atlantic and Pacific Oceans are barriers to long-distance dispersal, but have historically not always been so. There is clear evidence that *Rhizophora* at some time migrated from East to West across the Pacific.
6.   Recent tectonic events have added barriers. The best example is the formation of the Central American Isthmus CAI (Isthmus of Panama) about 3×10^6 MYA – separating Atlantic from Pacific mangrove populations and so isolating them genetically. This may also be the cause of a discontinuity in the genus *Pelliciera* (Fig. 4.1).

The biggest concentration of mangrove species is found in the Indo-Malayan region (Fig. 4.2); they decline in number in a western and eastern direction, so that East Africa has a diminished mangrove flora with the mangrove flora of Samoa (four species) as the eastern outlier of the IWP flora (Fosberg, 1975).

The numbers of species in the two regions can be expanded if the concept of "mangrove" is made broader, but because the floristic inventory favors the eastern group, the overall conclusion remains the same: there is a five-to-one disparity between the number of species in the two groups.

mangroves are ancestors of other flowering plants. Before elaborating biogeographic theories of mangrove redistribution, it is necessary to review briefly the relevant geographical and ecological facts.

1.   Mangroves are tropical and require climatic conditions appropriate for an ever-growing state. The distribution of mangroves in time will therefore depend on the distribution of hospitable tropical shores in time and space.

2.   Mangroves are restricted to intertidal regions. The minor incursion of mangroves into river mouths does not obscure overall constraints. Mangroves cannot cross land barriers, but must migrate along coasts or archipelagoes with continual climatic suitability.

3.   Mangroves are dispersed exclusively by sea currents, seemingly widely and efficiently. There are limits to their establishment, however, independent of dispersibility. Mangroves did not reach the Hawaian Islands, but have flourished there once introduced artificially, but considered an unfortunate anthropogenetic experiment. Mangrove propagules are frequently dispersed into regions where they cannot establish themselves or persist for any length of time (allochthonous occurrence).

4.   As a corollary of fact 3, the distribution of fossil mangroves does not necessarily represent their range of occurrence as growing plants (autochthonous occurrence). Propagules, pollen, and plant fragments are readily carried beyond the natural range (e.g., the London Clay Flora, Wilkinson, 1981).

5.   Distributions have changed with time. The fossil occurrence of *Nypa* is much more extensive than its present distribution (Tralau, 1964); the evidence comes from the wide occurrence of pollen (Fig. 9.1) and seeds. *Pelliciera* is a species the original wider range in the Caribbean of which has been reduced as a consequence of Pleistocene glaciation (Fig. 4.1). Consequently, its present range reflects part of a larger refugium for species in tropical America (Gentry, 1982). Even though this interpretation is based on fossil pollen, it is unlikely that the fossil occurrence is allochthonous because *Pelliciera* pollen is not capable of wide dispersal. Whether or not a species can migrate completely from one area to another without morphological change remains a matter for speculation, but is much supported by their morphological uniformity over great distances.

6.   Modern mangrove taxa can be of appreciable geological age (Muller, 1981). The oldest is *Nypa* (end of the Cretaceous, 69×106 years BP); records for *Pelliciera* and *Rhizophora* go back to the Eocene (30×10^6 years BP); other genera appear at progressively later periods. On this evidence, the possible previous range of mangroves older than 30 million years is dependent on the probable distribution of mangal climates as determined by plate tectonics and independent factors such as atmospheric carbon dioxide concentrations and solar input variations. Therefore, we have to sum several uncertainties in estimating past mangrove distribution. There has been a tendency, however, to interpret the present distribution of all mangroves in terms of tectonic events too ancient to influence relatively modern phytogeography.

From these generally accessible facts and principles, several often sharply contrasting hypotheses have been constructed to account for the present distribution of mangroves; for example, van Steenis (1962), Aubréville (1964), Chapman (1976), and Mepham (1983). In the early Cretaceous, there was an extensive tropical sea, the Tethys, separating the northern landmass of Laurasia from the southern Gondwanaland. The western margin of the Tethys became the Mediterranean by the juxtaposition of Africa and Arabia, although this sea is no longer hospitable to mangroves. (Paradoxically, the tropical seagrass *Cymodocea nodosa* has invaded the Mediterranean via the Suez Canal from the Red Sea, developing a sinister reputation as an invasive species.) The present Indo-Malayan concentration of mangroves is usually regarded as a relict of their origin somewhere on the southern margin of the continent Laurasia, on the eastern shores of the Tethys Sea. The continuance to the present of a rich mangrove flora in this area is consistent with a persistent equable climate. From this eastern center, migration had earlier proceeded via the western end of the ancient sea into the Atlantic, the future Caribbean, and eventually the Pacific side of South America. The total past and present distribution of *Nypa* could exemplify the results of this migration. The isthmus of Panama remained open until only three million years ago, the period of former continuity between the Atlantic and Pacific Oceans, sufficiently long to establish the western (but depauperate) mangrove flora. Evident divergence between the floras of the Atlantic and Pacific coasts of Central America in genetic terms has been forthcoming for *Rhizophora* in recent years (Takayama et al., 2013), this itself being preceded by the initial presumed transatlantic migration from West Africa. *Pelliciera* seems to be an example of a genus "trapped" in the Pacific by this process (Fig. 4.1). A similar genetic divergence has been reported for *Avicennia germinans* (Nettel and Dodd, 2007). This kind of genetic differentiation within species has also been reported latitudinally between northern and southern populations of *Rhizophora mangle* on the Brazilian coastline (Pil et al., 2011), as a result of southward migration. On the opposite (Pacific) coast, westward migration across the Pacific from the "trapped" AEP group introduced *Rhizophora mangle* to the Western Pacific in the form of taxon *R. samoensis*. Apart from *Rhizophora*, the other migrants from the original Tethyan stock were *Avicennia* (to diversify as its western group of species), *Laguncularia* (from some proto-Combretaceon ancestor), and possibly *Carapa* (a freshwater swamp species) and a close relative of *Xylocarpus*; all events proposed by Aubréville (1964) and Chapman (1976). The idea that the western group is an immigrant and therefore recent flora could account for its relative poverty, but the impoverishment could be geomorphological because of the limited availability of mangal sites. Morphological information is not helpful in this analysis, however.

This kind of cumulative evidence seems to dispose of the alternative opinion of van Steenis (1962); that is, that the migration from the Tethyan stock was primarily eastward across the Pacific; he claims that the shores of the Tethys were too dry to support mangroves. In his view, fossil mangroves are evidence of the dispersal ability of mangroves or their propensity to be carried as fragments in drift rather than an indication of the existence of a local mangrove community. This was also his explanation of the London Clay mangroves, part of the rich London Clay tropical flora (Chandler, 1957), and the creation of an Atlantic mangrove flora by migration in an easterly direction, before the creation of the CAI, achieving some secondary speciation in the Caribbean region. The only present evidence for van Steenis' idea of trans-Pacific migration would have to be the aforementioned *R. samoensis*. However, as Mepham (1983) points out, faunistic fossil evidence renders van Steenis' objections untenable and most authors would agree that van Steenis' arguments are not very parsimonious. McCoy and Heck (1976) do not accept that the center of diversity is equivalent to the center of origin but consider that plate tectonics alone account for the present biogeographic patterns. They reach this conclusion by considering the distribution of genera, whereas a consideration of species distribution raises unexplained problems.

Both hypotheses depend to a greater or lesser extent on trans-Pacific migration, which the present distribution of mangroves (and the current distribution of currents!) suggests is impossible in either direction. The long distances to be traversed seem beyond the range of dispersal of mangrove propagules and there are east–west countercurrents. Eastern and western mangal have therefore diversified in isolation. A common feature of all theories is that southerly extensions beyond the Cape of Good Hope and Cape Horn have always been impossible for mangroves. The present barriers of these two poleward extensions of Africa and South America presumably have always existed. Examples of more limited discontinuities of a geomorphological kind could be the Malay Peninsula, and on a small scale, the restriction of *Rhizophora apiculata* to the east coast of New Caledonia.

Genetic diversity. The introduction of these early ideas suggests that now available modern methods of genetic analysis could provide detailed resolution of the palaeographic history of mangrove floras. For example, the creation of the CAI that led to the separation of Atlantic and Pacific populations of the AEP group has resulted in appreciable genetic diversification (Sandoran-Castro et al., 2014). The Malay Peninsula, as a barrier to gene flow between the Pacific and Atlantic Oceans, has been examined through a study of nuclear and chloroplast genes by Ng et al. (2015). They discovered high levels of inbreeding said to be due to limited pollen and propagule dispersal. For *Rhizophora apiiculata*, Yahya et al. (2014) demonstrate in the Greater Sunda Islands a measure of "genetic clustering" attributed not only to limited dispersability but also to sea-current movement. Hence, broad scale mapping of mangrove distribution does not reveal the discontinuities of populations when viewed over short distances.

Examples studied. With these molecular techniques, advances are being made in the understanding of local genetic diversity among mangroves; an appropriate endeavor as mangrove populations are very extensive and range over enormous distances in an essentially linear fashion. How far can gene extension occur, for example in *Bruguiera gymnorrhiza*, which extends from East Africa to Samoa? Answers to such questions impinge on many disciplines, such as biogeographic history (paleobiogeography), the study of breeding mechanisms, phylogenetic relationships, and systematics. This is a major topic that is beyond the scope of this book, but references are included according to the kinds of genes studied that can provide an initial guide. Future studies will undoubtedly contribute to a better understanding of mangrove biology. However, additional observations can be made with simpler techniques, as with the study of mutation rates based on a recessive gene that produces albino seedlings (Lowenfeld and Klekowski, 1992; Klekowski et al., 1994a).

Microsatellites and other markers. These provide information about genetic diversity in populations and species. Examples are listed according to the taxon studied.

1.   *Avicennia marina.* Giang et al. (2003); Zolgharnein et al. (2010).
2.   *Bruguiera gymnorrhiza.* Islam et al. (2006a); Sugaya et al. (2003).
3.   *Kandelia* species. Giang et al. (2006).
4.   *Kandelia candel.* Islam et al. (2006b). See also Sugaya et al. (2002).
5.   *Lumnitzera racemosa.* Su et al. (2006).
6.   *Rhizophora mangle.* Arebeláez et al. (2011); Rosero-Galindo et al. (2002); Takayama et al. (2008).
7.   *Rhizophora stylosa.* Islam et al. (2004); Takayama et al. (2009).

Nuclear and chloroplast DNA. Can show genetic diversity in relation to early mangrove transoceanic dispersal.

1.   *Rhizophora* species. Kojit et al. (2013).

Distribution of Extant Species

Discontinuities. Most mangroves have a wide, but essentially continuous distribution (Spalding et al., 2010), although with a few exceptions. *Bruguiera hainesii* seems discontinuous between western Malaysia and New Guinea, possibly reflecting a lack of discrimination by collectors. *Pemphis acidula* is recorded in East Africa but is absent from a broad intervening area from South India to Sumatra, reappearing at about 137°E in eastern Malaysia.

Vicariants. Different species of one genus with non-overlapping ranges are found in *Aegialitis* and *Camptostemon.* In Australia and eastern Malaysia, the range of *Aegialitis annulata* is contrasted with that of *A. rotundifolia* in Myanmar, Bengal, and the Andamans, and provides a unique example where the total range of the genus is comparable to that of several other mangroves but the range of each of its two species is different (Fig. 4.3). This suggests that the present range of the species represents separation of an originally continuous population after geographical isolation. *Camptostemon* may reflect the same process, with *C. philippinensis* in Borneo and the Philippines and *C. schultzii* in northern Australia and New Guinea. In both examples, it is not known whether the species of the same genus are inter-fertile.

Floristic decline. The floristic richness of the central portion of the eastern group (longitude 135°E–165°E) is well documented in Figure 4.2. Of interest is the progressive decline in the number of species toward the Pacific. This is discussed by Fosberg (1975), who finds the absence of mangroves from eastern Micronesia difficult to account for, as there are suitable substrates and dispersal to them should be possible. Fossil mangrove vegetation is recorded in the northern Marshall Islands, and it has been suggested that sea-level changes could have eliminated previously existing mangal. In the genus

Table 4.2 Distribution of Mangrove Species in the Isthmian Region of Tropical America

Occurring on both Atlantic and Pacific coasts

Avicennia bicolor	*Rhizophora mangle*
Avicennia germinans	*Rhizophora* x *harrisonii*
Laguncularia racemosa	*Rhizophora racemosa*

Occurring on only the Pacific coast

Avicennia tonduzii (if distinct from *A. bicolor*)	*Pelliciera rhizophoreae*
Crenea patentinervis[a]	*Phryganocydia phellosperma*[a]
Mora megistosperma (if distinct from *M. oleifera*)[a]	*Tabebuia palustris*[a]

[a] Mangrove associates with limited distribution.
Source: after Gentry (1982).

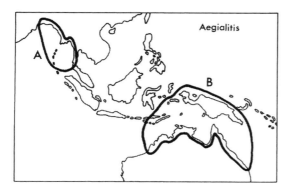

Figure 4.3. *Aegialitis.* Disjunct distribution of its two constituent species: *A. rotundifolia* (A) and *A. annulata* (B). (After van Steenis, 1949.) *A. annulata* is now known to range much farther south in western Australia (detailed distribution in Fig. 4.5).

Rhizophora, first *R. apiculata*, then *R. mucronata*, and finally *R. stylosa* disappear in an easterly direction (Fig. B.51). Why one species (*R. stylosa*) should be the most persistent in the Pacific while *R. mucronata* is the most persistent in the Indian Ocean is not at all apparent. Details of the distribution *of Rhizophora* are discussed in Part II.

Floristics of New World mangroves. The well-established disparity between the rich flora of the eastern mangal and the poor flora of the western mangal is somewhat tempered if one accepts more than the traditional number of mangrove-associate taxa in the western community. The increase in the number of New World mangrove flora has been proposed by Gentry (1982). This is set against a background of knowledge that suggests that much of the present distribution of mangroves in the New World is relict. Gentry argues that mangroves provide some evidence for the Chocó (the Pacific province of Colombia) as a refuge for tropical species during earlier glacial periods, as the mangal of the Pacific coast of Central and South America is floristically much richer than that of the Atlantic coast of northern South America (Table 4.2). However, his list of "true mangroves" includes species (as in his footnote) that lack true fidelity to the mangal and are better considered as mangrove associates.

Table 4.2 is based on what the author refers to as "rather casual personal observation and is perhaps not complete." In addition, Gentry lists *Conocarpus erectus*, *Amphitecna latifolia*, and *Muellera moniliformis* as "mangrove fringe species," whereas *Cassipourea elliptica*, *Pterocarpus officinalis*, and *Acrostichum aureum* are mentioned as mangrove components that are less specialized ecologically and also occur in other habitats. Gentry's argument is based on the concept that the present distribution of these species must have a phytogeographic, historical explanation rather than an ecological explanation. However, he also emphasizes the extreme wetness of the Chocó region, and on the basis of the earlier contrast made between the east (wet) and west (dry) coasts of Australia (Figs. 4.4, 4.5), it could be argued that the difference between Atlantic and Pacific mangrove floras is ecologically and ultimately climatically determined. Lot-Helgueras et al. (1975) describe mangroves near their limit of distribution in the Gulf Coast of Mexico, and they list many characteristic mangrove associates, including among the woody genera species in, for example, *Acacia*, *Clusia*, *Dalbergia*, *Ficus*, *Inga*, *Lycium*, *Mouriria*, *Pachira*, *Pithecellobium*, *Randia*, and *Ternstroemia*; none is restricted to coastal habitats. *Pachira aquatica* (Malvaceae: Bombacoideae) is used by Infante-Mata et al. (2014) to represent the inner limit of the mangal. The authors consider its level of salinity tolerance qualifies its status as a "high tide" mangrove, emphasizing that this title justifies a legally protected status.

Hybridization. Where, as is usual, different species of the same genus partly occupy the same range, it is of interest to know what are interspecific barriers and whether these are always maintained. Hybrids are known in *Avicennia*, *Bruguiera*, *Lumnitzera*, *Rhizophora*, and *Sonneratia*. These examples are discussed in detail when reproductive biology is considered.

A number of additional points need to be made before a synthesis can be attempted.

Incomplete Knowledge

Although it is possible to discuss the distribution of mangroves in general terms, precise information is often lacking and our present knowledge is often deficient in significant detail. For example, knowledge about the mangal flora of Queensland has been enormously extended by the detailed surveys carried out by John Bunt and Norman Duke of the Australian Institute of Marine Sciences. In little over 10 years, they increased the number of known species recorded for eastern Australia from 28 (Jones, 1971) to 45 (Bunt et al., 1982), an increase of 38 percent. Some of these represent new taxa (e.g., in *Avicennia* and *Sonneratia*); others are new records for Australia and considerably broaden the known range of species. The discovery of *Pelliciera rhizophorae* on the Atlantic coast of Colombia (Winograd, 1983) illustrates not so much a range extension as a distribution more in keeping with its fossil record. Alternatively, is it a consequence of human introduction?

Existing ranges of mangroves may be modified by human activities. Blasco (1977) estimates that species of Rhizophoraceae in western India have been virtually exterminated by overexploitation, as they do not coppice and cannot regenerate in the absence of

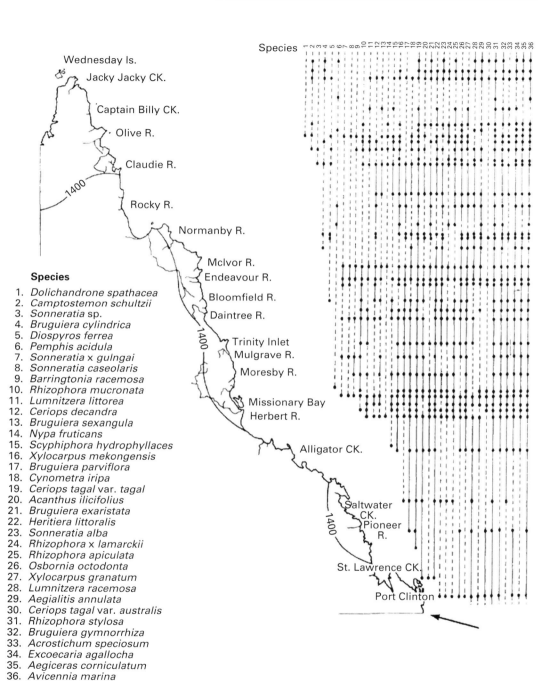

Species

1. *Dolichandrone spathacea*
2. *Camptostemon schultzii*
3. *Sonneratia* sp.
4. *Bruguiera cylindrica*
5. *Diospyros ferrea*
6. *Pemphis acidula*
7. *Sonneratia* x *gulngai*
8. *Sonneratia caseolaris*
9. *Barringtonia racemosa*
10. *Rhizophora mucronata*
11. *Lumnitzera littorea*
12. *Ceriops decandra*
13. *Bruguiera sexangula*
14. *Nypa fruticans*
15. *Scyphiphora hydrophyllaces*
16. *Xylocarpus mekongensis*
17. *Bruguiera parviflora*
18. *Cynometra iripa*
19. *Ceriops tagal* var. *tagal*
20. *Acanthus ilicifolius*
21. *Bruguiera exaristata*
22. *Heritiera littoralis*
23. *Sonneratia alba*
24. *Rhizophora* x *lamarckii*
25. *Rhizophora apiculata*
26. *Osbornia octodonta*
27. *Xylocarpus granatum*
28. *Lumnitzera racemosa*
29. *Aegialitis annulata*
30. *Ceriops tagal* var. *australis*
31. *Rhizophora stylosa*
32. *Bruguiera gymnorrhiza*
33. *Acrostichum speciosum*
34. *Excoecaria agallocha*
35. *Aegiceras corniculatum*
36. *Avicennia marina*

Figure 4.4. Distribution of mangroves in Queensland. Several species continue their distribution south of the Tropic of Capricorn (arrow); *Avicennia marina* extends to Corner Inlet, Victoria. Many areas receive over 1400 mm of rainfall annually. Place names are commonly rivers. (From an unpublished diagram by J. S. Bunt and N. C. Duke.) Minor elements unmapped are *Acanthus ebracteatus*, *Bruguiera* x *rhyncopetala*, and *Lumnitzera* x *rosea*.

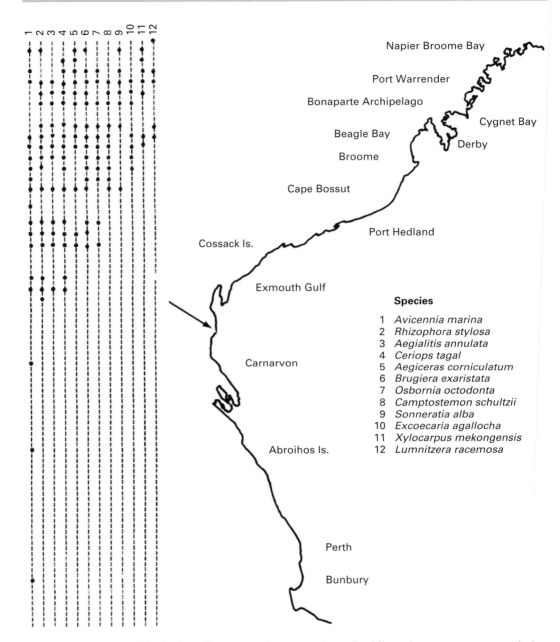

Figure 4.5. Distribution of mangroves in western Australia. All species except one are tropical (arrow, level of Tropic of Capricorn). Rainfall above 1000 mm per year only in the northernmost region (north of Derby). Place names are not distinguished by rivers. (From information in Semeniuk et al., 1978.)

seedling parents. Similarly, *Nypa* in the Ganges Delta has been largely eliminated. In contrast, several species have been established well outside their natural range, such as *Nypa* near Lagos, Nigeria, and in Central America. Such is the suggestive power of these apparent introductions, they have been discussed as natural trans-Atlantic modern migrations, an idea opposed by Duke (1991b). The limits of mangrove dispersibility, as we have seen, are shown by the Hawaiian islands, which naturally had no mangrove flora until *Rhizophora* was introduced, eventually to become regarded as an ineradicable invasive species, thereby demonstrating that Hawaiian coastlines are capable of supporting mangroves. *Bruguiera* and *Nypa* have also been introduced. The range of *Kandelia* in southern Japan may have been artificially extended as well.

Latitudinal Limits

Mangal is essentially a tropical vegetation type, with some outliers in subtropical latitudes, notably New Zealand, South Florida, South Africa, Victoria (Australia), and southern Japan (Fig. 4.2). The outliers are all a consequence of warm ocean currents moving from rich mangrove regions that provide a renewable source of propagules, plus a continuous coastline or island chain. At the limits of distribution, the formation is represented by scrubby, usually monotypic *Avicennia*-dominated vegetation, as at Westonport Bay and Corner Inlet, Victoria, Australia. The latter locality is the highest latitude (38° 45'S) at which mangroves occur naturally. The mangroves in New Zealand, which extend as far south as 37°, are of the same type; they start as low forest in the northern part of the North Island but become low scrub toward their southern limit. In both instances, the species is referred to as *Avicennia marina* var. *australis*, although genetic comparison is clearly needed. In Western Australia, *A. marina* extends as far south as Bunbury (33° 19'S). In the northern hemisphere, scrubby *Avicennia germinans* in Florida occurs as far north as St. Augustine on the east coast and Cedar Point on the west. There are records of *A. germinans* and *Rhizophora mangle* for Bermuda, presumably supplied by the Gulf Stream. In southern Japan, *Kandelia obovata* occurs to about 31 °N (Tagawa in Hosakawa et al., 1977, but initially referred to as *K. candel*).

Temperature Influence

The distribution of mangroves is clearly correlated with the sea-surface temperature, and their limits show a close correlation with the 24°C (75°F) isotherm; beyond that, the surface temperature is always colder than 24°C. Exceptions to this occur on the North Atlantic coast of America and Africa where the limit is closer to 27°C, and in southern Japan (see earlier). This suggests that *Avicennia* species in the western hemisphere are less tolerant of cold than the species in the eastern hemisphere. Mangroves, of course, endure air temperatures much below these values, but they tolerate very little or no frost and even temperatures around 5°C are inimical to the growth of most of them. Minimum sea temperatures may be lower than these average values for extended periods of the year in limiting regions.

The dependence of the limits of mangrove distribution on the sea temperature is clearly shown by the influence of cold currents with a northerly trend in the southern hemisphere. The southern limit of mangal in Pacific South America is 3° 40'S at Punta Malpelo, Tumbes, Peru, a latitude at which, in Southeast Asia, there are perhaps 30 mangrove species. This limited distribution in tropical America is conditioned by the cold Humboldt Current, which runs toward the equator, further accentuated by the arid coast. On the Atlantic side of South America, mangal occurs as far south as Florianopolis, Brazil. Similarly, the West African mangroves have a much lower latitudinal limit (12°S) at Lobito in Angola, under the influence of the South Atlantic cold current, whereas in eastern Africa, mangal extends south to the Nahoon River (33°S) in the Cape Province (Ward and Steinke, 1982). Some of these South African communities are artificially established, however.

The varying tolerances of different mangroves to presumed temperature ranges are well exemplified by the detailed information for eastern Australia (Fig. 4.4). The temperature effect is modified by aridity. Nevertheless, we should not assume a constant response to temperature by each species; this is hardly likely because of the broad range of mangroves and increasing knowledge about genetic diversity. A within-species variation in the chilling response has been demonstrated experimentally (McMillan, 1975; Markley et al., 1982) for *Avicennia* and *Rhizophora*. Samples of each species from poleward localities suffer less leaf damage at cold temperatures than samples from more equatorial latitudes. *Laguncularia*, on the other hand, showed a uniform chilling tolerance. Differential resistance could explain the skewed latitudinal distribution of *Kandelia* (Maxwell, 1995), which is asymmetric about the equator. Selection for cold tolerance may have occurred in northern populations that are now recognized as the distinct species *K. obovata* (Sheue, Liu et al., 2003). Transplant experiments seem to verify this hypothesis. The physiological basis for differential cold tolerance remains unexplained, but has been discussed by Markley et al. (1982).

Aridity

In addition to air and sea-surface temperatures determining absolute limits, the floristic abundance of mangroves is strongly conditioned by coastal aridity: the mangal is much richer in species along coasts that receive high levels of rainfall and heavy runoff and seepage. This is primarily related to the fact that many mangroves are typically estuarine and occur in greatest abundance in areas where there is extensive sedimentation. Abundant sedimentation provides a diversity of substrate types and nutrient levels higher than that of seawater. It is probably the maintenance of a balance between mineral nutrients and substrate salinity (Hanagata et al., 1999) that is relevant, rather than absolute nutrient levels.

A good example on a continental scale of this contrast between wet and dry coasts is found by comparing the mangrove floras of the (wet) eastern and (dry) western coasts of Australia. The contrasted floras are shown diagrammatically in Figures 4.4 and 4.5. At the same latitude (e.g., 20°S), Queensland has 18 species, Western Australia has only 4; and at the Tropic of Capricorn, the comparative figures are 8 and 10. This diagram also shows the progressive decline in the number of species at higher latitudes on both coasts.

This decline is not necessarily entirely due to decreasing temperature; the Queensland coast becomes drier in a southerly direction. More complete documentation of mangrove distribution in Australia is provided by Wells (1983), who similarly concludes that a major factor limiting their range is seasonal aridity.

Similar mangrove impoverishment is found in the dry areas of the Red Sea, which has only two or three species, the limit at 27° 14'N (Zahran, 1977). This impoverished mangrove flora is nevertheless in the area where mangrove trees were first described. The name *Avicennia*, after all, commemorates the Arabian philosopher Aba-bin-Senna. In other parts of the tropics, arid coasts may have few or no mangroves; examples are the Horn of Africa, northwestern Mexico, the Bahia region of Brazil, and parts of Western Australia.

An interesting microcosmic reflection of this correlation of mangrove distribution with climate is shown by the island of New Caledonia. There are 12 species of mangroves on the northeastern coast (*côte est*) (including 7 taxa in *Rhizophora* (Duke, 2010), but only 7 on the southwestern coast (*côte ouest*). The eastern coast receives the heaviest rainfall. Nevertheless, why a species as common as *Rhizophora apiculata* on the east coast should not be able to extend its range a few miles around the northern end of the island is a good measure of the mysteries governing mangrove distribution.

Mangrove Disturbance

We have discussed the biogeography of mangroves almost entirely in terms of the influence of natural events. However, the chief factor that currently modifies mangrove distribution is the activity of industrial man. Apart from the direct destruction of mangrove habitats, there is growing concern for the effects of pollutants like oil (Lewis, 1983) and toxic metals. The degradation of the mangal may produce an entirely new community, as with the appearance of a *Sesuvium–Iresine* association in place of the *Rhizophora–Avicennia–Laguncularia* association at Guanabara Bay, Rio de Janeiro, Brazil (Lacerda and May, 1982).

5   Shoot and Leaf Systems

The structure and especially the microscopic anatomy of mangroves have received considerable attention, first summarized by Chapman (1976). Information about form is necessary to an understanding of functional processes, particularly those that are specialized and adapt plants to mangal environments. The interpretation of observed structures in terms of function is, however, often based on assumption rather than experimental verification. Physiological investigations of mangroves is important because it is the suite of functional characteristics that allows mangroves to survive in tidal environments, but the literature is largely beyond the scope of this book. The present chapter, therefore, emphasizes crown structure and leaf morphology. Anatomy and wood structure is reserved for Chapter 6. The root is dealt with in Chapter 7. Water relations and salt balance are discussed in Chapter 8.

Tree Architecture

Vegetative survival and competitiveness of trees are dependent on a number of factors, including an efficient display of foliage and the ability to respond to environmental change and stress (e.g., mechanical damage). The result of the response to these factors can be translated into crown shape. The principles of the responsive processes are well understood, although an appreciation that crown form is a dynamic expression of deterministic and opportunistic processes has been slow in developing. A useful descriptive background for tropical trees is provided by Hallé et al. (1978), in which most principles are enunciated. The approach is essentially typological, which means that any set of trees in a particular environment can be compared with sets in other environments, using a convenient series of reference points defined in 23 "architectural models." Each model is named for a botanist who described a given example or "type." These models are abstract but dynamic; they constitute the deterministic component of form for each species; the visible expression of the architectural model of a tree at any time is referred to as its "architecture" (Figs. 5.1, 5.2). The sample size for mangroves is small, but the most commonly expressed model is Attims' model (Fig. 5.1A), which might suggest a limit to adaptive variation in architecture.

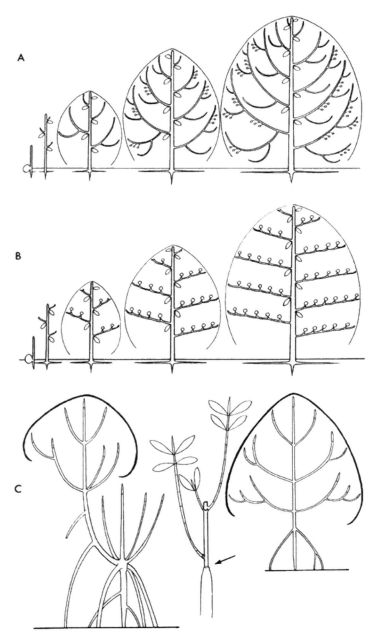

Figure 5.1. Common architectural tree models in mangroves. (After Hallé et al., 1978). A. Attims'
model, with continuous growth of the trunk repeated in the branches, which have lateral
inflorescences (e.g., *Rhizophora* spp.). B. Pettit's model, with rhythmic growth of the trunk; the
branches have terminal inflorescences and develop by substitution growth (e.g., *Lumnitzera
littorea*). C. Reiteration in *Rhizophora mangle*. Right: the tree conforming to its architectural
model (Attims' model, here with pronounced branch tiers); left: reiteration by reorientation of an
existing branch to form a reiterated crown; center: reiteration by prolepsis, significant only in
seedlings, and from adventitious buds on the hypocotyl (i.e., below the cotyledonary node –
arrow; see Plate 15).

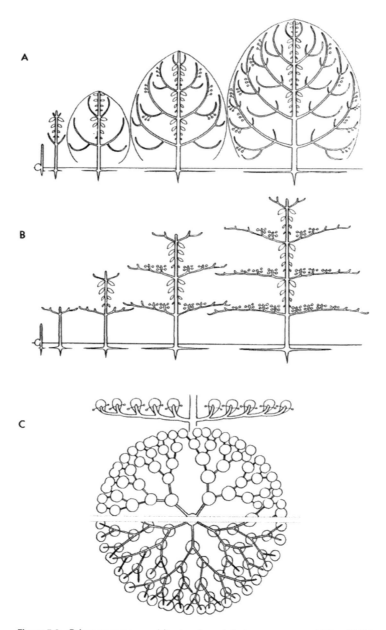

Figure 5.2. Other common architectural models in mangroves. (After Hallé et al., 1978).
A. Rauh's model, with rhythmic growth, undifferentiated branches, and lateral inflorescences
(e.g., *Xylocarpus* spp). B. Aubréville's model, with rhythmic growth of the trunk, branches
plagiotropic by apposition, and lateral inflorescences (e.g., *Terminalia catappa*). C. A single
branch tier in part B, from the side (top), from above (middle), and from below (bottom),
to show the distribution of leaf rosettes (circles) that can fill space in an economical and
photosynthetically efficient manner.

Architectural Models

The range of architectural models exhibited by mangroves is small; this may simply reflect floristic poverty, but as the observed models are structurally similar, a more likely explanation is that architecture is constrained by the environment. Hallé et al. (1978) comment on the high incidence of Attims' model in mangroves. It is found, for example, in all Rhizophoreae (except some species of *Bruguiera*). Attims' model is defined by the presence of a trunk with continuous growth, orthotropic branches with lateral flowers but with continuous or discontinuous branching. In summary, Attims' model is an all-purpose, generalized model, but with considerable plasticity in its expression, as illustrated in Plate 1B, D, F. The absence of pronounced rhythmic (episodic) growth seems to be correlated with the relatively uniform growing conditions for mangroves, but capable of appreciable seasonal modification, as shown for *Rhizophora* by Gill and Tomlinson (1971b) and charted extensively in Duke (2006). Inherent orthotropy of all axes allows for the rapid substitution of damaged leaders by adjacent existing shoots. In early stages, branch axes are readily converted into trunk axes, as is notable in *Rhizophora*, even though a branch axis can be supported initially by aerial roots (Fig. 5.1C). Branch expression is very opportunistic; vigorous trees may show continuous branching, but usually branches appear in loosely organized tiers.

Other architectural models found in mangal may be described as variants on the basic *Rhizophora* pattern. Rauh's model (Fig. 5.2A), which has rhythmic branching, is found in *Xylocarpus* and possibly *Heritiera*. Massart's model differs in that the rhythmically produced branches have a strong endogenous plagiotropy; *Myristica* is probably the best example. It is notable that these three genera are all elements of the back mangal. Furthermore, there may be a strong genetic imprint belonging to the families of these taxa.

Pettit's model (Fig. 5.1B) is common in species of the front mangal; it differs from Attims' model only in the presence of terminal inflorescences, so that the development of branches is sympodial by flowering. *Avicennia* and *Sonneratia* approach this, but much of their branching is independent of flowering, and corresponds to Attims' model. *Lumnitzera* shows contrasted architecture in its two species: *L. racemosa* (Attims' model) with lateral inflorescences, and *L. littorea* (Pettit's model) with terminal inflorescences. Here the difference in flower position may relate to contrasted pollinators, small insects that easily penetrate the crown in the former species, and birds that do not easily penetrate the crown in the second species. Elsewhere, this model is represented in coastal communities by *Morinda citrifolia* (Rubiaceae).

Aubréville's model (Fig. 5.2B) is well expressed in the large-flowered species of *Bruguiera* (e.g., *B. gymnorrhiza*). It differs from Attims' model (e.g., *Rhizophora*) in the pronounced plagiotropy by the abrupt apposition growth of its branches (Fig. 5.3). This "pagoda type" of earlier authors, is most clearly represented by the often-cultivated *Terminalia catappa*, closely analyzed by Fisher (1978). The developmental difference between Aubréville's and Attims' models is somewhat subtle; apposition growth is achieved gradually by branch complexes in *Rhizophora*, which are initially more or less orthotropic. *Terminalia* branching, whether by correlative control as in *Rhizophora* or by some endogenous mechanism as in *Bruguiera gymnorrhiza*, has been interpreted as

Figure 5.3. *Bruguiera gymnorrhiza* (Rhizophoraceae). A single sympodial branch complex in the lower canopy shows pronounced plagiotropy by apposition.

functionally beneficial in minimizing self-shading within a tier (Fisher and Honda, 1979a, b). The hexagonal pattern of structural units that supports the leafy rosettes itself represents a minimum path length arrangement (Fig. 5.2C).

Reiteration. Trees can only conform more or less precisely to their architectural model, depending on the circumstances in which they grow (i.e., they may or may not be "model conforming," as shown in Figs. 5.1A,B; 5.2A–C). They can deviate considerably from the idealized model because the architecture is modified by environmental influences (mechanical damage, predator attack, shading) that divert the normal course of development. The most usual response in the form of the branched tree to environmental perturbations is "reiteration," a partial or more usually complete repetition of the architecture of the tree (but not necessarily the root system). Reiteration may involve, for example, either regeneration of new trunk units from previously dormant meristems ("reserve meristems") or reorientation of existing axes (Fig. 5.1C, 5.4). These processes are the opportunistic component of form. This raises the question of the adaptive significance of tree form in mangroves: is relative uniformity of architecture compensated by marked adaptive features of reiteration? Regeneration of new units is the usual response to mechanical damage and depends on the availability of reserve meristems, which are developed very unequally in different mangroves. Axes can reorient in the absence of reserve meristems; this may be a response to mechanical damage but it also allows an existing axis to exploit an increased light level where there is a gap in the canopy. Extremes are represented by *Avicennia*, which coppices readily, and *Rhizophora*, which reorients branches readily. Both, however, show a remarkable plasticity of form, and the same architectural model can result in tall trees with a narrow crown in a closed canopy (Plate 1F, G), as well as the tiny but broad crown of a scrubby mangrove only a meter tall in transitional areas indicative of stress (e.g., Plate 1B). In contrast, *Nypa* (Fig. 5.5) is an example of a "tree" that is precisely model conforming in relation to its unique ability to branch terminally with a total absence of lateral

A

B

Figure 5.4. *Rhizophora mangle.* Contrasted growth forms to show architectural plasticity. A. Shrub form with dense crown of aerial roots; landward form in the Florida Everglades (cf. Plate 1B). B. Old exposed plant on the shoreline; Biscayne Bay, Florida. The parent trunk is lost, but replaced by numerous reiterated crowns and supported by a massive inverted "canopy" of aerial roots.

Figure 5.5. *Nypa fruticans* at mid-tide; Kuching River, Sarawak.

(axillary) vegetative meristems (Figs. 5.6, 5.7). Here there is no plasticity of form; the *Nypa* swamp has little diversity of aspect. Its uniqueness is expressed in the precise dichotomy of the horizontal "trunks." This feature is revealed most clearly in the grooved leaf base of the leaf subtending each branching event (Plates 17E, 18C).

Crown form, in summary, is the result of deterministic and opportunistic processes, but there seems to have been no opportunity for mathematical expression, thus little attention having been devoted to the subject of mangrove form by ecologists. There is scattered information in the classic texts by German botanists. Schmidt (1903) described bud structure and the development of branch complexes in mangroves in a remarkably complete way, but in a publication (in Danish) that has rarely been cited. Some summary information is provided in the examples of different models in Hallé et al. (1978). Gill and Tomlinson (1969) gave a basic description of *Rhizophora mangle*, presented in a phenological context (Gill and Tomlinson, 1971b). The later taxonomic descriptions in Part II of this book emphasize the dynamic aspects of vegetative morphology wherever possible; botanists neglect this kind of information at their interpretative peril. The tree throughout its vegetative life span continually unfolds and displays its leaves on the developing framework of axes. This vegetative activity is its most continuously expressed attribute. It seems unlikely that the processes involved and the resulting displays have no adaptive significance.

Elements of Crown Construction

The crown of a tall tree consists of a series of reiterated complexes reduced progressively along axes of higher order. The underlying architecture of each complex is more

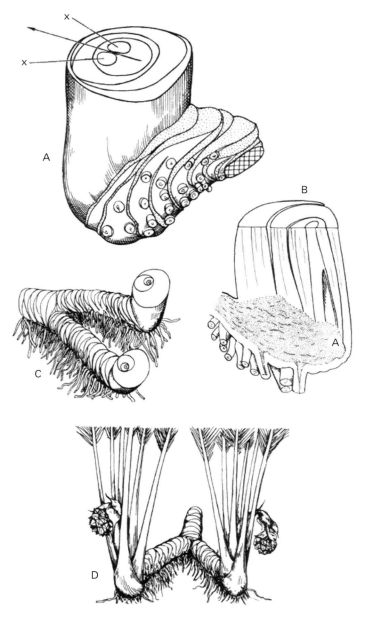

Figure 5.6. *Nypa fruticans* habit (diagrammatic, magnifications are approximate). A. Young plant (x1/4.) with older leaves removed to show late stage of first dichotomy. The growth direction of the daughter axes (X) diverges from the original growth direction (arrow). B. L.S. (longitudinal section) rhizome (x1/8), position of shoot apex at A. C. Older, dichotomized rhizome (x1/60). D. Rhizome from front with lateral (axillary) fruiting heads (x1/60). (After Tomlinson, 1971.)

Figure 5.7. "Resembling nothing so much as a series of overlapping cowpats": *Nypa fruticans*. Exposed old stem at low tide showing stem dichotomy; Kuching River, Sarawak.

or less evident, especially by comparison with strong leaders or saplings, which most clearly express the architectural model for the species. Within each complex, differentiated axes can be recognized; the main distinction is between orthotropic (erect) and plagiotropic (horizontal) axes, that is, basically between "trunk axis" and "branch axis." The branches themselves are the constituents of branch complexes, which may have developed sympodially either by substitution (replacement of determinate structures, e.g., terminal inflorescences) or by apposition (displacement of initially terminal shoots into a lateral position). This can be either an intrinsic architectural feature, as is striking in *Terminalia* (Figs. 5.2B, 5.2C), or as a progessive gradual loss of apical control. Where orthotropy is very pronounced, the shoot remains monopodial. The flower or inflorescence position is an important component of crown structure; it can influence vegetative development considerably, but the major mangrove taxa are largely characterized by lateral inflorescences. The chronology of branch initiation is also important. One may contrast syllepsis (the synchronous development of lateral and parent axes) (Fig. 5.8) with prolepsis (the delayed, nonsynchronous development of lateral compared with parent axis). Attims' model exclusively shows the former, presumably related to an essentially ever-growing condition.

Shoot morphology itself is determined by the primary leaf arrangement (phyllotaxis) and the degree of internodal extension. A contrast between long- and short-shoots is an essential component of the extended branch complex in *Bruguiera* (Aubréville's model), for example (Fig. 5.3). Bud morphology is often complicated. Many species have buds that would be described as "naked" because they lack specialized protective organs. Protection otherwise is afforded by such features as grooved petioles (*Avicennia*), hairs (*Conocarpus*), and varnish (*Cerbera*). Bud protection is the most obvious function of stipules, which temporarily enclose younger primordia. Phyllotaxis

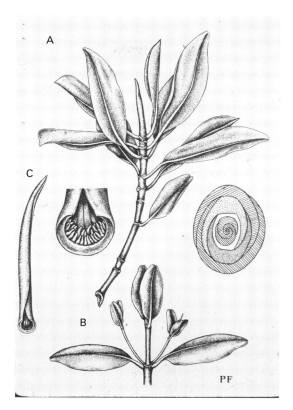

Figure 5.8. *Rhizophora mangle* shoot morphological detail. A Vegetative shoot, older leaves lost. B. Syllepsis illustrated. C. Single detached stipule. D. Inside base of stipule with colleters. E. Configuration of inrolled leaves and stipules in bud.

ultimately determines branch arrangement, because lateral meristems are almost always axillary. Branch periodicity is largely an expression of growth periodicity. Branch orientation is determined by a combination of primary and secondary effects; secondary orientation or reorientation may be the result of the development of reaction wood, an important but easily overlooked aspect of tree form (Fisher and Stevenson, 1981).

This set of structural features must be perceived in a chronological context; shoot growth and crown development must be studied developmentally. Studies of this type exist (Gill and Tomlinson, 1971b; Wium-Anderson and Christensen, 1978; Wium-Anderson, 1981; Duke et al., 1984). Some inferences about developmental processes can be made from structural characteristics; for example, the periodicity of shoot growth from variations in internode length. Proleptic and sylleptic branches, which contrast developmental differences, can be distinguished, after the event, by the presence or absence of basal bud scales (or their scars) on branches.

Bud morphology. Interpretating crown structure in terms of self-repeating units that maximize photosynthetic efficiency as far as light interception is concerned is but a preliminary step in understanding competitiveness in mangroves. Nevertheless, the characteristic dense canopy of tall mangrove seems understandable in terms of leaf arrangement, which in turn demands a study of bud structure.

Bud morphology in the tribe Rhizophoreae is distinctive and constant, and is discussed more extensively in the account of that family. In a detailed study of the whole of the Rhizophoraceae, Sheue et al. (2003) discuss stipule form and colleter structure, and conclude that the condition seen in mangrove Rhizophoraceae is the most complex and highly derived within the family, together with a characteristic fluid secretion. Each shoot ends in a bud with a limited number of primordial (usually three to five pairs) and associated stipule pairs. Each stipule pair stands below and encloses the associated leaves, but falls as the leaf pair expands. Phyllotaxis is a modified decussate arrangement (bijugate) with each pair at an angle less than 180° to the preceding pair (Fig. 5.9). This arrangement was pointed out by Schmidt (1903)

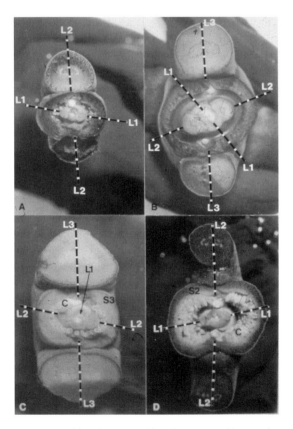

Figure 5.9. Rhizophoreae (Rhizophoraceae). Shoot apices with older leaves and enveloping stipules removed (all x3). A. *Kandelia candel*. B. *Bruguiera exaristata*. C. *Rhizophora mangle*. D. *Ceriops tagal*. Key to labeling: L1, youngest leaf pair (younger leaf primordia not visible by this technique were possibly ignored); L2 and L3, successively older leaf pairs; S2 and S3, stipules associated with leaf pairs L2 and L3; C, colleters. Lines show the dorsiventral planes of the first, second, and third pairs of leaves, estimated by eye, the method used for measuring angular divergence between successive leaf pairs. (From Tomlinson and Wheat, 1979.)

Figure 5.10. *Rhizophora apiculata* (Rhizophoraceae). Seedling from above to show bijugate phyllotaxis, which minimizes self-shading; left: part of aerial root with characteristic bark.

but not documented in detail until much later (Tomlinson and Wheat, 1979). The difference between bijugate and decussate is not trivial because the bijugate arrangement reduces shading and produces branch systems with a greater diversity of orientation compared with a decussate arrangement (Fig. 5.10). A feature of the vegetative buds of Rhizophoreae is the secretion of fluid of varying viscosity and emanating from the palisade of glandular colleters within each stipule base (Lersten and Curtis, 1974). The fluid fills space between the loosely packed primordia. Fluid secretions from colleters associated with either enclosed or exposed primordial also occur in *Aegiceras*, *Cerbera*, *Osbornia*, and *Scyphiphora*. *Aegialitis* has mucilaginous buds, with the mucilage accumulating in the cavity of the deep petiolar channel.

Vegetative Propagation

Reserve meristems. Complete regeneration of the crown in trees is possible, at least theoretically, after trunk damage or decapitation, from "resting" or "reserve" buds that develop as epicormic shoots or stump sprouts on the persistent trunk portion as an expression of the general process of *reiteration* (Hallé et al., 1978). The forester's term "coppicing" is often used, but this relates to regeneration from stumps left after felling, an entirely artificial circumstance. Mangroves have a limited capacity for regeneration from stump sprouts, but extremes may be contrasted.

Mangrove Rhizophoraceae are distinctive because they lose the ability to produce reserve meristems early. *Rhizophora* is the best-studied example (Gill and Tomlinson,

1969, 1971b) but other genera seem identical. *Rhizophora mangle* seedlings can regenerate after loss of the plumular apex by the development of adventitious buds on the hypocotyl, as described in detail later. A decapitated sapling will do the same from previously dormant axillary buds (Fig. 5.1C). In older branched trees, dormant axillary buds are produced at most nodes but they abort within two or three years, as shown by pruning experiments. Consequently, older woody axes lack reserve meristems. The inability of damaged *Rhizophora* to develop sprouts from older axes, which is familiar to mangrove sylviculturists (Watson, 1928), thus has a simple morphological basis.

Most other common mangrove genera (e.g., *Avicennia*, *Laguncularia*, and *Sonneratia*) retain reserve meristems and develop epicormic sprouts when damaged. Where mangal is clear-felled, however, regeneration of the forest from stump sprouts is not usual. One possible reason is that the delicate root–shoot relationship that supports the tree in its normal architectural configuration is severely disturbed beyond the regenerative ability of the tree.

Regeneration. Where temperate forests are disturbed or clear-felled, regeneration may come from seeds or some form of vegetative regrowth, usually in the form of stump sprouts ("coppicing"). In contrast, an examination of mangrove plants shows that they have little or no capacity for vegetative regeneration or vegetative spread, and no natural capacity for vegetative dispersal. This generalization may be taken as a conservative view and perhaps can be contradicted in some localities and by some examples. Blasco (1977), for example, suggests that *Avicennia* and *Excoecaria* in western India have persisted in the face of overexploitation because they coppice, whereas mangrove Rhizophoraceae have been eliminated through their lack of such ability. No mangrove produces detachable vegetative structures that indiscriminately lead to establishment at a distance. Chai (1982) reports that it is possible for uprooted clumps of *Nypa* to float away and become re-established. Experimental replanting of rhizome segments should be a method of verifying this observation. Craighead (1971, p. 17) reports that large clumps of *Rhizophora*, detached during a hurricane, can survive in a new location. This kind of vegetative dispersal is accidental and seems of little significance in mangrove distribution. Biological requirements for dispersal and establishment at a distance are met by floating seeds or seedlings; that is, by sexual means. Agamospermy (i.e., the production of seeds by nonsexual processes) is not reported for any mangrove.

Clonal development. In a few specialized forms, clonal spread is possible either by rhizomes or recumbent stems. *Nypa* (Fig. 5.5) spreads by the unique dichotomy of its rhizome apices, the simplest form of vegetative branching known in flowering plants. Clonal development of this type (Figs. 5.6, 5.7) accounts for the uniformity of the *Nypa*-type formation of Schimper (1891) and raises interesting questions of homozygosity, as it is not known to what extent the individual leafy shoots within a *Nypa* swamp originate sexually (from seed) or asexually (by branching).

Acrostichum is also rhizomatous. It has a potentially unlimited capacity for spread because a branch meristem is developed on the back of each leaf base in adult shoots (Fig. 5.11). However, most of these either abort or remain inhibited, because rhizomes branch little if at all. The vegetative spread of *Acrostichum* within mangal seems restricted by its rather specialized edaphic requirements. In Southeast Asia, for example,

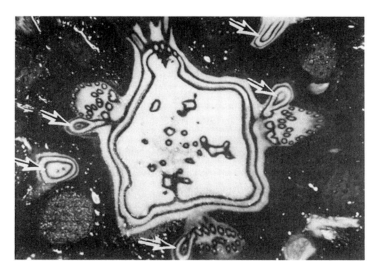

Figure 5.11. *Acrostichum* sp. (Pteridaceae). Surface view of planed rhizome to show bud or branch traces (arrows) at different levels, diverging from the base of each leaf. The section is cut obliquely, somewhat obscuring the phyllotactic pattern. (From a negative by AMJ.)

it characteristically occurs on the drier pinnacles of mounds thrown up by burrowing lobsters. In disturbed habitats, where these special requirements are more extensively accommodated, it may become quite weedy, in part by rhizomatous growth. In fresh-water swamps *Acrostichum* is long-lived and may form very large clumps, but still with few branches.

Acanthus species are somewhat sprawling plants, as they do not develop woody, strongly self-supporting axes. Spread of the recumbent stems is facilitated by the development of aerial roots.

Sonneratia has a limited ability to spread because its lower branches may recline under their own weight and root distally. Holbrook and Putz (1982), with *Sonneratia alba* in peninsular Malaysia, traced individual axes (clones) to a maximum length of 37 m (mean = 23.6, s.d. = 11.0). The direction of spread was always seaward, and it is suggested that this is the result of the rather exposed rocky coastline at the study site, in which seedlings could not be established, and landward spread is prevented by the inshore *Rhizophora* community.

Rhizophora itself has a marked capacity to extend its lower branches (Fig. 5.4) as a result of their continued horizontal growth supported by aerial roots (stilt roots). Varying claims are made about how important these branches are in the vegetative spread of the trees. Davis (1940) emphasized them as playing a major role in *Rhizophora mangle*, especially in a seaward direction, and consequently as having a major role in land building. Egler (1952) discounted this idea. It is possible that the morphological ability of branches to persist in this way after the parent trunk dies may account for the persistence of otherwise almost sterile colonies of *Rhizophora* hybrids. Architecturally, it is possible for *Rhizophora* generally to persist indefinitely in this way if upturned ends of branch axes dedifferentiate to become trunk axes, but this may only occur in saplings. The remarkable variability seen in the expression of growth form in *Rhizophora* is testimony to the combination of architecture and unusual root development.

It is appropriate to mention here that no mangrove spreads vegetatively by root suckers. Reports of this occurring from the aerial roots of *Rhizophora* result from an observer mistaking a branch for an old root.

Natural grafts. There are no reports of natural stem grafts in mature mangroves. Within the intertwining arching aerial roots of a *Rhizophora* forest, however, natural grafts among roots of the same species can develop (e.g., *R. mangle* in South Florida) and even between roots of different species (e.g., *R. apiculata* and *R. stylosa* in Queensland; J. S. Bunt, personal communication). The demographic significance of this is not known.

However, the grafting ability of *Rhizophora* at the earliest stage of development is pronounced; Larue and Muzik (1954) showed that seedling hypocotyls of *R. mangle* could be joined readily by artificial grafts in a variety of ways. The remarkable regenerative ability or plasticity of this genus at the seedling stage because of hypocotylar adventitious buds is contrasted with its adult inflexibility.

In summary, mangroves have a limited capacity for vegetative spread, *Nypa* being the most conspicuous exception. Regeneration, whether artificial or natural, almost always comes from seeds. Seedlings may have a more pronounced regenerative ability than adults, notably in the Rhizophoraceae. The management of mangroves depends on these biological restrictions on re-establishment; where mangroves are destroyed, seeds or seedlings are the normal mechanism for regeneration of the community. This strong propensity for the natural establishment of populations solely from seed is part of the "pioneer" status of mangroves.

Leaf Age and Phenology

Leaf age. We can illustrate the interdependence between shoot growth and photosynthetic ability by considering leaf age. How long do leaves persist on mangroves? What determines their longevity? What is the effect of their longevity? Representative values for *Rhizophora mangle* growing in the seasonal environment of South Florida illustrate some of the methods needed in any phenological study (Gill and Tomlinson, 1971b). Leaf longevity cannot be derived by casual observation, as the trees are evergreen; individual shoots must be marked and monitored at regular intervals. Values by Gill and Tomlinson are for three populations of shoots contrasted in their location, vigor, and reproductive status. The average leaf age was 6 to 12 months, with a maximum of 17. A number of factors account for this variation, but notably the time of year when a leaf is initiated. Those that expand in winter have a shorter life span than those that expand in summer. One reason is that growth is intrinsically continuous, but modified by the climate, so that leaves are produced faster in summer than in winter. In addition, leaf abscission is environmentally influenced, again with leaves falling faster in summer than in winter. This synchrony of leaf development and loss results in a fairly constant number of mature leaves (four or five) on each shoot. Leaves produced in winter are likely to be involved in the peak of leaf loss in summer. It must be emphasized that the leaf turnover rate can be different on different shoots, by virtue of the inherent continuous potential for growth in any shoot and the place of that shoot in the branch complex. This seems a general property of mangroves. Leaf longevity thus can be seen as a compromise between leaf retention that maximizes gain on the initial investment in biomass versus leaf abscission, which can be beneficial because in mangroves it is a method of salt elimination. Salts are not concentrated in older leaves, but seemingly accumulate via the mechanism of leaf succulence (Chapter 8). We can recognize interdependence among overall architecture (which determines shoot location), shoot vigor, leaf age, and maintenance of the salt balance.

Evergreens characterize not only mangroves but also mangal. Exceptionally, species of the back mangal may be briefly deciduous; the term "leaf exchanging" of Longman and Jenik (1974) seems appropriate here because leaf loss is followed almost immediately by expansion of a new leaf flush. *Terminalia catappa*, a frequently cultivated tree, is spectacular in this respect because leaf senescence throughout the crown results in what would be called "fall coloring" in a temperate tree. *Dolichandrone*, *Excoecaria*, *Terminalia*, and *Xylocarpus* are other examples. Front-mangal species are necessarily evergreen if leaves are important for maintaining the ever-present metabolic and physical processes involved in salt exclusion and stable water potentials. The balance may be delicate. As we have seen in *Rhizophora*, Gill and Tomlinson (1971b) indicate that leaf development and leaf loss are synchronized so that the shoot composition is constant. Defoliation of mangroves by hurricanes may be one factor in their death following such storms (root asphyxiation could be another). No experimental defoliations seem to have been attempted.

Mangroves are somewhat analogous to conifers at high altitudes and latitudes; that is, evergreenness is part of their survival kit in stressed environments, although for different reasons (maximizing assimilation versus maintaining stable salt balances). Additional stress cannot be accommodated by mangroves; therefore, they are excluded from climates with cold winters where the mangal is replaced by a treeless salt marsh.

Phenology. Phenology involves a study of the periodicity of production of plant parts (leaves, stems, flowers, and fruits). Stylized diagrams in Duke (2006) provide generalized information for individual species. The phenological observations of Gill and Tomlinson (1971b) on *Rhizophora* are preliminary and relate to a single species in a region with a markedly seasonal climate. A subsequent series of observations in the monsoon climate of southern Thailand (Christensen and Wium-Anderson, 1977; Christensen, 1978; Wium-Anderson and Christensen, 1978; Wium-Anderson, 1981) provides comparative data on more species. These authors again tagged individual shoots and measured the rates of leaf production and leaf loss for at least a year. Most species can be described as growing continuously but not necessarily with a uniform rate of leaf production. For *Rhizophora mucronata*, there was no seasonal pattern of leaf production and shedding, hence the species could be described as strictly "ever-growing." Shoot construction remained constant, given that the "mean leafing rate" (0.33, expressed as a percentage of the leaves produced in relation to the total number of leaves initially present per day) was not significantly different from the "mean shedding rate" (0.30 percent, expressed on the same basis). The average lifetime for the leaves was 330 days. Other species showed either a unimodal peak of leaf production (e.g., *Scyphiphora hydrophylacea*, *Lumnitzera littorea*) or a bimodal peak (*Rhizophora apiculata*, *Ceriops tagal*, and *Avicennia marina*). The unimodal pattern seemed related to the time of heaviest rainfall, whereas it was suggested that the bimodal pattern characterizes species with a more landward distribution that are subjected to least frequent tidal flooding. Overall, the average leaf lifetime varied from as short as 9 months (*Lumnitzera littorea*) to as long as 24 months (*Avicennia marina*).

The much longer leaf ages obtained in Thailand compared with South Florida may be explained partly by the deliberate selection of more vigorous shoots in some Florida populations. That the Thailand observations may be more generally applicable, however, is supported by the observations of Duke et al. (1984) in Queensland. These authors, in part, based their findings on the number of appendages (leaves and stipules) that fell into litter traps. This gives values for leaf loss, but because, in the Rhizophoraceae, there is a stipule for each leaf (in pairs), stipule fall can be used as a measure of leaf production. This provides a very objective means of assessment. Values were converted to rates of leaf production per 100 shoots per day, a more precise expression than used in the Thailand studies. The Queensland studies show some unimodal species (*Bruguiera gymnorrhiza*, *Ceriops tagal*) and some bimodal species (*Rhizophora stylosa*, *R. x lamarckii*), and it was concluded that *R. apiculata* was trimodal. These peaks correlate mainly with rainfall and temperature. Leaf age values fell within the range observed in Thailand.

All these regional studies considered flower and fruit phenology, as in most species flowers are lateral appendages and their development is intimately dependent on leaf production and shoot extension. The most extensive observations are by Duke et al. (1984) in north Queensland, where litterfall studies and generalized observations are combined. In summary, species may be described as having continuous flowering but with peaks of flower production (usually related to peaks of leaf production), or as having definite seasonality of flower production; that is, with flowerless periods that are usually short. Peaks of flower initiation (as in Duke, 2006) can be followed as populations so that values can be given for a reproductive "cycle." *Rhizophora apiculata* has the longest period, as it takes about two and a half years from the appearance of visible flower buds to the maturation and release of the seedling. The shoot morphology of this species accords well with these values, because flowers function beyond the oldest leaves on a shoot (Fig. B.53a, b). Viviparous seedlings consequently mature much beyond the leafy rosette. Other *Rhizophora* species have a shorter cycle (12 to 18 months), whereas *Bruguiera cylindrica* has the shortest (7 to 8 months). The reproductive phenology of *Rhizophora apiculata* is considered by Pandey et al. (2010) to have practical importance where propagules need to be harvested for rehabilitation projects.

The conclusion from these pilot studies is that although mangroves tend to be ever-growing and ever-flowering, there can be marked seasonality, and the values obtained for a limited region should not be too extensively generalized. Consequently, each region and climatic regime must be assessed empirically. The three sets of studies in Florida, Thailand, and Queensland provide excellent guides. All studies demonstrate a need for intimate knowledge of shoot morphology as a basic requirement.

6 Structural Biology

Gross morphology is necessary for an understanding of visible processes in an eco-logical context, but an observer is eventually drawn to further details, especially of a microscopic character, if functional processes are to be understood. This reduction is, of course, artificial but is useful in a descriptive context.

Leaf Morphology and Texture

The rather dull, uninspiring, and superficially unchanging aspect of mangal is largely due to the uniform leaf shape and texture of mangroves, coupled with the evergreen habit (Plate 8), although the fieldworker will soon find more subtle differences. Figure 6.1 presents the outlines of leaves or leaflet (*Xylocarpus*) of 18 species of mangrove (both New and Old World) to illustrate the uniform size and generally ovate to elliptic outline, with blunt to pointed apex and usually entire margins or with, at most, obscure crenulations. Even the leaflets of the compound leaves of *Xylocarpus* match this general pattern well. Nevertheless, there is sufficient variation that shape can be diagnostic; it can be used to distinguish species of *Avicennia* where the range of leaf form in the IWP species closely matches that of the AEP (Figs. B.14, B.15). Neverthe-less the white undersurface of the lamina in *Avicennia alba* makes identification obvious (Plate 8C). Leaf morphology is particularly uniform in the mangrove Rhizo-phoraceae, although there are differences in texture and average size. In *Rhizophora*, the mucronate versus non-mucronate shape of the leaf apex segregates species on a broad geographical basis. The hybrid between species of the eastern and western taxa (*Rhizophora* x *selala*) has an intermediate leaf tip morphology showing that leaf apex morphology is genetically controlled (Tomlinson et al., 1978). This feature seems trivial in functional terms, however.

Leaf texture is fairly consistently firm to almost coriaceous, but not rigid. The veins are obscure and not prominent; this is related to the usual absence of vein sheaths and vein sheath extensions. Succulence is a character that varies according to the degree of salinity and also leaf age. The anatomical basis for this is the differential expansion of mesophyll cells (Walter and Steiner, 1936), and is discussed later in this chapter. Constancy of leaf appearance is not matched by structural characteristics; leaf dry weight per unit area in different species, for example, varies over the range 6.4 to 21.6 mg.cm^{-2}, according to Johnstone (1981).

(a)

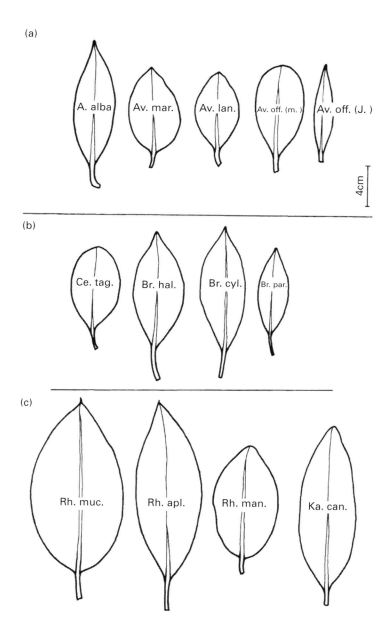

(b)

(c)

Figure 6.1. Mangrove leaf shape outlines with uniform size and shape; see also Plate 8.
A. *Avicennia*: from left to right, *A. alba, A. marina, A. lanata, A. officinalis* (mature), *A. officinalis*
(juvenile). B. Rhizophoreae: from left to right, *Ceriops tagal, Bruguiera hainesii, B. cylindrica, B.
parviflora*. C. Rhizophoreae: from left to right, *Rhizophora mucronata, R. apiculata, R. mangle,
Kandelia candel*. D. *Sonneratia*: left, *S. caseolaris*; right, *S. alba*. E. Top, *Xylocarpus
mekongensis* (one leaflet), *Excoecaria agallocha*; bottom, *Lumnitzera racemosa, Laguncularia
racemosa*.

(d)

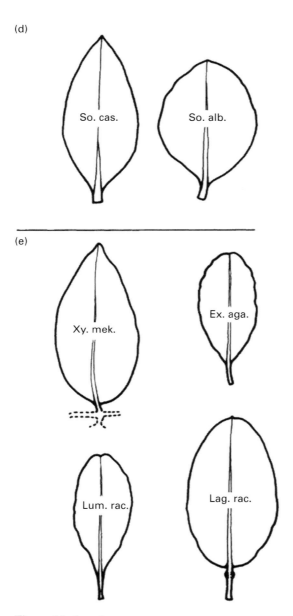

(e)

Figure 6.1. (*cont.*)

A conspicuous indumentum does not usually develop. Conspicuous waxy deposits do not occur. Sometimes young leaves are somewhat hairy in contrast to adult leaves (e.g., *Lumnitzera*). The most notable exception to the glabrous condition is *Avicennia*, with a dense palisade of short uniseriate capitate hairs (Fig. 6.2D; Plate 9E) covering the lower leaf surface and giving it a characteristic grayish or greenish-brown color.

Their function, along with other surface glands, is described in Chapter 8. *Pemphis* has an indumentum of pointed uniseriate hairs, whereas a scaly indumentum characterizes *Camptostemon* as well as the back mangroves *Brownlowia*, *Glochidion*, and *Heritiera*. Other surface features (Stace, 1965a) are so-called cork warts in *Rhizophora* (Plate 9C), seen as black spots on the lamina surface and which seem to be constellations of tannin cells. Salt glands, which in most instances function to maintain the salt balance, are described more appropriately in the physiological discussion in Chapter 8.

Secretory structures. Structures regarded as hydathodes have been described in mangrove leaves. These are areas either where the epidermis is interrupted or where there is an enlarged stoma ("water stoma" of Stace, 1965b) or where a modified epithelial tissue or epithelium is developed (Fahn, 1979). The presumed hydathodes subsequently become occluded, often with the development of a periderm. This may account for the cork warts of *Rhizophora*. It must be emphasized that there is no record of guttation from mangrove leaves, but this is not surprising because positive pressures in the xylem are hardly possible with plants rooted below seawater. Nevertheless, in some plants hydathodes function only in the early ontogeny of the individual leaf or leaflet, as in some ferns (Sperry, 1983). Consequently, in addressing the problem of the functional morphology of leaves, one should consider ontogenetic events related to the life span of the individual leaves. Structures on the lamina surface and margin that are prominent in the developing primordium may persist as inconspicuous structures in mature leaves. For example, in leaves with a slightly irregular outline, marginal indentations usually represent the site of structures that are conspicuous only in young leaves (e.g., in Combretaceae and Euphorbiaceae). Extrafloral nectaries are not recorded in strict mangroves, but the lateral petiolar glands of *Laguncularia* may have this function. Extrafloral nectaries certainly occur on the outside of the spathulate calyx of *Dolichandrone spathacea*.

Leaf Anatomy

The uniformity of leaf morphology is matched by features common to most mangroves, but still showing appreciable differences in their configuration (Plate 9). Most consistently is development of "colorless" (i.e., achlorophyllous) or "water-storage" tissue, short tracheids terminating vein endings (Plate 10G, H), and the marked absence of sclerotic vein sheaths. Sclereids of various shapes are, on the other hand, quite common (Figs. 6.2B, C), although leaf texture remains pliable because sclereids are isolated from each other. This can be appreciated by contrasting the stiff, almost brittle leaf of *Heritiera littoralis* and its well-developed bundle sheath extensions (Fig. 6.2A) with that of any true mangrove. Nevertheless, mangrove leaves have been assigned along a range of "sclerophylly" in varying degrees by Seneskide-de Lima et al. (2014). Contrasts and similarities can be found in the anatomy of the leaves of mangroves that belong to the same family. For example, mangrove Rhizophoraceae have diverse anatomical features, which is surprising in view of their uniform leaf shape and texture. The Combretaceae are much more uniform in this respect; *Laguncularia* and *Lumnitzera*

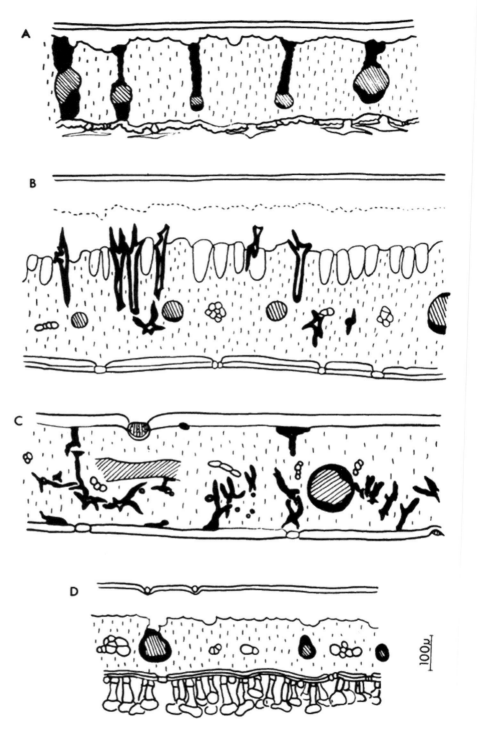

Figure 6.2. Leaf anatomy, diagrammatic, dorsiventral leaves. A. *Heritiera littoralis*, with well-developed bundle sheath extensions. B. *Rhizophora stylosa*, adaxial hypodermis with three distinct layers. C. *Aegialitis annulata*. D. *Avicennia marina*. Key to symbols: Lined area, assimilating tissue; black, bundle sheath sclerenchyma or sclereids; hatched, veins; open circles, terminal tracheid clusters.

are distinguished chiefly by their limited indumentum. We can ascribe their similarity to both a close systematic relationship and an adaptation to similar environments.

Despite this suite of common characters, leaves of different mangrove genera can be readily distinguished anatomically, as in the synoptic key ending this chapter. The characteristics are generic; specific differences are usually small so that, for example, there are no reported observations of structural features by which different species of *Avicennia* or *Rhizophora* can be distinguished. The most variable genus in this respect is probably *Bruguiera*.

Stomata. The stomatal structure in mangroves shows no high degree of specialization. Stomata are scarcely sunken or not sunken at all, but guard cells are somewhat thick walled, often with prominent or even elaborated cuticular ledges, suggesting some increased resistance to stomatal transpiration. *Nypa* has probably the most complex stomata, as the guard cells have several parallel ledges on the facing surfaces (Tomlinson et al., 2011). In other mangroves, restriction of non-stomatal water loss is imposed by the thickened outer epidermal wall, which is always strongly cutinized (Arzt, 1936). The cuticle is usually smooth, and together with the usual absence of hairs, increases reflectivity, as the "finish" of the leaf is glossy rather than matt. *Aegialitis*, however, is an example of a genus with a sculptured cuticle.

Mesophyll. The mesophyll in mangrove leaves is thick, largely because of well-developed non-assimilating layers, so that the strictly assimilatory tissue may make up a relatively small fraction of the total leaf thickness. This contrast is emphasized in Figures 6.2 and 6.3, in which assimilating tissue is dotted and colorless tissue left clear. The non-assimilating tissue is "colorless," because it has no or few chloroplasts, but this description is not entirely appropriate because tannin is common; the cell contents are then brown. The illustrations in Plate 9 are stained in various ways so that differences are artificial and often quite dramatic. Polarized light, as in Plate 9A, emphasizes sclereids. "Water-storage tissue" is a descriptive term commonly used, but this implies a function that is not proved.

Leaves may be described as dorsiventral (Fig. 6.2) if there is a marked dissimilarity between the upper and lower halves, with the colorless tissue represented by a thick adaxial hypodermis, or as isolateral (Fig. 6.3) when the opposed surfaces of the leaf are about equal with the colorless tissue occupying the middle region of the mesophyll. In thick dorsiventrally symmetrical leaves, there can be a concentration of palisade layers toward each surface. In the vascular tissue, there is the frequent development of groups of enlarged terminal tracheids at vein endings (Plate 9G, H); this is found in all common genera, but not in all species of *Bruguiera*. Branched sclereids are not uncommon in the mesophyll and are abundantly developed in *Aegiceras*, *Rhizophora*, and *Sonneratia*, and especially in *Aegialitis*, where they constitute about 15 percent of total leaf tissue (Plate 9F). Sclereids in *Pelliciera* are elongated, fiber-like cells. In many genera, two or more kinds of sclereid can be distinguished according to size, shape, and location; for example, *Aegiceras* (Rao, 1971). Both sclereids and tracheids may be involved in capillary water storage (Zimmermann, 1983), and sclereids may provide mechanical support to leaves with diminished turgor, or discourage herbivores. Most of the distinctive structural features shared by mangroves probably have functional explanations, although these are not always directly demonstrable. Epidermal features relate to transpirational control and salt glands, colorless mesophyll layers to the maintenance of a salt balance, and lack of extensive sclerenchymatous structural support to high turgor. The suggestion regarding the capillary storage function of terminal tracheids and sclereids needs experimental verification.

   Ideoblasts. The complex leaf anatomy is further reflected in various internal structurally (and presumably functionally) specialized cells that are isolated and not associated to form a distinct tissue; that is, they are ideoblastic in distribution. Among these features (in addition to sclereids already mentioned) are oil cells (e.g., in *Osbornia*), mucilage cells (subhypodermal in *Rhizophora*, hypodermal in *Sonneratia* [Plate 9D]), crystalliferous cells (e.g., druse crystals are abundant in the Rhizophoreae), tannin cells (abundant in most genera and often concentrated in certain tissues, such as the hypodermal tissue of Rhizophoraceae), laticifers (e.g., in *Excoecaria*). The precise chemical composition of the contents of these cells is rarely known in any detail; most functional attributes are entirely conjectural. The student of mangrove anatomy is advised to approach this subject empirically, because freehand sections of fresh leaves give a clearer impression of microscopic structural diversity than a lengthy search of the literature. There is considerable variation within individual species; and most authors have looked at only limited samples. Leaves of different ages should also be examined.

Synoptic Key to Common Mangrove Genera based on Leaf Anatomy

1A.   Leaves dorsiventral, stomata restricted to lower surface; colorless tissue, if present, hypodermal (Fig. 6.2). 2

1B.   Leaves isolateral, stomata equally abundant on both surfaces; colorless tissue, if present, not hypodermal (Fig. 6.3). 14

2.   Leaves dorsiventral, with enlarged terminal tracheids (except *Bruguiera* and *Kandelia*)

2A.   Hypodermis of compact colorless (non-chlorophyllous) cells below one or both surfaces; sclereids and salt glands present or absent 3

2B.   Hypodermis of colorless cells not developed; branched stellate sclereids abundant, salt glands conspicuous (Plate 9F). . . . . *Aegialitis*

3A.   Hypodermis below both surfaces; abaxial hypodermis always least conspicuous. 4

3B.   Hypodermis restricted to adaxial surface. 10

4A.   Indumentum on lower epidermis of three- or four-celled uniseriate capitate hairs forming a dense palisade (Fig. 6.2D; Plate 9E, G). . . . .*Avicennia*

4B.   Indumentum usually absent; if present, not palisade-like. 5

5A.   Adaxial hypodermis up to seven cells deep, differentiated into outer two to four layers of tabular cells, inner layer(s) of anticlinally extended cells (Fig. 6.2B; Plate 9B); branched sclereids usually well-developed (Plate 9A); cork warts present (Plate 9C). . . . . *Rhizophora*

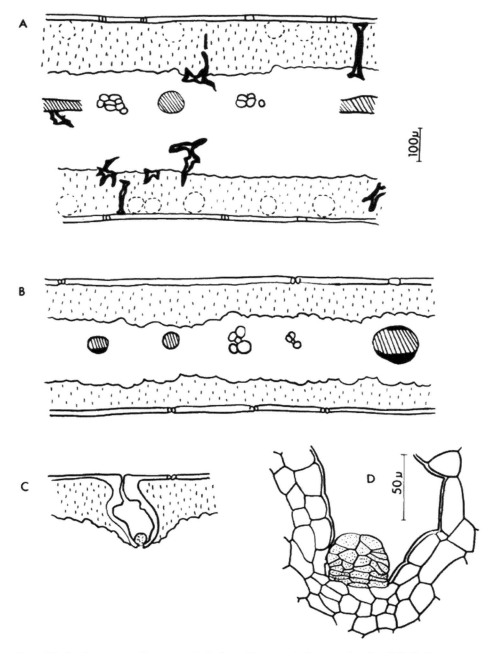

Figure 6.3. Leaf anatomy, diagrammatic isolateral leaves. A. *Sonneratia alba*, T.S. B–D: *Laguncularia racemosa*. B. T.S. C. Gland in pit on adaxial surface. D. Detail of presumed gland (salt-secreting?) at base of pit.

5B.   Adaxial hypodermis not differentiated into distinct layers; sclereids present or absent; cork warts absent. 6

6A.   Sclereids or fiber sclereids well developed, conspicuous. 7

6B.   Sclereids absent or inconspicuously developed 8

7A.   Sclereids stellately branched, thick walled; salt glands present (Fig. 8.4C, D); secretory cavities in mesophyll . . . . *Aegiceras*

7B.   Sclereids fiber-like, thin walled; salt glands and secretory cavities absent. . . . . *Pelliciera*

8A.   Terminal tracheids well developed. . . . . *Ceriops*

8B.   Terminal tracheids not well developed. 9

9A.   Mesophyll somewhat isodiametric, with assimilating palisade-like mesophyll layers below each hypodermis. . . . . *Kandelia*

9B.   Mesophyll dorsiventral, mesophyll palisade tissue present only adaxially . . . . *Bruguiera*

10A. Fibrous bundle sheath of most veins extended to one or both surfaces; indumentum of conspicuous peltate scales on abaxial surface (Fig. 6.2A). . . . *Heritiera*

10B. Bundle sheath not forming fibrous extensions to surface; surface scales absent. 11

11A. Salt glands present, especially on upper surface (Fig. 8.4E, F). . . . . . *Acanthus*

11B. Salt glands absent. 12

12A. Mesophyll somewhat bifacial with palisade-like layers toward each surface; laticifers present but not conspicuous in spongy mesophyll. . . . . . . *Excoecaria*

12B. Mesophyll dorsiventral; laticifers absent 13

13A. Terminal tracheids enlarged, conspicuous; bundle sheath not fibrous, not crystalliferous; adaxial hypodermis not enlarged. . . . . *Scyphiphora*

13B. Terminal tracheids not enlarged, bundle sheath fibers developed, often with crystals; adaxial hypodermis anticlinally enlarged in succulent leaves . . . . *Xylocarpus*

14.   Leaves isolateral, with enlarged terminal tracheids (except *Pemphis*). 15

15A. Enlarged spherical multicellular oil cavities, with lysigenous development, conspicuous and abundant below each surface . . . . *Osbornia*

15B. Enlarged oil cavities absent, hypodermal secretory structures, if present (*Sonneratia*), unicellular and not developing lysigenously. 16

16A. Leaf mesophyll relatively uniform, not conspicuously differentiated into central and peripheral layers; capitate hairs on both surfaces. . . . . *Conocarpus*

16B. Leaf mesophyll conspicuously differentiated into central colorless and peripheral assimilating layers; capitate hairs absent. 17

17A. Leaves with conspicuous slender, unbranched unicellular hairs. . . . .*Pemphis*

17B. Leaves glabrous (or unicellular hairs only in early development). 18

18A. Sclereids conspicuous, abundant (Fig. 6.3A); hypodermal mucilage cells common and conspicuous (Plate 9D). . . . . *Sonneratia*

18B. Sclereids absent or infrequent, hypodermal tannin or mucilage cells absent. . .19

19A. Lamina simple (Fig. 6.3B); "salt glands" in epidermal depressions sparsely present (Fig. 6.3C, D). . . . . *Laguncularia*
19B. Similar, but salt glands absent. . . . .*Lumnitzera*

Wood Anatomy

The anatomy of mangrove woods has been studied extensively, in part because of their economic importance and in part to correlate wood structure with physiological specialization toward the mangrove habitat (Panshin, 1932; Jansonnius, 1950; and for the Rhizophoraceae alone, van Vliet, 1976). An overview is presented in Plate 10. The following generalizations seem appropriate:

Mangroves usually lack growth rings. There has been a search for annual growth rings as a measure of tree age. The topic is reviewed by Robert et al. (2011, 2014), who conclude that, although growth rings do exist, their use in determining growth rate and tree age should be evaluated case by case. An example is in the wood of *Diospyros ferrea*, reported by Duke et al. (1981), which produces microscopically visible growth rings at an average rate of seven rings every four years. This unusual frequency seems determined by irregular trends in seasonal rainfall. Superficial examination of other mangrove woods often suggests the presence of growth rings, but these seem related to variations in the density of chemical wall deposits and not to microscopic cellular discontinuities as a result of variations in cambial activity. Similar discontinuities probably account for the claim by Davis (1940) that, in Florida, *Rhizophora mangle* produces annual growth rings. Estrada et al. (2008) refer to growth rings in *Laguncularia racemosa* as "obscure microscopically," but still refer to them as annual. They are not visible in Plate 10B.

Avicennia has conspicuous growth rings that relate to the anomalous alternation of bands of xylem and phloem tissue (Fig. B.11), but again these are not produced seasonally (Gill, 1971). *Aegialitis* also has successive cambia, producing alternate rings of xylem and phloem.

The best evidence for annual growth rings is for *Rhizophora mucronata* in coastal Kenya (Verheyden et al., 2004). Here there is a regular annular contrast between rings of high vessel density, formed during the short dry season, compared with low vessel density formed during the long rainy season. The rings are visibly contrasted as dark and light bands, but there is no cellular discontinuity; that is, growth rings are defined as "indistinct." On this basis, age determination of trees could be attempted, but with reliable variability dependent on site. Attention is drawn to similar results for *Rhizophora mangle* in Brazil (Ménezes et al., 2015).

Mangrove woods have narrow and often densely distributed vessels (Schmitz et al., 2006) (Yanez-Espinosa et al., 2001). This is a general feature of the examples in Plates 10A–E. Here a non-mangrove member of the Rhizophoracea, *Pellacalyx axillaris*, is included to contrast the much wider vessel diameter in a terrestrial species (Plate 10F). Panshin (1932) found that the tangential diameter of vessels in mangroves was in the range of small (i.e., narrow) to very small (usually less than 100 μm, rarely exceeding 150 μm), when compared with the total range of vessels in woods (range 50–250+ μm diameter). These results were confirmed by Jansonnius (1950), who suggested that vessel density was in proportion to the frequency of inundation. The extreme concentration is found in *Aegiceras*, with vessels all narrower than 50 μm and up to 150 per square millimeter. *Pelliciera* also has narrow vessels.

The high density of narrow vessels can be related on theoretical grounds to the fact that high tensions (negative pressures) have been measured in the xylem of mangrove stems (Scholander et al., 1964, 1965). These high values, in turn, are related to the high osmotic potential of seawater, which mangroves must overcome initially to absorb water, and to the high insolation, which promotes transpiration. Frequent high tensions in the xylem increases the likelihood of cavitation within vessels during extreme stressful conditions. Because such embolisms are contained within a single unit (the vessel), an element of safety is introduced if there are a large number of conducting units, as it is reasonable to assume that embolism is a random event, independent of the number of units, but directly dependent on vessel diameter (Zimmermann, 1983). However, given that the conductivity of a capillary is proportional to the fourth power of its radius, from the Haagen–Poiseuille equation, a slight decrease in diameter therefore brings about a proportionally much greater increase in resistance to flow (the reciprocal of conductivity). In other words, if one substitutes, for reasons of safety, narrow vessels for wide ones, maintaining a uniform flow rate requires the replacement of each capillary of unit diameter with 16 capillaries of one-half unit diameter (Zimmermann, 1983). A direct measure of relative conductive capacities can be obtained by comparing the sum of the radius raised to the fourth power of all conducting elements. The narrowness of vessels in mangrove woods (for safety, as it reduces the loss of conducting units by embolism) is therefore understandably compensated by an increase in their numbers, hence their density. Jansonnius gives vessel densities in mangrove woods (vessel number per unit cross-sectional area) from two to 10 times the average value for a wide range of woods, which is expected by these theoretical considerations. The reverse trend toward few wide vessels is found in lianes, which have relatively narrow stems because they are not self-supporting, and which cannot accommodate large numbers of vessels. However, this is not a safe system; lianes do not occur in the strict mangal. This discussion should also continue with vessel length, if short vessels are also safer than long ones, but there are no data.

It is to be emphasized that the previous discussion relates to stem anatomy only. The aerial roots of *Rhizophora* retain narrow vessels but rather diffusely distributed (Plate 10H). This may be understandable because many roots supply a single trunk.

Vessel elements in mangrove wood. These usually have simple perforation plates, with the notable exception of the mangrove Rhizophoraceae (tribe Rhizophoreae), which have scalariform perforation plates with few (two to seven) but broad thickening bars (Fig. 6.4). Scalariform perforation plates also distinguish the mangrove Rhizophoraceae from their terrestrial relatives (van Vliet, 1976). Because mangroves with and without scalariform perforation plates in their wood grow close together, it is not easy to provide an adaptive explanation for this structural peculiarity of the tribe Rhizophoreae. The simplest explanation, but not a very satisfactory one because it is not a functional explanation, is that this tribe is monophyletic and evolved from an ancestor that had scalariform perforation plates in its wood. Scalariform perforations were retained in

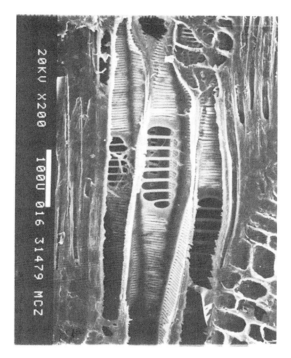

Figure 6.4. *Bruguiera gymnorrhiza*. Scanning electron micrograph of radial longitudinal surface of a wood sample to show scalariform perforation plates that characterize wood of the mangrove Rhizophoraceae.

descendants because they are adaptively neutral. An alternative explanation might be suggested if a function for perforation plates could be discovered. *Dolichandrone spathacea* is also exceptional; it has simple perforations in most vessels, but in addition, some unusual reticulate perforation plates.

Other wood structural features. A feature of wood anatomy especially in mangrove Rhizophoraceae is the high fiber content (Plate 10G). This accounts for the high density of such wood and contributes to its success as a fuel and in construction. Unfortunately, this high fiber content also allows it to be harvested for paper manufacture in a highly destructive way.

Variation in other wood characteristics relates mostly to the systematic position of a species. This suggests that there are no very precise adaptive constraints that relate anatomy to habit; that is, the structure of the mangrove wood says as much about the systematic position of a species as does its ecological location. This allows Panshin, for example, to construct keys for the identification of the mangrove woods he studied, often distinguishing closely related species. Some caution should be exercised in using these keys, because quantitative differences are often used and might refer only to the rather limited range of samples studied by Panshin. A more comprehensive evaluation using qualitative characters might be attempted and would be useful to ecologists and paleobotanists.

7 Root Systems

Aerial Roots

A feature of more highly specialized mangroves is that they develop some part of the root system exposed to the atmosphere, at least at low tide. The most obvious are the looping above-ground roots of *Rhizophora*, which impart the characteristic physiognomy to much of the mangal (Plate 1G). That these structures are truly roots has been well established, based on careful developmetal study (Gill and Tomlinson, 1971a). That they can be anything else, such as a "rhizophore" homolog, is a persistent idea (de Menezes, 2006), despite all evidence to the contrary. This unusual example of the root system of trees in tidal forests is built into the definition of "aerial root" given by Gill and Tomlinson (1975): "a root which is exposed to the atmosphere at least part of the day." The latter qualification is not necessary for most aerial roots in tropical plants, which are continually exposed, as in the initial descending roots of epiphytic aroids (Huggett and Tomlinson, 2010). Ascending or above-ground parts of aerial roots are common in trees of tropical swamp forests where they are continually exposed, that is, not subject to diurnal changes in water level. Descriptions and illustrations of such root systems can be found in Ogura (1940), Corner (1978), and Jenik (1978). In the more seaward mangroves, however, aerial roots are covered at high tide. This turns out to be important to their functioning. Physiologically, the need for the development of an above-ground part of the root system in trees in the tropics is related to the anaerobic nature of substrates in swamps and the need for atmospheric oxygenation of the absorbing system as a whole, rather than oxygenation via the absorbing root surface alone. The presumed mechanism is discussed in further detail later. The relative restriction on development of the within-substrate ("underground") portion of the root system is demonstrated by Ong et al. (2004) for *Rhizophora apiculata*; root biomass, as a percentage of total biomass above ground, is 10 to 20 percent versus only 5 percent below ground. In general, *Rhizophora* root penetration within a soft substrate is limited to about 1 m (Fig. 7.4). A survey of root systems in mangroves is provided by Troll (1930, 1943), based largely on his own extensive field studies.

In describing root systems, most descriptions are naturally concerned with the architecture plan of what is observed, rather than the actual layout developed under natural condition. By analogy with shoot systems, the distinction can be made between the "architecture" and "reiteration" (Hallé et al., 1978). The latter process understandably has largely remained unexplored in subterranean systems, but aerial systems, as in mangroves, can clearly overcome this limitation. Here the initial descriptions include diagrams of "architecture," as in Figures 7.2–7.5, whereas Figure 7.6 is a plot of an actual system. Many of the processes that are visible in the reiterative aspects of *Rhizophora* roots can be demonstrated to be normal features of underground roots in all plants, but are scarcely dealt with in the morphological literature. Mangroves thus become informative of approaches still to be deployed universally in plant root systems. There are several types of aerial roots in mangroves. Only general features that demonstrate common characteristics are described here.

Stilt Roots

This refers to the branched, looping aerial roots that arise from the trunk and lower branches of *Rhizophora* (Plates 1F, 11D, and schematically in Figure 7.2A). The description of them as "stilt" roots applies to them in depauperate trees (Plates 1B, 13H), whereas they function as flying buttresses in mature trees (Plate 13A), because the trunk at the base of the tree has limited secondary growth and characteristically it becomes obconical with age. Most secondary thickening at the base of the tree occurs in the aerial roots rather than the trunk. Méndez-Alonzo et al. (2015) have studied the biomechanics of these flying buttresses. Their support can account for the relative absence of leaning stems, although they are not produced radially in relation to prevailing winds. A suggested initial benefit is the ability to grow taller in early stages of development, that is, to be more slender than a free-standing tree. In Rhizophoraceae, stilt roots of the same type are developed to a limited extent in *Bruguiera* and *Ceriops* in the sapling stage, close to the stem base, becoming shallow buttresses in old trees. Stilt roots occur sporadically in other mangroves and are a common feature, for example, of *Avicennia* species, notably *A. alba* and *A. officinalis* (Fig. 7.1; Plate 13E). In other *Avicennia* species (e.g., *A. germinans*), aerial roots growing from the trunk seem to be partly a response to wounding. The structure and development of the root system of *Rhizophora* have been investigated in detail by Gill and Tomlinson (1971a, 1977). They emphasize (following Docters van Leeuwen, 1911) the opportunistic nature of the spread of the roots, because branch roots that extend the system arise either at a damaged apex or behind the apex when the axis becomes rooted distally.

Pneumatophores

Erect roots that are some form of direct upward extension or appendage of the subterranean root system are called "pneumatophores" in a descriptive sense; the term has no precise morphological connotation. Quite different developmental types can be recognized, each characteristic of a genus or group of species. The essential features are shown in Figure 7.2. In this type of root regeneration of new cable roots can come from the base of erect pneumatophores (Plate 13D). This provides a mechanism for relocating the cable system at a different level if the substrate depth increases,

Figure 7.1. *Avicennia officinalis* (Avicenniaceae). Aerial stilt roots, a common feature of this species; Ponggol, Singapore (see Plate 11D).

by sedimentation, or decreases, by erosion. Troll and Dragendorff (1931) regard this adaptability to be of prime importance ecologically.

Pneumatophores in **Avicennia** *and* **Sonneratia.** The pneumatophores are erect lateral branches of the horizontal roots buried in the substrate. The first-order roots ("cable roots") originate at the base of the trunk and extend horizontally for long distances through the substrate, whereas the visible parts of the root system are pencil-like erect branches spaced at regular intervals along the first-order roots (Fig. 7.2B). In *Avicennia*, pneumatophores are of limited height (usually less than 30 cm) and develop little secondary thickening. In soft substrates, the extent of the cable component is mapped by the location of the visible pneumatophores (Plate 11E). The early literature on *Avicennia* is summarized in Chapman (1940, 1947).

In *Sonneratia*, early descriptions include those of Goebel (1886) and notably Troll and Dragendorff (1931), represented by Figure 7.3. Here the pneumatphores have a much longer period of development and undergo secondary thickening, thus they become quite tall (Plate 11B), exceptionally up to 3 m (Plate 12F). These mature structures are mechanically rigid, their material properties such that they resist tidal bending, as measured by Zhang et al. (2015). The pneumatophores usually remain unbranched, but are capable of branching, either when damaged, or as a response to sedimentation. The surface texture in these two types differs: in *Avicennia*, the surface remains smooth and spongy; in *Sonneratia*, a characteristically flaky bark develops in younger roots, whereas in older roots the bark is smoother. In both genera, the roots include chlorophyll in the subsurface layers.

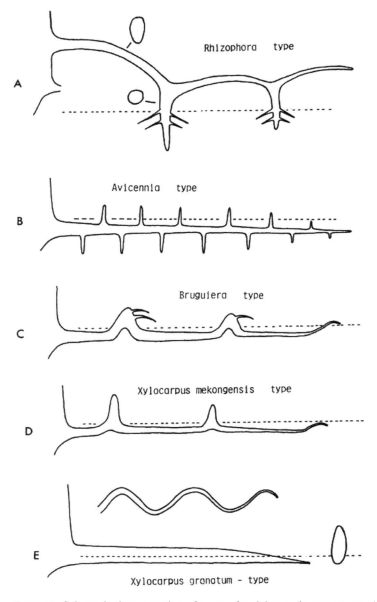

Figure 7.2. Schematic demonstration of types of aerial roots in mangroves. All have developed from left to right. Dotted line, substrate level. Some or all of the roots may be inundated at high tide.

Pneumatophores in **Laguncularia.** This genus develops erect, blunt-tipped pneumatophores under certain circumstances, rarely exceeding a height of 20 cm (Jeník, 1970). They are more frequently branched than in *Avicennia* and *Sonneratia*. The pneumatophores seem facultative in their development; therefore, populations in some localities lack them.

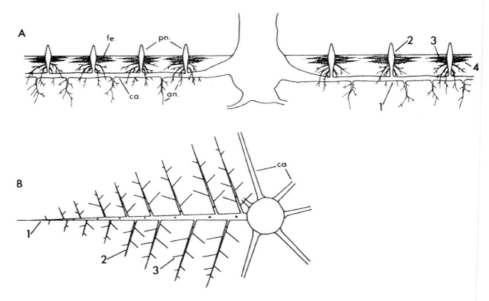

Figure 7.3. Root system of *Sonneratia*. Numbers refer to branch orders. A. Profile schematic view; the ultimate- (fifth-) order branches are too fine to appear. B. Plan schematic view. Specialized root structures are referred to as cable roots (ca), pneumatophores (pn), anchoring roots (an), and feeding roots (fe). Order number does not determine the type of root (e.g., pneumatophores can be second-or third-order erect branches; second-order branches may be pneumatophores, anchoring roots, or high-order cable roots). Note that second-order cable roots must intersect extensively close to the trunk if all roots developed according to this scheme. (After Troll and Dragendorff, 1931.)

Root Knees

Root knees in **Bruguiera** *and* **Ceriops.** In these genera, the horizontal roots periodically reorient as they grow through the substrate away from the parent tree so that the tip forms a pronounced loop before continuing its horizontal growth. This process is shown in sequence for both genera in Plate 12B, D. At the site of the loop, eccentric secondary thickening occurs mainly on the upper side so that a blunt, knoblike structure is raised above the substrate surface. Branching of the root system is largely restricted to these "knees." The branches both proliferate new horizontal roots and form subsurface anchoring roots. The system collectively is very distinctive and its features are evident where some substrate erosion has occurred (Plates 11A, 12C). By the repeated development of loops, a single horizontal root develops a series of additional outgrowths at regular intervals. The size, shape, and frequency of root knees are characteristic for each species, although somewhat variable according to substrate condition.

Pneumatophores in **Lumnitzera.** The root knees of *Lumnitzera* are rather inconspicuous and seem to be derived mainly from laterals that execute looping growth but with little secondary growth. They may be thought of as structures intermediate between pneumatophores and knees.

Root knees in **Xylocarpus.** Localized erect outgrowths of the upper surface of the horizontal root system are a feature of *Xylocarpus mekongensis* (Fig. 7.2D). Here the structures are secondary and produced entirely by localized cambial activity on the upper surface of the horizontal root (Plate 12E). This should be contrasted with *Bruguiera* and *Ceriops*, where the knee is initiated as the result of primary growth. Root knees in this species of *Xylocarpus* may grow to a height of 50 cm. The knobby aerial structures of *Camptostemon* roots correspond closely to this type (Troll, 1933a).

Plank roots and buttresses. This category includes a number of comparable types in different species. *Xylocarpus granatum* (Fig. 7.2E) is contrasted with *X. mekongensis* because the horizontal root becomes extended vertically by eccentric cambial activity throughout its length, but without forming pneumatophores. As these roots extend, they become sinuous in their course; thus the result in the mature tree is a series of wavy, plank-like structures growing away from the base of the trunk (Plate 11C). When expressed profusely, such a root system becomes an acute impediment to foot traffic (Plate 13C).

Smaller plank buttresses occur in species of *Heritiera* (e.g., *H. fomes* and *H. littoralis*) as narrow surface outgrowths of the base of the trunk. These are transitional in form to the plank buttresses of the trunks of many rain forest trees.

This transition leads to a discussion of the distinctive method of buttress development in a number of mangroves, notably *Pelliciera* and *Ceriops*. Aerial roots of *Pelliciera* form fluted buttresses from aerial roots initiated in early stem development (Plate 13G), but with gradual fusion of adjacent, but originally independent roots (Plate 19A). A similar process is initiated in seedlings of *Ceriops* (Plate 13F) and also eventually forms buttresses of fused plank-like construction (Fig. B.60). In these two examples, the submerged root system is highly contrasted, unelaborated in *Pelliciera* but developing pneumatophores in *Ceriops*.

Mangroves without Aerial Roots

The following mangroves normally lack any elaborated aerial part to the root system: *Aegiceras, Aegialitis, Excoecaria, Kandelia, Osbornia, Scyphiphora,* and *Nypa*. However, certain structures may be developed to aerate the root system. The base of the stem in *Aegialitis* is enlarged and fluted, with a very spongy texture. *Kandelia* tends to develop aerial roots in some limiting environments (Hosakawa et al., 1977).

In *Nypa*, the petioles of individual leaves function like giant pneumatophores growing close to the stem apex, facilitated by a somewhat spongy texture of their ground tissue and complimented by the spongy ground tissue of the rhizome itself (Chomicki et al., 2014). After the distal portion breaks off, this aerating function continues in the persistent leaf base. Breakage, which sheds the distal portion of the leaf, always occurs at a precise level on the petiole via an abscission zone just below the rachis, and hence at a quite consistent height.

Mangal associates do not develop any aerial roots, with the exception of the palms *Phoenix*, *Raphia*, and *Oncosperma*, which usually develop slender pneumatophores. This condition is characteristic of many palms that grow in wet sites (e.g., *Metroxylon* and *Mauritia*).

Root Functions

The preceding description largely refers only to the aerial portion of the total root system, which has received much emphasis because it forms such a conspicuous element of mangrove forests. If one examines *in toto* the root system of the more specialized mangroves, three basic structural and therefore functional components can be recognized, even though these components may have a different morphological origin in different species. The equivalence is shown by different kinds of shading in Figure 7.4. It is the existence of these three components, rather than just the aerial portion, that is the most distinctive feature of mangroves. The functional correlations between the various parts of the root system therefore need emphasis.

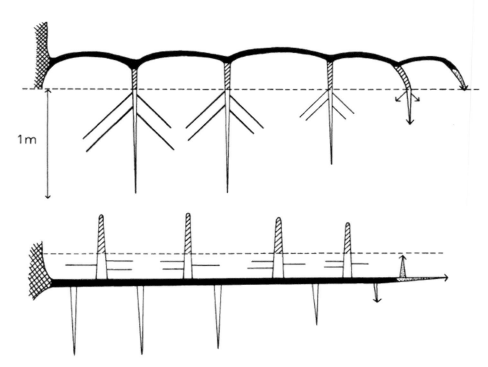

Figure 7.4. Simplified representation of two major contrasted types of root system in mangroves (top, *Rhizophora* type; bottom, *Avicennia* type) to show common structural features. Key to symbols: cross-hatched area, trunk; black area, cable root; hatched area, pneumatophore; white area, anchoring roots; fine lines, absorbing roots; stippled and arrows, growing regions; dashed line, substrate level. (After Gill and Tomlinson, 1977.)

Aerating component. The erect part of the root system, which, in all specialized mangroves, projects above the root system, relates to gas exchange. As we have seen, this part is produced differently in different root types: the columns of the looping root system of *Rhizophora*; the pneumatophores of *Avicennia, Laguncularia,* and *Sonneratia*; the root knees of *Bruguiera, Ceriops,* and *Xylocarpus mekogensis*; and the plank extension of the roots of *Xylocarpus granatum*.

Absorbing/anchoring component. This component has a dual function: it forms the basis of anchorage for the whole system and is the site of absorption. In most mangroves, this is the result of the development of appendages that grow downward from the horizontal system. They are direct first-order laterals in *Avicennia* and *Sonneratia*, for example, but second-order laterals from the region of the root knee in *Bruguiera* and *Ceriops*. In *Rhizophora*, each unit is the direct extension of one loop, which penetrates the substrate (Fig. 7.4, top panel). The absorbing and anchoring components can be morphologically segregated, because the more massive downward-directed root appendages provide the main anchorage; the finer ultimate orders of the branching system are the immediate sites of absorption, as with the capillary rootlets of *Rhizophora*. This system is the equivalent of that which is initiated by the seedling (Fig. 7.5).

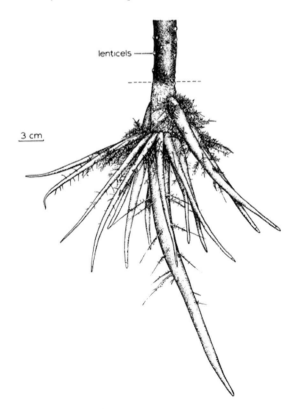

lenticels

3 cm

Figure 7.5. *Rhizophora mangle.* Submerged portion of the root system developed from the anchored aerial root; note the lenticels on the "column" above the substrate level (dotted line). (From Gill and Tomlinson, 1977.)

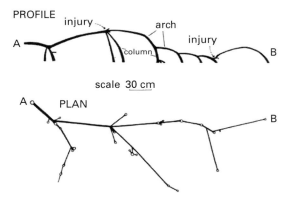

Figure 7.6. *Rhizophora mangle.* Sector of the aerial root system in profile and plan, growing from the trunk (A) to the most recently attached but still unbranched segment (B). Open circles, insertion of column into substrate. Branching of cable roots may be due either to injury (a form of "reiteration") or subsequent to rooting. The branch insertion distinguishes between two regions referred to as "column" and "arch." (From Gill and Tomlinson, 1977.)

Cable component. This is the extended horizontal part of the root system that unifies the aerating and absorbing/anchoring part of the root system. In *Rhizophora*, this system is above ground and is a direct result of extension of the primary aerial roots (Fig. 7.6) by branching processes that can be seen as reiteration (opportunistic). In all other mangroves, the cable system is underground (Fig. 7.4, bottom panel). In both examples, the proliferation of the root system and its extension may be mostly opportunistic, as in the earlier comment about *Sonneratia*. It should be emphasized that the absorbing or anchoring and the cable components exist in root systems of all plants; the aerating component is the frequent additional component in swamp or mangrove forest species.

Integrated Root Functions

Disregarding the finer details of root architecture, which have diverse morphological origins, recognition of the three basic components allows an overview of the root architecture in functional terms. This involves absorption (of water and mineral nutrients), aeration, and transport.

The sites of absorption are the ultimate branches of the descending roots; their immediate parent roots, the more massive portion of the descending system, have the primary mechanical function of attaching the whole system to the substrate. In mangroves, this absorptive system is characteristically shallow, its depth presumably limited by the lack of available oxygen for deep-growing roots. Localized sites of aeration are provided by that portion of the root system that extends into the atmosphere, and there has to be transport from aerated to unaerated regions. Commonly, there is a close structural association between ascending and descending components (e.g., *Rhizophora, Bruguiera, Sonneratia*), such that the pathway for oxygen diffusion is minimized.

The ascending and descending components are linked by the cable system, initially simply a developmental necessity, but finally providing the transport system so that water and nutrients can move from distal sites of absorption to the trunk (in the xylem), while assimilates needed for metabolic processes are moved from the trunk to the distal parts of the root system (in the phloem). Viewed in this way, the root system of mangroves can be appreciated as an integrated whole. The special requirements for existence in an anaerobic environment are met. The further specialization for absorbing fresh water from seawater by the exclusion of salt is a micromorphological mechanism at or close to the absorbing surface that does not necessitate major morphological modification.

Gas Exchange

The general concept that the above-ground component of the root system in mangroves ventilates the buried portion for respiratory purposes continues to be supported by experimental methods. That the composite root system is permeable to the mass flow of gases, with atmospheric exchange occurring through lenticels in the pneumatophores (aerial vents or "chimneys"), is evident from the extensive internal gas space system that develops largely by the enlargement of intercellular spaces supplemented by cell breakdown. Continuity through the gas space system can readily be demonstrated by blowing air through the root. Resistance can be accounted for, in part, by known anatomical constrictions, as at the level of attachment of branch roots to the main axis in *Avicennia* (Chapman, 1940). Morphological analysis has demonstrated a uniform structural system of vents, absorbing roots, and connecting roots, even though there are several contrasted developmental procedures for achieving this structural set. The spacing of vents can be highly deterministic, as in *Avicennia*, where pneumatophores are spaced regularly and are 15 to 30 cm apart (Plate 11E). The total extent of these vents is enormous. Scholander et al. (1955) calculated that a single tree of *Avicennia* only 2 to 3 m tall had well over 10 000 pneumatophores. As they point out, such an extensive structural specialization requires some functional explanation.

Development of the air-space system is a possible combination of those termed *lysigenous* and *expansigenous* by Seago et al. (2005). The resulting gas space can be measured in a cut length of root by weighing it before and after artificial waterlogging under a partial vacuum. The weight difference can be converted to a volume difference, representing the volume of displaced gas, provided there is no significant shrinkage or swelling of tissue. For example, values of about 40 percent of total root volume have been measured (Gill and Tomlinson, 1971a). Purnobasuki and Susuki (2004) defined root porosity as the percentage of air spaces to total root cross-sectional area, measured digitally, in *Sonneratia alba*, and found values up to 60 percent. This refers to the primary stage of development of all root categories with a wide cortex; secondary growth reduces this value considerably. Even so, Sun et al. (2004) found intercellular spaces in the secondary xylem rays of pneumatophores in the same species, and discussed this feature in terms of aeration or water storage. The root wood is unusual in that it is composed largely of vessels.

Because diffusion is inadequate to account for the extent of gas exchange of the requisite volumes and over the distances involved, some mechanism for mass flow has been sought. Westermaeir (1900) suggested that alternate compression and relaxation of roots, caused by tidal fluctuation, could work a simple pumping or exhalation and inhalation mechanism. However, Scholander et al. (1955) considered the pneumatophores insufficiently compressible. Furthermore, measurements showed that the gas pressure in the roots was, for the most part, reduced at high tide, rather than increased, as would be required by Westermaeir's compression hypothesis. Scholander and his colleagues suggested, and to a large extent substantiated, an alternate theory dependent on changes in the partial pressure of different gases in the internal gas space system during the tidal cycle.

These authors showed that the reduced pressure occurred at high tides when the pneumatophores were completely covered with water. Reduced pressure is the result of the respiratory consumption of oxygen when the lenticels are closed and the relatively high solubility in water of released carbon dioxide. The maximum suction (negative pressure) generated was equivalent to about a 55-cm water column. The vents are not flooded by the hydrostatic pressure of the water above them, because the suction is insufficient to overcome the surface tension of the wet intercellular spaces at the lenticel surface. Lenticels remain continuously open because they constantly produce new cells, forming a very friable tissue. This need for continuous cell activity should be seen as a component of the endless metabolism that restricts mangroves to the tropics. The suction developed is proportional to the length of time the pneumatophores remain submerged because respiration is continuous. When the pneumatophores eventually become uncovered as the tide falls, the pressure imbalance between the external and internal atmosphere is restored as air is drawn into the root. The proposed mechanism operates only in aerial roots that are regularly inundated. In the higher reaches of the mangal where the tidal influence is minimal, the mechanism cannot operate, and roots are presumably aerated (or oxygenated) by diffusion. Nevertheless, Scholander et al. showed a slight fluctuation in the gas composition of two roots of *Avicennia* above high tide. Aeration of above-tide roots may be less critical if the substrate itself is better aerated. Some mangroves (e.g., *Laguncularia* and possibly *Lumnitzera*) are facultative in their development of pneumatophores; this may be controlled by the extent to which the substrate is anaerobic.

Strong supporting evidence for the tidal suction mechanism comes from the observation that the relative concentrations of carbon dioxide and oxygen change in the manner required ("molar gas reduction"). Oxygen concentrations can change as much as 10 percent during the tidal cycle in pneumatophores that become inundated. Artificial and prolonged sealing of the lenticels by vaseline or wax reduces the oxygen concentration to very low values; the roots become asphyxiated. This may cause widespread mangrove death after storm tides, where high water levels may be impounded for abnormally long times. At this point, the reader is likely to ponder this statement in relation to projected sea-level rises brought on by global warming.

Scholander et al. (1955) emphasize that ventilation of the root system through the lenticels of the pneumatophores is essentially a continuous process and is modified only by the tide in the significant ways they measured. Simple diffusion remains an important component of the process, as is suggested by the observation that the anchoring and descending absorbing roots are relatively shallow and do not penetrate the substrate much beyond a meter, the limit shown in Fig. 7.4.

The mechanism of root ventilation cannot be discussed in terms of an isolated root segment; the root system as a whole is highly integrated. Further experiments by Scholander and his associates show that a single pneumatophore can serve as a port of gas entry for an extended portion of the system from which other pneumatophores have been removed. The unrelated observations of Dacey (1980, 1981) of what were termed "internal winds" in the yellow water lily (*Nuphar luteum*) are noteworthy. An internal cycling or mass flow of gas was demonstrated from young, newly emergent leaves down into the rhizome and back to the older leaves via the long petioles. The mechanism depends on the development of a positive pressure in young leaves as they are heated by the sun, and the relative impermeability of the leaf surface to gas exchange. The pressure generated is vented via the internal gas space through older leaves, which have much lower resistance to gas exchange. Apart from eliminating respiratory carbon dioxide, the mechanism circulates methane (CH_4) absorbed by the roots from the sediment. Because the internal wind is a simple consequence of the physical properties of the system, although of benefit to the plant, it could operate in mangrove roots if they developed some differential permeability. However, emergent roots do not present a large surface area to the sun. Nevertheless, somewhat similar conditions could develop where submerged and emergent pneumatophores are attached to different ends of the same cable root.

Technical methods. Because the apparatus used by Scholander et al. is simple and easily reproduced, diagrams are included here. Gas is sampled from the substrate portion of the root with a hypodermic needle, and a distant plunger is inserted into a root that is first excavated and then reburied (Fig. 7.7). Samples can be drawn off at

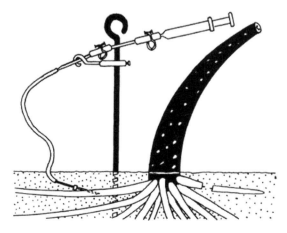

Figure 7.7. Apparatus used by Scholander et al. (1955) to extract gas samples from intact roots of mangroves.

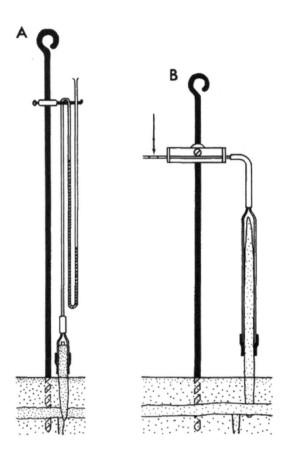

Figure 7.8. Apparatus used by Scholander et al. (1955) to measure gas pressures in pneumatophores of *Avicennia*: A. Manometer attached tightly to a cut pneumatophore records pressure changes compared to atmosphere. B. Sleeve fitted tightly around an intact pneumatophore measures changes in rate of gas exchange via the recording bubble (arrow).

regular intervals. For measuring changes in gas pressure, a simple water manometer is attached tightly to the stump of a pneumatophore, the woody cone of which is greased to make it airtight, so that gas exchange occurs only through the aerenchymatous cortex (Fig. 7.8, left). For measurements of volume changes, the apparatus in Figure 7.8 (right) is attached to an entire pneumatophore in such a way that the collar is airtight but the enclosed lenticels are not obstructed. The horizontal calibrated tube has a kerosene indicator drop.

Alternative theories. The general conclusion that aerial roots have a demonstrable ventilatory function seems well founded, but has not always been supported. Troll and Dragendorff (1931) present views that differ with this standard interpretation. Working primarily with *Sonneratia*, they made extensive observations on the respiratory mechanism and concluded that the major function of the root system is to adjust the level of the absorbing and anchoring system to changes in the substrate level, especially those due to progressive sedimentation. The method of development of the root framework undoubtedly permits the necessary degree of plasticity, and instances of adjustment to depth are easily seen. However, the root construction seems overelaborate for the adjustments necessary. Most mangrove root systems are not subjected to regular or extensive changes in substrate levels. Similar pneumatophores occur in freshwater swamp-forest species that are rarely subjected to sedimentation. Terrestrial trees adjust to changes in substrate levels quite frequently, but without developing specialized tiering mechanisms. It may be simplest to recognize that structures that develop primarily in relation to root aeration are also beneficial in allowing depth adjustment.

Root Anatomy

The anatomy of mangrove roots has been extensively investigated in relation to development and function, but also requires a comparison of different branch orders that clearly have dissimilar functions. Summaries are provided by Troll (1943) and Chapman (1976); individual articles of particular relevance are by Goebel (1886), Schenk (1889), Brenner (1902), Liebau (1914), Chapman (1940, 1947), and Baylis (1950). Most information is available for the above-ground portions of root systems, as these are relatively accessible.

In all roots, a basic pattern of root-like characters can be recognized: root cap, endogenous origin of lateral roots, exarch protoxylem, and alternate primary phloem and xylem strands. A common modification is an enlarged, polyarch stele with a wide parenchymatous medulla (Figs. 7.9, 7.10). Sometimes some normal root features are

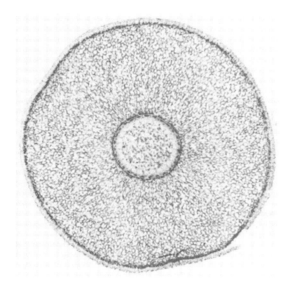

Figure 7.9. *Rhizophora mangle.* T.S. of subterranean root to show broad lacunose cortex, narrow medullated stele (x12). No trichosclereids are developed.

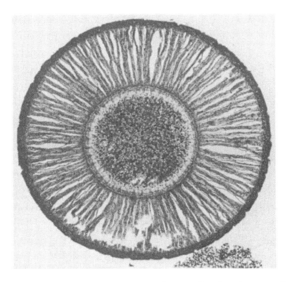

Figure 7.10. *Bruguiera gymnorrhiza*. T.S. of underground anchoring root to show well-developed aerenchyma in cortex and wide, medullated stele (x10).

obscured and have been misinterpreted. In *Rhizophora*, superficial observation suggested that the aerial root had bicollateral primary xylem and phloem strands and endarch protoxylem, leading to the suggestion that these organs combined features of stem and root or were even novel organs, comparable to the "rhizophore" of *Selaginella* (Pitot, 1958; de Menezes, 2006). This mistake is easily avoided by a more dynamic approach. Developmental study (Gill and Tomlinson, 1971a) shows that the aerial root of *Rhizophora* is entirely normal in the early stages of vascular differentiation, with exarch protoxylem (Fig. 7.12A), but protoxylem appears mesarch and even endarch as development proceeds (Fig. 7.12B). This unusual development is correlated with an extended region of elongation behind the apical meristem proper, whereas underground roots have a short zone of apical elongation. That the aerial root is actively growing over a distance of about 10 cm is reflected in its plastic or rubbery distal texture. The suite of characters developed by the aerial roots of *Rhizophora* is possible because (1) they are not mechanically constrained by the substrate, and (2) they are not directly absorbing organs. The initial exarch organization in this sense is almost atavistic. Aerial roots, therefore, are appropriately modified for life above ground. They have a well-developed periderm, trichosclereids are abundant, and secondary thickening is pronounced. When the same roots reach the ground, they undergo a remarkable transformation that adapts them to a subterranean existence, illustrated here by the contrast of aerial (Fig. 7.11A) versus buried (Fig. 7.11B) root apices. As the root extends into the substrate, the extension zone becomes short, primary vascular differentiation is normal, the phellogen is little developed, trichosclereids are absent, and secondary thickening is limited. In addition, the ground tissue becomes decidedly spongy.

   The contrasted suite of characters developed by the aerial root of *Rhizophora* before and after it becomes buried develops in varying combinations in the roots of other

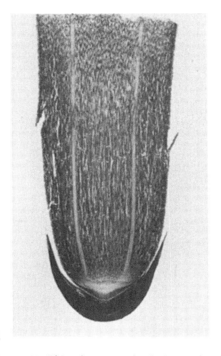
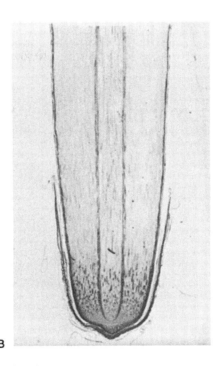

A   B

Figure 7.11. *Rhizophora mangle*. A. Apex of aerial root in LS. Note pronounced root cap (x8). B. The same root after it becomes submerged. Note the marked change in shape of the root cap and the absence of tannin and trichosclereids.

mangroves, where there is a more consistent differentiation between roots of different orders. A phellogen-producing cork is present in above-ground root parts. In young pneumatophores of *Sonneratia*, the cork exfoliates in a characteristic way. The cortex is often lacunose; in *Avicennia*, the cells of the middle cortex become lobed, creating enlarged intercellular spaces, and they are supported by bands of wall thickenings. Similar thickening bands occur in some Rhizophoreae. The aerenchyma of the roots of this and other genera developmentally probably belong largely to the "expansigenous" type described by Seago et al. (2005), and may be compared to the studies of Purnobasuki and Suzuki (2004, 2005) in *Sonneratia alba* and *Avicennia marina*.

Secondary thickening in the roots of mangroves is roughly proportional to the order number, so that cable roots have most wood and ultimate orders none. There are many exceptions to this generalization. Pneumatophores of *Avicennia* have limited secondary growth; those of *Sonneratia* are made up of largely secondary tissue. Root knees in *Bruguiera*, *Camptostemon*, *Ceriops*, and *Xylocarpus granatum* are composed of mostly secondary wood. Lenticels are important components of the aerial roots of mangroves for aeration. They are especially conspicuous in Rhizophoreae, on the columns of the root system of *Rhizophora* (Fig. 7.5) and in the root knees of *Bruguiera* (Plate 12D). In *Ceriops* (Plate 12B), the surface layers of the knee exfoliates bark extensively. Aerial roots of mangroves may also undergo extensive secondary changes with age

A

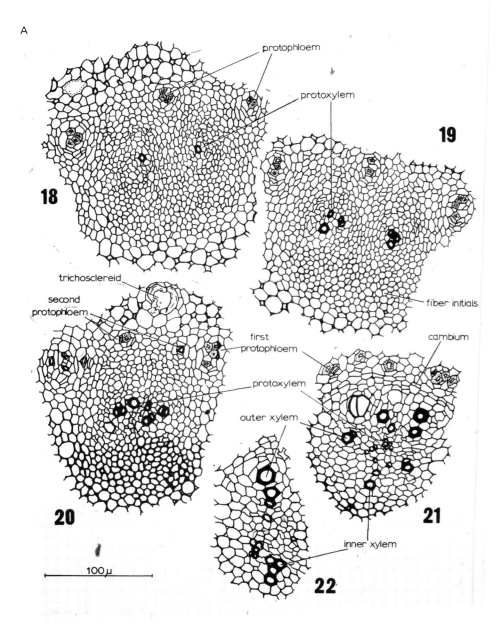

Figure 7.12A, B. Vascular development in the aerial root of *Rhizophora mangle*. (After Gill and Tomlinson, 1971a). A. Early differentiation of conducting tissues. Protophloem and protoxylem originate on alternate radii, correct for a root. Protoxylem is exarch at this stage, correct for a root. B. In later differentiation, protoxylem is disrupted and later xylem elements appear endarch, as in a stem. This condition arises as a consequence of the considerable length over which extension growth occurs (up to 20 cm). These observations eliminate any notion that this organ, despite its peculiarity, is anything but a root.

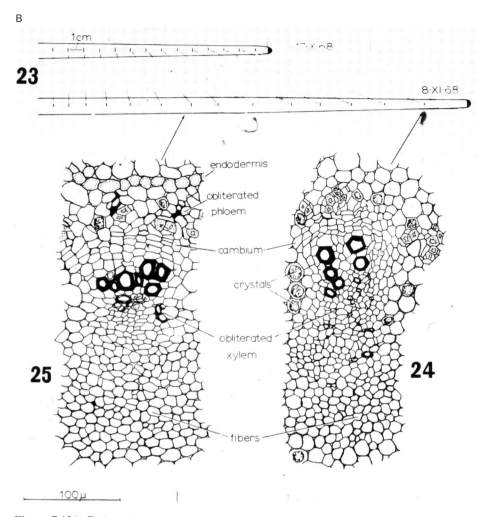

Figure 7.12A, B. (*cont.*)

related to their unusual milieu. Roots may undergo alternate submersion and exposure, adjust to internal enlargement, and effect gas exchange when exposed. Apart from the normal cork, extensive secondary cortical tissues are often produced internally by the phellogen.

Less is known about the anatomy of higher-order roots. In *Rhizophora*, there is a progressive reduction in diameter and complexity with each order of submerged roots; the ultimate order or "capillary rootlet" has a very simple organization. It has a diarch stele and a narrow cortex and the epidermis develops no root hairs (Attims and Cremer, 1967). The absence of root hairs seems to characterize all strict mangroves; it is a feature common to different kinds of aquatic plants. Knowledge of the structure of the epidermis and endodermis of the ultimate branch orders of roots is important, as these are the most likely sites for the ultrafiltration mechanism that makes life in seawater possible. The study by Moon et al. (1986) is the most important to date. They used fluorescent tracer dyes to contrast apoplastic versus symplastic pathways into the xylem sap. An apoplastic barrier is provided by a superficial exodermis and the interior endodermis. These are both provided with a suberized or ligno-suberized wall layer, as can easily be demonstrated by the use of routine histochemical tests. In view of the absence of root hairs, the endodermis is the most likely layer controlling entry of nutrients, because it would become the effective absorbing layer and also the site of salt exclusion. An important emphasis is that disturbed roots differ physiologically from undisturbed roots, although this facet of root biology is not always appreciated.

Fouling Mechanisms and the Rhizosphere

The highly visible part of the aerial root system of mangroves, most pronounced in *Rhizophora* (Plate 1G) has led to the consideration of how it may become occluded by epibiontic organisms, as a natural event. This may have either beneficial or detrimental effects. Snails move about and probably have little effect (Plate 2A). Most obvious is the colonization by oysters, leading to the canard of "oysters growing on trees" (Plate 2B, C), but in some communities they are of economic importance. However, when excessive, this biota can lead to mass mortality in hypersaline conditions (Orihuela et al., 1991). More significant but less obvious is the activity of microorganisms in the rhizosphere of ultimate rootlets, that is, the region immediately surrounding the root surface. An association with nitrogen-fixing cyanobacteria may be seen as beneficial, although it largely adds fixed nitrogen to the ecosystem as a whole (Sheridan, 1991, 1992). Although the biota of the mangrove substrate leads to anoxia, it can be mollified by oxygen supplied externally through the root system (Nickerson and Thibodeau, 1985; Thibodeau and Nickerson, 1986). Another aspect is the biogeochemical relationship between the rhizosphere and trace elements (Lacerda et al., 1993). These intimate associations are clearly disrupted by toxic wastes in seawater, most spectacularly by oil-spills, which receive national attention. In mangal communities, mangroves are likely to be asphyxiated by occlusion of lenticels on aerial roots. Finally, it is suggested that high concentrations of polycyclic aromatic hydrocarbons may increase mutation rates (Jiménez et al., 1985; Whitten and Daminik, 1986; Klekowski et al., 1994b, c).

8 Water Relations and Salt Balance

High salt (sodium chloride) concentrations are inimical to the growth of most plants (glycophytes) either because they increase the water potential of the soil to levels beyond which the plant can absorb water through the roots, or more usually because absorbed salts are toxic. This is a major problem in irrigated soils where salts may accumulate. Knowledge of the salt-tolerance mechanism is therefore of commercial significance (Greenway and Munns, 1980). In contrast, halophytes, including mangroves, develop a tolerance to soil salinity because they both maintain a high cellular water potential and are relatively insensitive to salt toxicity. The mechanisms of salt tolerance are complex and variable and involve factors such as ionic potentials across membranes, osmotic relationships, enzyme activation, and protein synthesis. For example, the toxicity of Na^+ anions at high concentrations is a result of their effect on enzymes. An extended early discussion of the subject is provided by Waisel (1972) and Flowers et al. (1977).

A comprehensive discussion of all halophytes and their adaptation to saline conditions includes mangroves as a special case in which the salt concentration throughout the tree is added to a discussion of hydraulic architecture (Reef and Lovelock, 2015). How this affects osmolality and ionic composition has been measured experimentally in xylem sap (López-Portillo et al., 2014).

Some direct evidence for adaptation to a saline environment comes from the progressive development of leaf succulence in most mangroves, as described later. Nevertheless, mangroves have a fairly low limit to their salt tolerance compared with strict, herbaceous halophytes, which are usually much more succulent but not very tall. A landward transition to salt marsh is common next to mangal. Where salt persists at very high concentrations, no plants grow and a "salt desert" results. The interrelation between mangal and salt marsh in the tropics is discussed extensively in Chapman (1977a).

Salt and Plant Growth

It has been suggested that mangroves have a requirement for sodium chloride; they have been described as "obligate halophytes" (Wang et al., 2011) that do not grow well in the absence of salt. Does this, therefore, invalidate results from mangroves grown experimentally in fresh water? This question was discussed rather inconclusively by Benecke and Arnold (1931). Flowers et al. (1977) cite evidence that enzymes in vitro are inhibited by salt concentrations only half of that in the leaf cells of halophytes. This is evidence for a nonobligatory relationship between plant and salt, as it shows that enzyme systems neither require salt nor are particularly resistant to it. Stern and Voigt (1959) obtained better growth of *Rhizophora* seedlings in salt solutions of low concentrations compared with those in fresh water, suggesting a requirement for salt, but this seems to be contradicted by later research (Hanagata et al., 1999). Connor (1969) found that optimum growth of *Avicennia marina* occurred in culture solutions with a sodium concentration about half that of seawater. A similar study on the same species also indicates a requirement for salt in the development and function of the shoot system (Ngyen et al., 2015). A special example is seen in *Sonneratia*, where it was shown that two species differed in their salt sensitivity, *S. alba* being more salt tolerant and *S. lanceolata* more salt sensitive. This difference in salinity optima, that is, the salt concentration at which each species is most successful, can account for their different ecological status, *S. alba* occurring more seaward to *S. lanceolata*. In Connor's studies, higher than normal concentrations of calcium and potassium depressed growth. This was specifically an ion effect and not due to osmotic potential alone. Cell mechanisms can involve other osmolitica that can maintain turgor. However, we note that in our own experiments on seedlings of Rhizophoraceae, they were all grown in fresh water (Tomlinson and Cox, 2000).

Most mangroves seemingly grow very well in fresh water, and some penetrate considerable distances inland along riverbanks where water is permanently fresh and tidal fluctuations are small or absent (Plate 1A). Salt water is denser than fresh water; hence, mangroves may exist inland in a lens of fresh water above one of salt water, as in the Florida Everglades. It is usually assumed, therefore, that mangroves are excluded from terrestrial communities by competition with other kinds of plants that do not carry the burden of features that are adaptive only in stressed environments. Mangroves toward their ecological and geographical limits become stunted; for example, in the transition from salt- to freshwater marsh communities in the Florida Everglades (Plate 1B). We do not know the reason for this stunting, especially where water might seem to be readily available. More obviously, the decline in the stature of mangroves along gradients of increasing salinity (Soto and Jiménez, 1982) shows the increasing stress of high salt concentrations. The simplest conclusion is that mangroves tolerate salt to a limited degree. The usual figure quoted is 90 0/00 of NaCl on the basis of the maximum salinity of soils in which mangroves will grow. Soto and Jiménez (1982) suggest values up to 155 0/00 with annual averages of 100 0/00. Hanagata et al. (1999) give details of how changing salinities affect growth responses of different kinds, such as photosynthesis and transpiration.

Sap Ascent

The overall water relations of mangroves are paradoxically compromised by their growth in abundant seawater because they have to deal with an external osmoliticum (sodium chloride), which inhibits water intake. Numerous observers (such as Ashton, 2014) have commented that this is not easy by virtue of the small number of taxa that successfully overcome the limitations of the mangal, compared with species-rich terrestrial moist forests. Discussion has focused on the effect of constant or periodic presence of salt water upon such physiological processes as gas exchange, hydraulic architecture, photosynthesis, respiration, and stomatal function. Here we touch only briefly on complex topics. The reviews by Ball (1996) and Hanagata et al. (1999) provide useful introductions.

The water relations of mangroves (Fig. 8.1) can be described in terms of the water potential (Ψ) of the tissues of the plant and of the environment, where in any system, the water potential is the algebraic sum of the pressure potential and the osmotic potential ($\Psi = \Psi_p + \Psi_\pi$). In a plant cell, the water potential is determined by the osmotic potential of the cell sap and the turgor pressure of the cytoplasmic membranes against the cell wall. By convention, the water potential for pure water is established as zero; seawater, at a concentration of three percent sodium chloride (NaCl) has a negative pressure (tension) of about 25 bars. In effect, the plant rooted in seawater has to generate an additional hydrostatic pressure of about 25 bars as compared with a plant rooted in fresh water. Mangroves also raise water against a hydrostatic gradient to heights of up to 30 m. The process can be understood in fairly straightforward physical terms, although the detailed structural features that provide the physical mechanism are not well known. Values expressed in "atmospheres" or "bars" in the older literature can be directly transcribed into SI (Standard International) units (MPa), now used by plant physiologists, by the appropriate conversion factors.[*] Simple integers are used in the examples for easy comparison.

Observations and experiments by Scholander (1968) demonstrate that the root system effectively functions as a partial to almost complete ultrafiltration system, and that the water potentials of the leaf cells are always lower than that of seawater. Under these conditions, if leaf tissue were in direct contact with seawater, it would absorb water to an extent determined by the water potential difference. In the living tree, under conditions of no transpiration (stomata closed), that is, no xylem flow of water to the leaves to replace water lost by transpiration, there is an additional hydrostatic pressure in the column of water in the xylem equal to –1 bar per 10 m to be overcome. This hydrostatic pressure occurs in all trees. Figure 8.1 shows the essential relationships with representative values for a mangrove rooted in seawater.

Under conditions of transpiration (stomata open), work is done to move the continuous water column and requires a pressure gradient. The pressure in the crown is decreased by transpiration. Water molecules move along a steep gradient from the leaf (at a low negative water potential determined by the osmotic potential and turgor pressure of the leaf cells) to the atmosphere (at a very low negative water potential for relatively dry air) via the open stomata. Transpiration is controlled collectively by the stomata. These interpretations and generalizations are supported by our knowledge

---

[*] 1 bar = 0.1 MPa; 1 atmosphere = 0.1013 MPa.

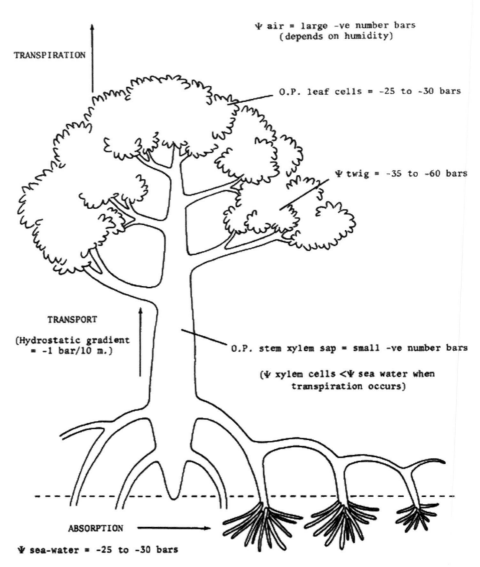

Figure 8.1. Common values for osmotic potential (O.P.) and water potential (Ψ) involved in water balance of a mangrove tree rooted in seawater. All values are either calculated or measured. For water to move from sites of absorption to sites of transpiration, a negative hydrostatic gradient must be maintained. If Ψ seawater exceeds Ψ xylem cells, water should flow back, into the sea.

of the morphology and structure of mangroves, and the measurements of Scholander and his colleagues.

Sperry et al. (1988) provide a model study of the vulnerability, to embolisms, of conduits (specifically vessels) in the xylem, caused by high tensions in xylem sap. They appropriately contrasted *Rhizophora mangle* with a terrestrial moist tropical rain forest member of the Rhizophoraceae, *Cassipourea elliptica*, in terms of vulnerability and wood structure (Plate 10). The terrestrial species of the family have longer, wider, and thinner-walled vessels than the mangrove (Plate 10F), with *Rhizophora* resistant to higher xylem pressures (–6 MPa) than its terrestrial relative. It is significant that the mangrove has scalariform perforation plates (Fig. 6.4) versus the simple ones of *Cassipourea* and *Pellacalyx* (Plate 10F). This difference accords well with the idea that scalariform plates minimize air bubble size in embolized vessels and facilitate refilling of cavitated vessels.

Structural Modifications and Transpiration

Water loss is controlled most immediately by stomatal closure. There is some controversy as to whether mangroves have high or low rates of transpiration. High rates would reflect the high insolation of the mangal environment and the continual availability of water, even though it may have to be absorbed against negative substrate water potential. Low rates would reflect the general physiognomy of mangrove plants, which suggests that they conserve water as "physiological xerophytes"; that is, they grow in an environment of low water availability entirely because the substrate water potential would tend to extract water from root tissues. Such a designation is artificial; the mangal environment is unique and mangroves have distinctive physiological mechanisms to adapt to it. The controversy about the rate of transpiration is genuine, because it depends on the true direct measurement of a functional process. Early estimates of low transpiration rates in mangroves (von Faber 1913, 1923) were contradicted by later workers (Walter and Steiner, 1936). Measurements involving detached shoots in potometers are artificial, however, because hydraulic relations are disturbed. This is not entirely eliminated by alternative methods that measure water loss at the surface of attached leaves by porometry. Modern psychrometric methods measure water potential from the equilibrium value of the relative humidity of air in a closed chamber enclosing a leaf disk.

Scholander et al. (1962) measured transpiration rates indirectly by the ingenious method of comparing the amount of salt excreted at the leaf surface with the concentration of salt in the xylem sap (Fig. 8.2). This gave values (Table 8.1) significantly lower than those that typically occur in terrestrial trees (ranging from 10 to 55 mg.dm^{-2}. min^{-1}.). Scholander suggested that the effect of permanently sequestering salt by transferring it to leaf or other tissue would constitute a larger error for nonsecreters than for secreters (as defined later); hence, the figures for the latter group are likely to be on the low side.

Structural modifications for the leaves of mangroves that seem to be related to these presumed low values include a thick cutinized outer epidermal wall (to reduce cuticular transpiration), somewhat elaborated stomata (indicating precise control of stomatal aperture), well-developed colorless mesophyll or hypodermal layers (for temporary water storage), terminal tracheid clusters (possibly for direct apoplastic response), and the frequent presence of ideoblastic sclereids (for maintaining leaf shape under wilting conditions). The general succulent or coriaceous texture of mangrove leaves has led to

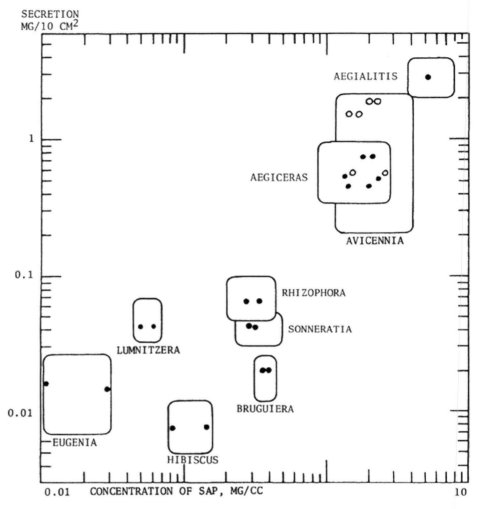

Figure 8.2. Secretion of sodium chloride in various mangroves and two terrestrial trees, over nine daylight hours, as related to salt concentration in the xylem sap (range of values within the rectangles). (After Scholander et al., 1962.)

their comparison to the leaves of xerophytes, with a resulting predisposition to create functional explanations without supporting evidence.

Hydrostatic Pressure Measurements

As we have seen, even in a steady, nontranspiring state, mangroves must maintain a higher water potential in their xylem elements if water is not to be lost to the substrate through the roots by "back filtration." The development and application of the pressure

Table 8.1 Rate of Transpiration from Rate of Salt Secretion

Species	Rate (mg.dm^{-2}.min^{-1})
Salt Secreters	
Aegiceras	2.5
Aegialitis	5
Avicennia	6.5
Nonsecreters	
Sonneratia	1.5
Rhizophora	2.5
Lumnitzera	6.5
Hibiscus	6.5
Eugenia	7.5

Source: Scholander et al. (1962).

bomb technique, primarily by Scholander (Scholander et al., 1965), greatly facilitated measurements of hydrostatic pressure in mangrove xylem. Only indirect methods can be used, because xylem water under negative pressure is in a metastable state in the undisturbed tree. With this technique, a detached shoot is enclosed within a sealed pressure bomb and a measured positive pressure is progressively applied until leakage from the cut xylem begins. This critical positive pressure is assumed to represent the negative pressure that was in the xylem at the time the shoot was detached, as it restores water to the previous volume state. The method is best used comparatively. Scholander found very low values for mangroves, −30 to −60 bars, but the values always exceeded the negative water potential of normal seawater. Under these circumstances, back filtration would not occur. In progressively more saline conditions (when Ψ for seawater becomes increasingly more negative), however, the ability of the mangrove to absorb water decreases. The upper limit seems to be a value of 90 0/00 NaCl, given that even the most efficient mangroves are excluded, as a number of distribution with salinity correlations have shown. The work of Scholander has been challenged to some extent by Zimmermann et al. (1994) in the description of gel-like substances with hygroscopic properties in the cell wall of vessels, purporting to facilitate flow. How they might inhibit cavitation is not clear.

Coping with Excess Salt

Scholander (1968) has discussed the physiological status of plants that grow in mangal in an article titled *How Mangroves Desalinate Water*, which describes the necessary process in direct terms. Salt in high concentrations in plant tissues is toxic as an ionic effect, presumably on enzymes (Parida and Das, 2005). All mangroves exclude most of the salt in surrounding seawater. They can be further divided into two groups (Figs. 8.2, 8.3): those that secrete salt (secreters) and those that do not secrete salt (nonsecreters) (Table 8.1). An interesting phylogenetic statement can be made here,

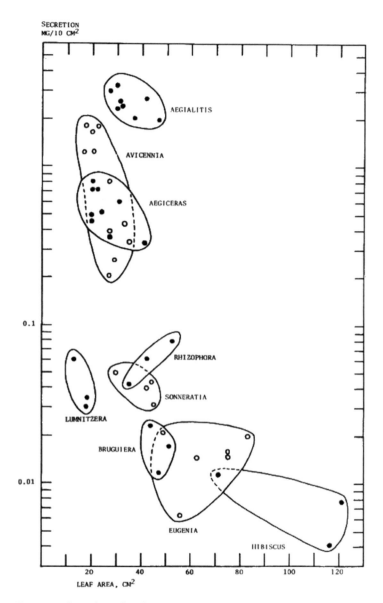

Figure 8.3. Secretion of sodium chloride from leaves of several mangroves and two terrestrial species (*Eugenia* and *Hibiscus*) over nine daylight hours. [After Scholander et al., 1962.]

because Hanagata et al. (1999) show the independent evolution of these contrasted mechanisms, with secreters being later divergent taxa in angiosperm phylogeny. In salt secreters, such as *Aegialitis*, *Aegiceras*, and *Avicennia*, the NaCl concentration of xylem sap is relatively high, but still about one-tenth of the concentration of salt in seawater. Salt is, therefore, only partially excluded at the roots. The absorbed salt is primarily excreted metabolically via salt glands, which are described later. The voided salt in solution can crystallize on the leaf surface by evaporation (Plate 8A; Fig. B.12d), and may blow away, or be washed off by rain. In nonsecreters, such as *Bruguiera*, *Lumnitzera*, *Rhizophora*, and *Sonneratia*, xylem sap has a salt concentration less than one-hundredth of that of seawater, but still almost 10 times more concentrated than that of non-mangrove plants such as *Eugenia* and *Hibiscus*. Salt exclusion is more efficient here, but the small amount absorbed must accumulate and has to be disposed of.

Although nonsecreters have no specialized mechanism for actively secreting salt, they could lose some salt through the leaf surface, possibly by cuticular transpiration. Scholander was able to leach a salt solution from the leaves of plants in this category with mineral elements in the same proportion as they occur in the xylem sap. An additional mechanism for the elimination of salt in all mangroves is simply by loss of parts, notably leaves. However, there is no indication that salt accumulates preferentially in older leaves; that is, there is no active transport mechanism toward senescing leaves. This would be expected from simple physical considerations, because any localized imbalance of the water potential by unequal osmotic potential would generate pressure gradients.

Scholander demonstrated experimentally that the salt separation process must occur at or near the root surface and is mediated by physical processes alone, because it is not inhibited by poisons or high temperatures like a metabolic process. Most convincing of all was Scholander's ability to reverse the process; that is, to move water out of the root into the substrate by simply applying a high hydrostatic pressure to a cut root. No detailed information about the structure of the ultrafiltration mechanism is known. Presumably, it occurs either at the absorbing root surface (epidermis) or at the root endodermis; the latter region might be the most likely because the ultimate absorbing roots (e.g., the capillary rootlets of *Rhizophora*) lack root hairs. This indicates that the absorbing area is reduced in comparison with a terrestrial plant. However, compensation for this loss of absorbing ability may be possible if absorbing rootlets live a long time or are produced in especially large numbers. No quantitative estimates are available.

Salt Glands and Salt Secretion

The most distinctive trichome developed in certain mangrove leaves is the structure that secretes solutions of certain ions, mainly Na^+ and Cl^-. These form a general class of secretory structures called "salt glands" by Fahn (1979), which also occur in other halophytes. The particular kind found in mangroves is referred to as "multicellular glands" by Fahn. Salt glands occur in *Acanthus*, *Aegiceras*, *Aegialitis*, and *Avicennia* (Fig. 8.4), all of which are mangroves that are salt secreters. The salt evaporates and crystallizes in a conspicuous manner in dry conditions, to be lost either by wind or gravity, sometimes in such quantities as to coat adjacent dry ground (Hanagata et al., 1999). Salt glands are abundant on leaves of these plants but are not necessarily equally frequent on upper and lower surfaces. Other mangroves may have epidermal structures

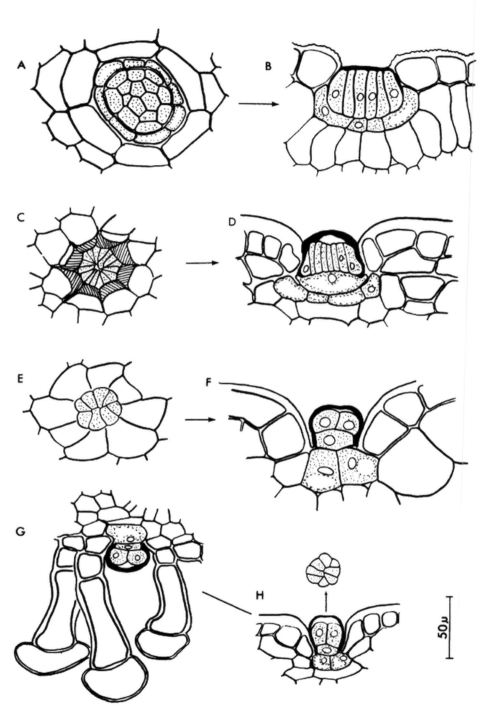

Figure 8.4. Salt glands in mangroves. A., B. *Aegialitis annulata*. C., D. *Aegiceras corniculatum*. E., F. *Acanthus ilicifolius*. G., H. *Avicennia marina*. All are from adaxial epidermis except (G), which is from abaxial epidermis.

that somewhat resemble salt glands, but where a secretory ability has not been demonstrated precisely, as in *Laguncularia* and *Conocarpus* (Fig. 6.3C,D).

The salt glands of *Avicennia* have been much investigated; accounts by Fahn and Shimony (1977) and Shimony et al. (1973) summarize the relevant early literature and present detailed structural, developmental, and physiological information. These authors describe the hairs of *A. marina*, but the information undoubtedly applies to all species of this anatomically uniform genus. Salt glands are scattered in individual shallow pits on the upper leaf surface and much more densely within the abaxial indumentum (Fig. 6.2D; Plate 9E), where they are not sunken but still obscured by the palisade of three- to four-celled capitate nonglandular hairs (which impart the distinctive appearance to the leaf undersurface in *Avicennia*). Each salt gland (Fig. 8.4G, H) consists of two to four basal vacuolated cells at the level of the epidermis, a single stalk cell with an almost completely cutinized wall, and a radiating series of at least eight terminal cells that have a thin, minutely perforated cuticle, separated apically from the noncutinized cell wall to leave a shallow cavity. The stalk and terminal cells are nonvacuolate, and the dense cytoplasm is rich in ribosomes and has a large nucleus and numerous organelles such as mitochondria, all features that suggest high metabolic activity. Plasmodesmata intimately connect the cells of the gland.

Developmentally, glandular and nonglandular hairs are similar until the three-celled stage, when the appearance of the short middle (stalk) cell distinguishes the future gland. This early similarity led Fahn and Shimony to homologize the two kinds of hair and suggest that the glandular hair is phyletically derived from the nonglandular type. The homology is supported by the thick cuticle of the stalk cells in both types of hair.

The precise mechanism of salt secretion is not understood, but it does require energy and can be stopped by metabolic inhibitors. Salts move symplastically into and along the hair, down a decreasing ionic gradient (Shimony et al., 1973), and eventually into the terminal cells, where they are voided into the subcuticular cavity, and hence, via the cuticular pores, to the leaf surface where they visibly accumulate in solution (Scholander, 1968). Reabsorption via the apoplast is prevented by the leaf cuticle, and especially the cutinized wall of the basal cell.

The structure of salt glands in other salt-secreting mangroves is surprisingly consistent in view of the fairly remote systematic affinity of the several families involved, providing evidence for evolutionary convergence. Salt glands all have basal (collecting) cells, a single cutinized stalk cell, and a capitate group of terminal cells (Fig. 8.4A–F). The ultrastructure in each taxon, insofar as it is known, seems similar. Comparable experimental evidence to that found for *Avicennia*, supporting a secretory function, is available for *Aegialitis* (Atkinson et al., 1967).

Looking at other organs in mangroves, one discovers that glandular hairs with similar structure may have different secretory functions. Evidence for this is found in mangroves that have glands on the leaves, which secrete salt, and identical glands in the corolla that are presumably responsible for nectar secretion. The flowers of *Avicennia*, *Acanthus*, *Aegialitis*, and *Aegiceras* are all pollinated by insects and offer nectar as the main floral reward. There are very similar epidermal leaf glands in the back mangrove *Dolichandrone*, but these are not known to secrete salt. On the other hand, capitate glandular hairs are common in the Bignoniaceae; the association here therefore seems to be systematic.

In other mangroves, salt secretion is less clearly established. In *Laguncularia race-mosa*, for example, Biebl and Kinzel (1965) described three types of secretory structures on leaves. The smallest are simply depressions of the leaf surface, with a rather irregular group of densely cytoplasmic cells prominent at the base of the pit (Fig. 6.3C, D). These structures occur in infrequent but variable numbers. In *Pelliciera*, glands on the inner margin of the in-rolled leaf are secretory, but contrary to the report of Collins et al. (1977), the exudate is not salty.

Photosynthesis and Salinity

Photosynthetic rates and stomatal responses to changes in both salinity (either short-term or long-term) and humidity, were measured by Ball and Farquhar (1984a, b) in *Avicennia* and *Aegiceras*, in order to quantify the effect of increased salinity in depressing carbon assimilation. *Aegiceras* was shown to be more sensitive to high salinity than *Avicennia*. The short-term salinity experiments show that there is some "mesophyll resistance" to small increases in salinity; that is, the species does not respond immediately. An important conclusion in these experiments was that even when leaf metabolism was affected, water loss in relation to carbon gain was always minimized so that, in one sense, photosynthesis was not less efficient in high salinities even though the carbon gain was reduced. These authors, together with Hanagata et al. (1999), provide a useful discussion of how salinity affects photosynthetic capacity.

It seems appropriate to mention at this point the experiments by Walsh et al. (1979) on the toxicity to heavy metal ions of *Rhizophora mangle* seedlings. At the concentrations applied (up to 500 $\mu g.g^{-1}$ soil, 250 μg for lead), seedlings were more tolerant of lead and cadmium than of mercury. The authors suggest several mechanisms, such as formation of insoluble sulphides at the root surface, ion exclusion at the root surface, and internal detoxification, but favor the first.

Leaf Succulence

A fleshy texture and a high water content usually develop in mangrove leaves with an increase in leaf thickness with age. In *Laguncularia racemosa*, for example, Biebl and Kinzel (1965) measured a four-fold increase in leaf thickness from the youngest to oldest leaves along a shoot. Such an increase is probably representative for many species. Anatomical changes resulting in increased succulence always involve an enlargement of existing colorless cells, not cell division. Cell enlargement may be localized or generalized; Walter and Steiner (1936) provide illustrations of the changes. In dorsiventral leaves, as in *Rhizophora*, the hypodermis alone expands; in isolateral leaves, the central mesophyll layers enlarge, as in *Laguncularia* and *Sonneratia*; or the whole leaf tissue becomes succulent, as in *Pemphis*. In *Xylocarpus*, a single palisade layer expands considerably. Biebl and Kinzel (1965) measured a decrease in the number of stomata per unit area with increasing succulence, presumably because of the increased surface area of the epidermis.

Leaf succulence in mangroves seemingly has a simple explanation in terms of salt balance. The osmotic potential of the leaf cells of mangroves is high, as measured in the epidermis by plasmolytic methods (Walter and Steiner, 1936), or cryoscopically in expressed sap (Scholander et al., 1964). A high osmotic potential is essential if mangroves are to draw water from the sea with its high negative water potential (Ball, 1996; Hanagata, 1999). However, several authors note that the salt concentration of mangrove leaves remains constant and independent of leaf age. We have also mentioned that measurements of the salt content of xylem sap demonstrate incomplete salt exclusion at the roots. Therefore, mangroves accumulate salt, but as we have seen, this differs in amount for either salt secreters or salt excluders. In the former, where salt exclusion is less efficient, accumulation is compensated by salt secretion via salt glands. In the latter, where exclusion is more efficient, accumulated salt is sequestered largely in leaves and can be voided only when the leaves eventually fall. If salt concentration is constant and independent of leaf age, salt must accumulate by an increase in the volume of the leaf cells, inducing succulence. This has also been mentioned as being understandable in terms of water relations, because different salt concentrations would complicate the balance of water potential in adjacent leaves.

An exception to this generalization is found in developing seedlings in viviparous mangroves, in which a salt concentration lower than that in leaves has been measured (Kipp-Goller, 1940; Pannier, 1962). This has led to the development of theories that explain the viviparous condition in terms of differential salt tolerance, discussed further in Chapter 10. Nevertheless, the permeability barrier that allows this differential salinity is poorly understood.

Leaf succulence in mangroves, in these simple terms, may therefore be accepted as a consequence of life in an environment that provides ample water at the expense of some compensation for that water's high salinity. Further evidence comes from the observations of Camilleri and Ribi (1983) that leaf thickness (presumably also succulence) in *Rhizophora mangle* is correlated with soil salinity, as leaves were thicker in sites of constant high salinity. These authors deny that succulence (which they term "formation of water storage tissue") is a simple function of leaf age, but they also fail to cite Biebl and Kinzel's results.

These simplistic conclusions avoid the developmental question of how succulence is induced in conditions of high salinity; that is, how cell volume is increased. The answer is remote from our present level of understanding of the control of developmental mechanisms in plants.

Selective Ion Absorption

Of the mineral cations essential to the nutrition of plants, potassium is required in the largest amounts. The salt exclusion mechanism of mangroves, therefore, must be selective, it must have a sufficient discriminatory capacity to absorb ions in appropriate concentrations from external solutions of lower potassium concentration, and there must be a preferential selection of potassium in competition with high concentrations of sodium. Rains and Epstein (1967), studying the leaf tissue of *Avicennia*, provide evidence for two mechanisms: one operating at low and the other at high potassium concentrations but resistant to interference by sodium. The net result is a preferential absorption of potassium, and it is suggested that the mechanism occurs in roots as well as leaves.

Resistance to high salt concentrations is not complete, however, because Ball and Farquhar (1984a) have measured decreased potassium concentrations at high salinities in *Avicennia* and *Aegiceras*. This is relevant to their studies of photosynthetic responses to high salinities, as potassium ions are involved in the stomatal mechanism, and indirectly, the potassium concentration could affect gas exchange.

Conclusion

This discussion of the growth of mangroves in a saline environment has taken us well away from the simple description of mangrove features, which is the main purpose of this book, but even a brief summary with a strong historical component serves to emphasize that structural description is a necessary precursor to functional elucidation. The topic of the water relations of mangroves will continue to exercise the interest and ingenuity of botanists as an essential requirement for ecological understanding.

9 Flowering

As a consequence of vivipary and other seedling peculiarities, interest in the reproduct-ive biology of mangroves has been centered on seed biology, which will be discussed later. Although floral mechanics is described in some detail in the individual accounts and the words "visitor" and "pollinator" have been used interchangeably, it is only recently that genetic isolating mechanisms and the degree of heterozygosity have been measurable using the techniques of molecular genetics. Ge and Sun (1999), in a study of *Aegiceras corniculatum* (Myrsinaceae), provide a model protocol for experimental study of the pollination mechanism and genetic consequences of insect pollination, with emphasis on the additional influence of gene migration via long-distance dispersal. The requirement for pollination seems strict because agamospermy (seed set by unpol-linated ovules) is unknown in mangroves.

Breeding Mechanisms

The simple analysis of sexual differentiation and floral mechanisms in Table 9.1 shows a trend toward outbreeding mechanisms based initially on the analysis by Primack and Tomlinson (1980). They contrast mangroves with trees in terrestrial lowland forests and relate differences to the pioneering propensities of mangrove taxa. In the strict mangrove species, dioecy is rare (4 percent) and monoecy uncommon (11 percent) and most taxa are hermaphroditic. In dioecious organisms, outbreeding is obligate. It may be promoted in monoecious organisms, especially where sexual segregation is reinforced by lack of synchrony in the maturation of male and female flowers. This occurs in *Nypa* where the inflorescence is strikingly protogynous. Each vegetative shoot supports only one functional inflorescence at a time, with the terminal female flowers becoming exposed and receptive well before pollen is released on lateral spikes by the numerous male flowers. *Nypa* is an example of extreme dimorphism between male and female flowers. Nevertheless, this indication of potential outbreeding should be tempered by the appreciation of rhizomatous habit of this palm, such that gene interchange is still most likely within a clone. Therefore, this example is unique.

   In *Xylocarpus*, the structural difference between male and female flowers is slight (Fig. B.41) and was not appreciated by early botanists, although functional differences between them is evident at the time of fruit set. Female flowers produce sterile pollen

Table 9.1 Breeding Mechanisms and Distribution of Sexual Types in Common Mangroves

Genus	Perfect	Monoecious	Dioecious	Outbreeding Mechanisms
Rhizophora	+			Possibly weak protandry, self-compatible (?)
Bruguiera	+			Pollination may favor outcrossing
Ceriops	+			Pollination may favor outcrossing
Kandelia	+			
Avicennia	+			Protandrous
Sonneratia	+			
Lumnitzera	+			
Osbornia	+			
Pemphis	+			Heterostyly plus self-incompatibility
Aegiceras	+			
Aegialitis	+			
Scyphiphora	+			Protandrous
Pelliciera	+			
Camptostemon	+			
Nypa		+		Protogynous
Heritiera		+		
Xylocarpus		+		
Laguncularia[a]			+	Outbreeding obligate
Excoecaria[a]			+	Outbreeding obligate

Generic summary: hermaphrodite 74%, monoecious 16%, dioecious 10%
Specific summary: hermaphrodite 85%, monoecious 11%, dioecious 4%

[a] Probably not consistently dioecious.

and the ovules in male flowers are nonfunctional. Morphological similarity presumably facilitates pollen transfer, because visitors (mainly bees) make no discrimination between the two sexes; the reward in both flower types is nectar. Trees are monoecious, but usually with only a single female flower in each panicle.

Among mangroves with perfect flowers there is clear evidence for an outbreeding mechanism in *Pemphis* (Lythraceae), which has floral dimorphism (heterostyly) associated with self-incompatibility (Fig. B.36). Otherwise there is evidence for varying degrees of dichogamy, usually protandry, as in *Avicennia* (Fig. B.13a, b, c, d) and *Scyphiphora* (Fig. B.63), whereas weak protandry is recorded for the mangrove Rhizophoraceae. In many of these species, there is limited evidence for self-compatibility based either on a few controlled pollinations or on observation of the fruiting of isolated individuals or of heavy fruit set in natural populations. Theoretical arguments would support the idea that mangroves, if they are primarily colonizing species, would retain the need for self-fertility if they are to establish populations in isolated localities, as argued by Primack and Tomlinson (1980). However, more direct confirmation of this hypothesis is needed; for example, abundant fruit set could be evidence of efficient pollination as well as of self-compatibility.

Pollination Biology

This subject refers to the actual mechanism of pollen transfer from one flower to another. Mangroves are almost exclusively pollinated by animals, and the classes of flower visitors are remarkably diverse. Mangroves must interact with each other or even compete with each other for pollinators; the evidence, particularly for the mangrove Rhizophoraceae (Table 9.2), which has been studied in comparative detail, suggests that the interaction minimizes the competition for pollinators so that the "pollinator resource" (the total available population of pollinating agents) is used in the most efficient manner possible. First we must deal with one conspicuous exception to our generalization; *Rhizophora*, which seems to be pollinated mainly by wind. Subsequently, pollination is best described in terms of flower visitors, that is, presumed pollinators. This itself leads into a topic that is largely avoided here, that is, floral scent chemistry. Azuma et al. (2002) have summarized what is known, and not unexpectedly, show considerable diversity. *Nypa*, for example, is particularly rich, with more than 25 chemicals identified in its flowers; it is visited by a diversity of small insects (Hoppe, 2005), with an emphasis on drosophiloid flies (Essig, 1973).

Pollinator resource and interaction. The diversity and generalized nature of the pollinators that are effective in mangroves allow two basic conclusions: first, the spectrum of pollinators is broad so that no plant is highly dependent on one specific pollinator; and second, plants are specialized only to the extent of being associated with a given class of pollinator. These conclusions have a reasonable explanation in the wide geographical range of mangroves, so that they are not dependent on a specific pollinating agent with a limited geographic range. Furthermore, because each mangrove adapts primarily to a generalized type of pollinator, competition for the available pollinator resource is reduced.

The primary division is between nocturnal and diurnal flower visitors; secondary divisions involve adaptation to different classes of pollinators or pollinators distinguished by size as well as by behavior. This is most clearly seen in the mangrove Rhizophoraceae, where each genus or group of species within a genus is adapted to a given class, with *Rhizophora* "opting out" of competitive interaction because of its primary dependence on wind as the pollen vector (Table 9.2). A particular feature of this

Table 9.2 Pollinators and Pollen-discharge Mechanisms in Rhizophoreae

Genus or Subgroup	Flower Visitor or Vector	Pollen Discharge
Rhizophora	Wind	Nonexplosive
Bruguiera		
Large-flowered species	Birds	Explosive
Small-flowered species	Butterflies, etc.	Explosive
Ceriops		
C. tagal	Moths	Explosive
C. decandra	Small insects	Nonexplosive
Kandelia	Small insects	Nonexplosive

group of related species is that, in some of them (*Bruguiera* and *Ceriops tagal*), the mechanism of pollen discharge is explosive and highly specialized, yet is related to visits from three classes of pollinators: birds, butterflies, and moths. *Ceriops decandra* and *Kandelia candel* are much more generalized in their method of pollination and seem to lack a highly elaborated mechanism. The subject is presented in greater detail in the description of each taxon (Part II). A feature common to all mangrove Rhizophoraceae is the precocious dehiscence of the anther so that pollen is released within the unopened flower bud. This feature may say something about the common ancestor of these diverse flower types.

Wind Pollination

The chief evidence for wind pollination in *Rhizophora* is a knowledge of the floral mechanism in comparison with that of related genera in the tribe Rhizophoreae and prolific pollen production. Evidence is presented more completely in the detailed description of the group. Here, it is only summarized.

1. Method of pollen release: dispersed by petal hairs after stamens dehisce in the bud (most species) or by a "pepper pot" mechanism (*R. apiculata*).
2. High pollen/ovule ratio* as much as an order of magnitude higher than related genera.
3. Light powdery pollen (but also characteristic of other Rhizophoreae!).
4. Absence of an attractive odor.
5. Absence of abundant pollinator reward (other than pollen).
6. Inverted flowers.
7. Ephemeral pollen presentation time (stamens and petals fall within 12 hours).
8. Screened flowers set fruit.

Arguments against *Rhizophora* being pollinated by wind are the absence of an elaborated stigma suited to catching wind-borne pollen and the frequent records of flower visitors (bees) that may visit the flower for pollen. In addition, there is some indication that flowers produce an exudate (nectar?) after the floral organs have been lost. Chai (1982) has produced evidence that *Rhizophora* sets fruit when animal visitors are denied access to flowers enclosed by a fine mesh bag that does not exclude wind-borne pollen. *Rhizophora* appears to be self-compatible, however, and there is therefore the possibility of self-pollination. Again, stigmas may not become receptive until after pollen is released (weak protandry). Much more experimental work is needed.

---

* P/O ratio, the ratio of pollen produced per ovule, is characteristically high in wind-pollinated plants and is an indication of a low efficiency of pollination. *Nypa*, which is pollinated by insects, probably must have a very high P/O ratio, suggesting inefficient pollination but facilitating outcrossing in an organism that spreads clonally by rhizomes.

Animal Pollinators: Nocturnal

Bats. *Sonneratia* is a well-known example of a flower visited by bats (Marshall, 1983). The flowers open in the early evening, when the calyx expands to expose the attractive mass of extended stamens with powdery pollen (Plate 6D). Flowers produce nectar in some quantity from a basal disc. The stamens mostly fall from the flower by the following morning (Plate 6E), so that the flowers are ephemeral. In west Malaysia, bats are known to fly as far as 50 km from inland roosting sites to feed on *Sonneratia*. Bats are not the only flower visitors, however; there are also records of hawk moths, another important group of mangrove pollinators.

Hawk moths. (Sphingidae). Primack et al. (1981) have recorded hawk moths visiting *Sonneratia* in Queensland. These animals (i.e., insects) are important alternative pollinators for the genus because they may occur in areas where there are few or no bats. That *Sonneratia* is pollinated exclusively by bats is implied in the literature, but for many mangroves, flexibility in the pollinating agent is very important in view of their wide distribution. *Dolichandrone spathacea* is also probably pollinated by sphingid moths. Its flowers open at night (Plate 7K) and are ephemeral. The floral tube is very long (up to 20 cm, Fig. B.19a) and the nectar can be reached only by an animal with a long tongue. No visitor records exist, however.

Moths. Small moths seem to be the most likely pollinators of *Ceriops tagal*; its flowers are small and inconspicuous but have white petals, and a sweet scent develops in the early evening. Tomlinson et al. (1979) record moths as flower visitors in Queensland.

Animal Pollinators: Diurnal

Birds. Birds are common as visitors to the mangal, although a study in north Australia suggests that they are largely insectivorous rather than pollinators (Mohd-Azian et al., 2014). The large-flowered species of *Bruguiera* (e.g., *B. gymnorrhiza*) are well adapted to bird visitors and there are several independent records of them (Gehrmann, 1911; Davey, 1975; Tomlinson et al., 1979). The flowers are recurved and typically point backward into the crown of the tree (Plate 22G); this posture facilitates an approach by a perching bird. Nectar is produced in abundance and held in the deep floral cup. In *B. gymnorrhiza*, the calyx is red, a color attractive to birds. The bird probes the base of the flower, exploding the petal pouches, and pollen is thrown onto the head of the animal, which is likely to transfer pollen to the stigma of a subsequently visited flower. Cross-pollination may be favored by weak protandry because the stigma may not become sensitive until two or three days after the flower opens. Pollen is not usually discharged in the absence of a flower visitor. Birds as a class are flower visitors to *Bruguiera*, as there are records of sun birds in some parts of its range (e.g., East Africa and Sarawak) and of honey-eaters in Queensland. *Lumnitzera littorea* (red flowers) is described as bird pollinated in contrast to *L. racemosa* (white flowers), which is visited by small insects (Plate 6A, C).

Bees. These are the most common flower visitors in the mangroves and have been observed on flowers of *Acanthus*, *Aegiceras*, *Avicennia*, *Excoecaria*, *Rhizophora*, *Scyphiphora*, and *Xylocarpus*. In South Florida, "mangrove honey" is made, a major constituent of which is said to be the nectar from *Avicennia germinans*. Mangrove honey is widely produced in India (Spalding et al., 2010). Most bee flowers are rather small, although they are commonly aggregated, with a short corolla tube; the petals are typically white, pink, or lilac; nectar is produced in small quantities; the flowers characteristically have an attractive scent, and they may last for several days. *Excoecaria* is exceptional but has elaborate glands on its spicate unisexual inflorescence (Fig. B.27). The kinds and sizes of bees vary considerably. *Nypa* is visited by small trigonid bees. A large bee is required to extract nectar from the flower of *Acanthus* (Fig. B.2d, e), and the pollination mechanism depends in part on the forceful separation of the flower parts (Primack et al., 1981).

Butterflies. Records of butterflies visiting *Bruguiera parviflora* have been provided by Tomlinson et al. (1979), and the flowers in this and other small-flowered *Bruguiera* species seem well-adapted to this kind of small visitor (Plate 23G). The flowers are directed outward, facilitating visits by an insect approaching the shoot. Crab spiders may predate such visitors.

Generalized small insects. Some mangrove flowers, like *Ceriops decandra* and *Kandelia candel*, seem to be pollinated by a generalized population of flying insects of suitable size. Visitors of this kind, apart from bees and butterflies, are wasps and flies. There seem to be no reliable records of mangrove flowers being visited by beetles, which are considered to be "primitive" pollinators and typically associated with large flowers or inflorescences of very generalized type. Such "beetle" flowers are absent from mangroves.

Vector Dependence

Thus far, we have dealt with pollination in mangroves primarily from the point of view of the needs of the plant, and we have stressed their partitioning of the pollinator resource. It is equally important, however, to consider the system from the point of view of the pollinators. A feature of the plant–pollinator interdependence in mangroves is that most of the pollinators live outside the community and the mangroves share them, or even compete for them, with plants of adjacent terrestrial communities. One can argue that the pollinators are less dependent on the plants than the plants are on the pollinators. Close scrutiny shows that there may be seasonal interdependence in certain areas that works to the advantage of the mangroves, whereas mangroves may flower seasonally (Duke, 2006), hence are not necessarily a constant resource.

Sonneratia visited by *Macroglossus minimus* in the Malay Peninsula provides a well-investigated example (Start and Marshall, 1976; Marshall, 1983). One of its major pollinators in the Klang area is the bat *Eonycteris spelea* (the cave fruit bat), which eats the nectar and pollen of the flowers. Each flower in *Sonneratia* functions for just one night. The bat roosts inland in large caves, such as those in the limestone area known as Batu near Kuala Lumpur. The bats have a range of up to 50 km in their feeding flight, and *Sonneratia* (especially *S. alba*) is one of a number of favored flowers, which also include the commercially important durian (*Durio zibethinus*). We do not know what proportion of the bats' feeding is on *Sonneratia*, but it is clear that this mangrove species forms a link in the chain that ensures a good durian crop, because the durian has a limited flowering (and hence, fruiting) season and the bats need alternative food resources. In return, *Sonneratia* is indirectly dependent for its successful pollination on the ability of bats to sustain themselves elsewhere during seasons when *Sonneratia* overlaps little with durian, so that again competition is minimized. The bat–mangrove–durian interrelationship represents interdependence between mangroves and terrestrial plants that may be quite common. It is discussed extensively in Ashton (2014).

Another example is provided in parts of north Queensland where species of *Bruguiera* that are pollinated by birds (*B. gymnorrhiza* and *B. exaristata*) flower most abundantly in the dry season, at which time terrestrial vegetation may support few flowers that are suited to nectar-eating birds (e.g., honey eaters). Under these circumstances, the mangal may become the main food source for birds that otherwise feed extensively in terrestrial plant communities. *Bruguiera* is less directly dependent than terrestrial plants on rainfall for growth, as it is rooted in tidally inundated areas; it can take maximum advantage of the availability of bird visitors, thus a mutual dependence results. If this interpretation is correct, the mechanism of control of flowering is presumably selected for in relation to some climatic trigger, because flowering obviously cannot be stimulated by flower visitors.

Where several species of *Avicennia* grow together, there is evidence of nonsynchrony in flowering times, which might minimize the competition for pollinators (probably bees), and at the same time, spread the availability of nectar over a more extended period. In Malaysia, for example, Watson (1928) comments (in a letter in the Herbarium of the Singapore Botanic Garden) on a visit to an area where the three common Malayan species (*A. alba*, *A. marina*, and *A. officinalis*) grew together, so that each species was at a different stage of the reproductive cycle. Mature flowers in abundance were present in only one species at one time. This may be part of the reproductive isolating mechanism for these species, which have very similar flowers and may well be served by the same class, if not the same species, of pollinator. Bees and wasps represent a group of pollinators that nest in mangroves (Plate 3E), and some populations are therefore more completely dependent on mangal for their existence than is usual in plant–animal interaction in mangroves.

Pollen

The study of the pollen grains of mangroves may appear to be a somewhat arbitrarily selected topic, as it deals with pollen from an unrelated group of plants, but it has several important applications (Vezey et al., 1988; Tyagi and Singh, 1998). Some examples are shown in Figs. 9.1–9.4). First, fossil pollen is the most extensive and systematically reliable source of information about the ancestry of individual

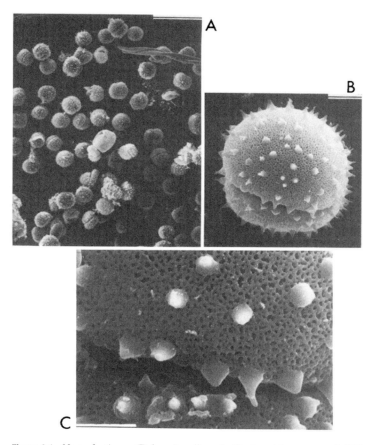

Figure 9.1. *Nypa fruticans* (Palmae) pollen. A. Grains at low power (x200, scale = 10 μm).
B. Single grain (x1500, scale = 100 μm), oblique view "resembling a rather forbidding
hamburger." C. Detail of circumpolar sulcus and sculpturing (x5000, scale = 250 μm). Pollen is
sticky and can be transported only by an animal vector.

mangroves. Available information has been reviewed by Muller (1981). Because extant
mangroves are only in coastal areas, their pollen record can be interpreted as a record of
the existence of mangrove vegetation, provided it has not been widely redistributed
before fossilization. The dispersibility of mangrove pollen is therefore relevant to the
study of fossil mangal. Graham (1977) compares the distribution of *Pelliciera* pollen
in Tertiary deposits with the present distribution of the genus to demonstrate that the
range has contracted in recent geological time (Fig. 4.1).

Second, pollen structure is related to the floral mechanism, and its analysis may be an
important adjunct to a study of floral biology and may give some indication of the
pollen vectors, as in the study of Mancina et al. (2005), who found *Tilipariti* pollen in
bat stomach contents. The large grains of *Scyphiphora* and the sticky grains of *Nypa*
(Fig. 9.1), for example, can clearly be transported only by animal agents. The abundant
dry, light pollen of *Rhizophora* is part of the evidence suggesting pollination by wind,

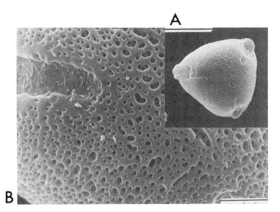

Figure 9.2. *Rhizophora apiculata* (Rhizophoraceae). Pollen morphology via scanning electron microscopy. A. Single grain (x200, scale = 10 μm). B. Detail of sculpturing (x10 000, scale = 2 μm). Pollen in the mangrove Rhizophoraceae is quite constant in size and morphology, yet the group includes both wind-pollinated and animal-pollinated species.

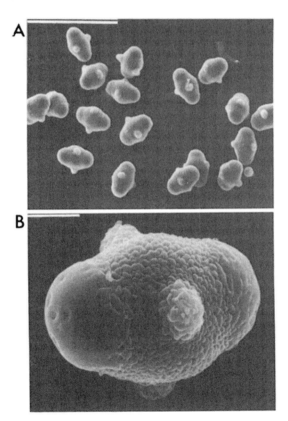

Figure 9.3. *Sonneratia caseolaris* (Sonneratiaceae). Pollen morphology via scanning electron microscopy. A. Grains at low power (x350, scale = 100 μm). B. Single grain (x2000, scale = 10 μm). Grains are large and appropriate for an animal-pollinated species.

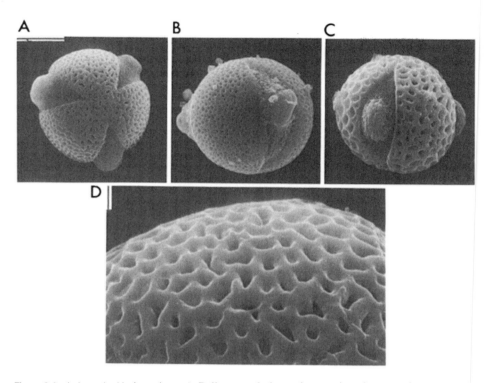

Figure 9.4. *Avicennia* (Avicenniaceae). Pollen morphology via scanning electron microscopy. A–C. Whole grains (x2000, scale = 10 μm). D. Detail of wall sculpturing from part A (x10 000, scale = 1 μm). A. *A. marina* var. *australis* (New Zealand). B. *A. germinans* (Florida). C. *A. marina* var. *eucalyptifolia* (Queensland). Size and morphology seem quite constant in the genus, but wall sculpturing is variable.

and the likelihood of wide dispersal in marine sediments has to be considered when fossil *Rhizophora* pollen is examined (Fig. 9.2). On the other hand, a study of the floral biology of *Bruguiera*, *Ceriops*, and *Kandelia*, which have pollen structurally similar to that of *Rhizophora* but in limited quantities, provides evidence that they are not wind pollinated. This produces a paradoxical situation of pollen with seeming considerable dispersal ability, but not associated with the pollination mechanism.

Third, pollen abortion can give evidence of hybridization, an application found initially in the work of Muller and Hou Liu (1966) on *Sonneratia* (see also Wright, 1977).

Fourth, there is limited application of information about pollen structure in the recognition of the source of honey, which is sometimes derived from the nectar of mangroves (notably *Avicennia*).

Fossil pollen. In his summary of the appearance of angiospermous families in the geological record based on their identifiable pollen, Muller (1981) documents the existence of the genus *Nypa* in the very late Cretaceous (Maestrichtian, 69×10^6 years BP). He interprets this as an invasion of the mangrove environment by the

Table 10.2 Germination Type and Seed Size in Mangrove Associates

Genus	Type of Propagule	Vivipary$^{(++)}$ or Cryptovivipary$^{(+)}$	Seed Length (cm)
Hypogeal Germination			
Cerbera	One-seeded fruit	—	7–5
Terminalia	One-seeded fruit	—	7–5
Heritiera	One-seeded fruit	—	6–5
Calophyllum	One-seeded fruit	—	4–3
Epigeal Germination			
Cynometra	One-seeded fruit	—	3–2
Dolichandrone	Seed	—	1
Acanthus	Seed	—	1–0.5
Conocarpus	Seed	—	0.4–0.2

Lumnitzera, *Nypa*, *Pelliciera*, *Scyphiphora*, and *Terminalia*. In *Xylocarpus*, the testa is thick and corky. Special adaptation is particularly notable where the method of dispersal is different between mangrove and terrestrial relatives. In *Heritiera spp.* and *Terminalia catappa*, for example, seeds are dispersed in water, whereas many other members of these genera have winged seeds suited for wind-assisted dispersal. In *Dolichandrone*, the corky seed wing differs from the papery wings suited for wind dispersal, which are otherwise characteristic of many Bignoniaceae. The genus *Osbornia* is difficult to place in the Myrtaceae because its fruits are neither capsular nor fleshy, whereas the family as a whole has been subdivided (at least biologically) on the basis of this character. Where the seedling is the propagule, the cortex of the hypocotyl is aerenchymatous, thus providing buoyancy in varying degrees (e.g., mangrove Rhizophoraceae, Fig. 10.5).

Vivipary

Definitions. The development of the embryo in seed plants is a continual process, normally interrupted by a longer or shorter period of seed dormancy, with tropical species tending to have the briefest dormancy; they are said to be "recalcitrant" (Berjak and Pammeter, 2006). Thus, there is a potential continuum of developmental possibilities. In the extreme condition of "true vivipary" (Fig. 10.3), the embryo that results from normal sexual reproduction has no dormancy but grows first out of the seed coat and then out of the fruit while still attached to the parent plant (Juncosa, 1982a, b); this organ (propagule) is thus the seedling and neither fruit nor seed. Vivipary occurs sporadically outside of mangroves, as summarized by Elmquist and Cox (1996), but development of the seedling on the parent tree in this way is a feature of several mangroves and is found in the whole tribe Rhizophoreae of the family Rhizophoraceae. Why vivipary should be so common in mangroves is a topic that has fostered plenty of discussion but no generally accepted explanation. A new one is offered as follows.

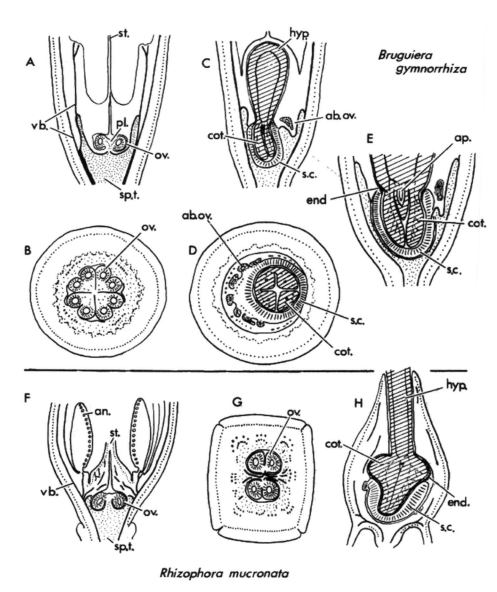

Rhizophora mucronata

Figure 10.3. Structural features of vivipary in mangrove Rhizophoraceae. A.–E. *Bruguiera gymnorrhiza*: A. L.S. flower. B. T.S. ovary at level of placenta. C. L.S. fruit with developing seedling still enclosed within ovary. D. T.S. fruit at level of cotyledons of an almost mature seedling. E. the same, with details of seedling apex. F.–H. *Rhizophora mucronata*: F. L.S. flower. G. T.S. flower at level of placenta. H. L.S. fruit with apical part of emerged seedling. Key to symbols and abbreviations: dotted line, layer of secretory cells (laticifers); hatched area, embryo; ab. ov., aborted ovule; an., anther; ap., plumule; cot., cotyledon; end., remains of endosperm; hyp., hypocotyl of seedling; ov., ovule; pl., placenta; s.c., seed coat; sp.t., spongy tissue; st., stigma; vb., vascular bundle. The remains of the endosperm in *Bruguiera* scarcely protrude from the seed, whereas in *Rhizophora* they are extruded well beyond the seed. (Redrawn from Kipp-Goller, 1940.)

Tan and Rao (1981) describe what they refer to as vivipary in *Opiorrhiza* (Rubiaceae), a plant found in wet places. Seedlings simply germinate within the undetached infructescence and are not specialized. Premature germination is known in some cultivated fruits, as in *Citrus* where it is usually associated with some form of cleavage polyembryony. Adventitious vegetative buds can develop on inflorescences in such plants as *Agave* and *Poa alpina* without involving sexual processes. These nonsexual or parasexual processes are often referred to as "false vivipary" and are not uncommon in flowering plants.

True vivipary needs to be studied as part of the continuum of normal seed development and not as an isolated phenomenon, as explanations often involve a comparison with terrestrial species. The absence of a dormant phase in the development of the embryo of *Rhizophora mangle* has been documented by Sussex (1975), who uses the example to describe the processes that govern embryo development in seed plants generally. The reviews and classifications, by Ng (1978, 2014) and De Vogel (1980), of germination processes in tropical woody plants are valuable for comparative information. These authors confirm that rapid germination of the seed (i.e., with limited dormancy) is common in tropical rain forest trees and that the seed life span is usually short (a few weeks). Rapid germination of this type does not, however, normally involve exposure of the seedling before the fruit falls. In a more advanced state found in some mangroves, cryptovivipary exists, in which the embryo emerges from the seed coat or even the fruit before the latter abscises (Carey, 1934). Examples are *Aegiceras*, *Avicennia*, *Nypa*, and *Pelliciera*. In the palm *Nypa*, the plumule emerges from the fruit before this is released, and may even assist abscission (Plate 18D, E). In the mangrove Rhizophoraceae, although true vivipary represents the most advanced condition, there are significant differences between the genera so that comparative study is possible. In other marine ecosystems, true vivipary occurs only in the seagrasses *Amphibolis* and *Thalassodendron* (Black, 1913; Tomlinson, 1982b). Vivipary is unknown in non-mangrove halophytes.

Explanations. The association between vivipary and the littoral habitat has occasioned much comment, with many authors attempting to account for the correlation in adaptive terms. Morphological, ecological, and physiological explanations have been developed and comparative studies have offered explanations in evolutionary terms (Juncosa, 1982a, 1984a,b). Embryological study has revealed a surprising diversity of systematic characters (Tobe and Raven, 1988). Inland members of Rhizophoraceae and its putative relatives have a range of germination types, from epigeal (e.g., *Cassipourea*) to hypogeal, either of which could represent the ancestral state, as it is reasonable to assume that the mangrove Rhizophoraceae had a terrestrial ancestor within the same complex that gave rise to the present-day family. Most of the evidence suggests that the Rhizophoreae have modified epigeal germination.

An explanation of vivipary might therefore account for its origin among other types of germination, its high incidence in mangroves (but not in other halophytic habitats), its convergent development in unrelated groups, and its functional or adaptive significance. The structural relation between flower and fruit is shown for *Bruguiera* and *Rhizophora* in Figure 10.3A–E, F–H, respectively. This illustrates the massive

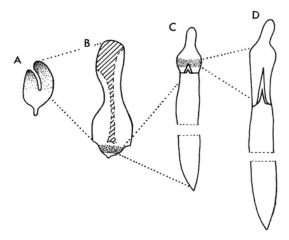

Figure 10.4. Embryo development in *Rhizophora*, schematic; steps are about half an order of magnitude different in size. Stippling represents region of maximum growth (cell division and expansion). A. Growth in cotyledons. B. Growth in hypocotyl. C. Second growth phase in cotyledons. D. Mature embryo. Hatched area in (B) represents the median sinus of cotyledons.

rearrangement of floral tissues as the embryo develops. Details of embryonic development are shown in Plate 21 for *Rhizophora* alone. Figure 10.4 shows how regions of active growth (stippled) shift in an embryo, in embryos at four successive stages.

Developmental Features

An extensive analysis of systematic differences in seedling anatomy within mangrove Rhizophoraceae is provided in the comparative study of Tomlinson and Cox (2000). This was an approach adopted by Wilkinson (1981) in a study of *Ceriops* seedlings that provided information by means of which fossil seedlings could be identified. The following are more detailed accounts of well-studied taxa.

Rhizophora species. A study of *Rhizophora mangle* by Cook (1907), with further detail by Juncosa (1982b), can illustrate the distinctive features of embryo development and germination in the genus leading to the conspicuous appearance of the hypocotyl of the seedling (Plate 21A). In *Rhizophora*, the cotyledon develops as a cylindrical body ("cotyledonary collar"), which remains in the fruit at the time the seedling is released (Plate 14C). Of the four ovules in each ovary, normally only one becomes an embryo (Fig. 10.3F–H), but there is a low incidence of two- or even three-seeded fruits, where two or three ovules develop. Each ovule has two integuments. Endosperm development is initially free-nuclear and then cellular, but normal for this type in angiosperms. Plate 21 presents some microscopic detail. The embryo is initially attached to the integument at the micropylar end by a long multicellular suspensor whose basal cells disintegrate (Plate 21F). This allows the embryo to move deeper into the endosperm. These features are considered by Juncosa to be unique among the angiosperms. Transfer cells are conspicuous at the interface between embryo and endosperm (Wise and Juncosa, 1989). Intrusive growth of the endosperm envelops the embryo, forces the micropyle open, and carries the embryo out of the integument (Plate 21F). This is curious because if "germination" refers to the extrusion of the embryo from the seed, it is growth of the endosperm, rather than of the embryo, that initiates the process in this form of vivipary. This process is illustrated in Plate 21B, in which the undifferentiated white tissue of the embryo is contrasted with the surrounding dark endosperm, now largely enlarged outside the original micropyle. This becomes the anchor-shaped tissue projecting well beyond the embryo (Plate 21C).

Within the embryo itself, the cotyledons are initiated on the top-shaped embryo as a single torus opposite the remains of the suspensor (Plate 21G, H), which only later becomes lobed so that there is, at first, virtually no indication of a pair of cotyledons. Further growth occurs apically in the cotyledonary body (Fig. 10.4A), but the most active growth is soon restricted to a diffuse intercalary hypocotylary meristem below the cotyledon, but separated from it by a region of differentiated cells (Fig. 10.4B). Growth of the hypocotyl (Figs. 10.4B, C) thus extends the seedling until the radicle end protrudes beyond the fruit (Plate 21D). The much enlarged embryo by now has absorbed the endosperm. A final stage is the further elongation of the cotyledons (Fig. 10.4C), the basal abscission zone of which becomes exposed outside the fruit (Fig. 10.2B). The radicle apex is enclosed by embryonic tissue and its own root cap.; it remains almost wholly inactive and usually plays no part in establishing the seedling.

The fruit itself (Fig. 10.2D) enlarges to its final size, achieved as the seedling is first visible externally (Fig. 10.2C–F). The hypocotyl shows evidence of a geotropic response as it may become curved (Plate 21A). Lateral roots, visible on older seedlings as apical protuberances, arise early in seedling development and delineate the morphological root at the end of the hypocotyl. Three pairs of leaves and associated stipule pairs are produced while the seedling is still on the tree; the first pair of leaves (after the cotyledons) aborts and persists only as minute vestiges on the detached seedling, but their associated stipule pair is well-developed and protective of the enclosed plumule.

Other species of *Rhizophora* have not been studied in the same detail, but the seedling size at maturity varies and is somewhat diagnostic. *Rhizophora mangle* has short seedlings (20 to 30 cm long), whereas in *R. mucronata* seedlings as long as 70 cm are common (Plate 14B). Juncosa emphasizes that the *Rhizophora* seedling is distinctive in the timing of meristematic activity within the embryo, rather than in any major structural modification. It seems appropriate to view the *Rhizophora* seedling as having modified epigeal germination, as both types are characterized by an elongated hypocotyl and epigeal germination is the dominant type in other mangroves.

Histologically, the *Rhizophora* seedling is well protected, which seems appropriate in view of its long exposure both on the tree and during dispersal. There is an incipient phellogen toward the radicle end, usually sufficiently developed to mask the superficial chlorophyllous tissue. The aerenchymatous ground tissue is abundantly tanniniferous and there are numerous trichosclereids in the cortex and medulla. Clusters of stone cells occur in the outer cortex. A peculiar feature of the epidermis is the virtual absence of stomata so that gas exchange may occur only through the numerous lenticels; their

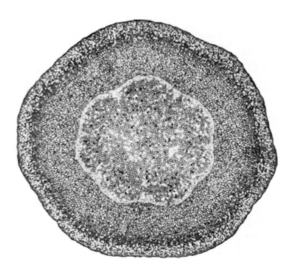

Figure 10.5. *Bruguiera cylindrica*. T.S. seedling hypocotyl with lacunose cortex; wide medullated stele with limited peripheral vascular tissue (x10).

development has been described by Roth (1965). In *R. mucronata*, the lenticels are warty and prominent.

   Bruguiera exaristata. Juncosa (1984b), in his study of *Bruguiera*, placed it between *Cassipourea*, a terrestrial member of the Rhizophoraceae with epigeal germination (Juncosa, 1984a), and *Rhizophora* in terms of the degree of specialization in embryo development. The same growth phases as in *Rhizophora* are recognized; that is, apical growth of the cotyledons followed by hypocotylary extension (Fig. 10.3A–E). The ovule has a very narrow nucellus. The suspensor is wide and short to a degree considered aberrant in the family, but it does not become detached from the integument as in *Rhizophora*. Transfer cells are less well developed than in *Rhizophora*. Two or even three discrete cotyledons are recognizable as the embryo differentiates, and their apical growth lengthens the embryo. The radicle apex is protected by a specialized tissue referred to as the coleorhiza, which begins to develop before the tip emerges from the fruit. The second growth phase, mainly localized in the upper hypocotyl, causes "germination" in the sense that the radicular end of the embryo is extruded from the seed and ultimately the fruit. Juncosa emphasizes that this process corresponds exactly to that in plants with epigeal germination except that the seed is still within the attached fruit. Unlike *Rhizophora*, there is no extrusion of the endosperm (Fig. 10.3C, E). The first plumular leaves are poorly developed. A distinctive feature of this species is the development of buds in the axils of the cotyledons. The histology of the seedling is less elaborate than that of *Rhizophora*; only the inner cortex is aerenchymatous and sclereids and trichosclereids are absent (Fig. 10.5).

   Bruguiera differs from all other Rhizophoraceae in that the fruit remains attached to the seedling and is dispersed with it (Fig. 10.1D, E). Less obvious is that the cotyledons of *Bruguiera* develop flattened, fundamentally laminar structures early and only later become solid and fleshy. In *Bruguiera parviflora*, the cotyledons remain laminar and provide further evidence for an epigeal ancestry.

Additional evidence for placing *Bruguiera* between *Cassipourea* and *Rhizophora* is the intermediate complexity of the cotyledonary vasculature, with *Cassipourea* the least and *Rhizophora* the most specialized. In this respect, a feature of the Rhizophoraceae that Juncosa emphasizes as occuring in few other angiosperms is the late differentiation of provascular tissue, after cotyledonary initiation. This might be a preadaptation for vivipary. *Rhizophora* can be interpreted as the most specialized in its vivipary compared with terrestrial relatives, because modifications associated with vivipary occur much earlier in development.

These extended considerations indicate the approaches that are needed to place mangrove vivipary in an evolutionary context. They show that the range of morphological variation in viviparous embryos is considerable, so that trends of increasing specialization can be established. Further extensive comparison with extant terrestrial relatives might well allow the process of evolutionary specialization to be reconstructed in detail.

Comparison with non-viviparous seedlings. Ng (1978, 2014) distinguished four main seedling types in his study of Malayan forest trees. The least common (only eight percent of his total sample) he describes as the "Durian type," in which the hypocotyl is extended, thus raising the cotyledons above the ground. These remain unexpanded within the seed coats, however, and are shed as a separate unit. *Rhizophora* is included in this category because of morphological similarities. Ng (1978) emphasizes that the Durian type is not very efficient, because the cotyledons are raised by the hypocotyl to no apparent advantage as they do not become photosynthetic, but we will see shortly how, in fact, it can explain the whole purpose of vivipary in mangroves. The common feature to be emphasized is indeed the elongated hypocotyl. In preparation, we can accept that vivipary in the mangrove Rhizophoraceae most likely evolved directly from an ancestor with epigeal germination, as this is the usual condition in terrestrial Rhizophoraceae.

The more elaborate subdivision of seedling types by De Vogel (1980) not only emphasizes the degree of specialization of mangrove Rhizophoraceae, but also contrasts non-viviparous mangrove and mangrove-associate species. Mangrove Rhizophoraceae are segregated as a separate type ("*Rhizophora* type"), which also includes *Aegiceras*. *Cynometra ramiflora* exemplifies a type with fleshy, food-storing cotyledons that persist at ground level, as the hypocotyl remains short. A peculiarity is the extended plumular axis above the cotyledons, which initially produces a series of scale leaves. *Pelliciera* of the New World apparently conforms to this type (Plate 19D); semi-epigeal and semi-hypogeal seem equally appropriate designations (Ng, 1978). *Barringtonia* is also made a distinct type by De Vogel; either it lacks cotyledons or they are vestigial, and the hypocotyl becomes fleshy as a food-storage organ within the seed so that only radicle and plumule eventually protrude . The plumule is again elongated and initially bears only scale leaves (Fig. B.35). This peculiar morphology seems primarily an adaptation that allows the seed to germinate when buried quite deeply. The long plumular axis grows upward until it finds the light. Mangroves with hypogeal germination correspond to less specialized types in De Vogel's classification. There is some minor subdivision of epigeal types; mangroves like *Avicennia* (Sloanea type) and *Sonneratia* (Macaranga type) are contrasted on the basis of thickness of the cotyledons. A perusal of this often extensive literature should point the way to a better understanding of seedling biology of tropical species generally.

Dispersal and Establishment

Dispersal

There is much discussion in early botanical literature about the ability of mangroves to disperse and subsequently establish their propagules (Guppy, 1906, 1917; Ridley, 1930). Clearly, mangrove propagules are adapted to float, at least initially. What then prevents them from occupying all suitable pantropical sites? Is it lack of a suitable substrate rather than an inability to disperse sufficiently broadly that prevents mangroves from reaching some remote islands? *Pemphis acidula*, for example, seems to be the only mangrove on the more remote atolls of southeastern Polynesia (Brown, 1935), because it behaves more like a beach or strand plant. Here lack of suitable growing conditions seems to exclude other (strict mangrove) species. On the other hand, even though the Hawaiian Islands have no native mangrove flora, imported seedlings of *Rhizophora* and *Bruguiera* have flourished and built synthetic mangal. Dispersal seems to be the factor that limits natural establishment on this isolated island group. The question is of phytogeographic as well as ecological interest because the present distribution of mangroves (Chapter 4) suggests that, ancestrally, mangroves may have crossed the Pacific at least once in geological time, but can no longer do so. The marked contrast between the eastern and western groups of mangroves shows that dispersal and subsequent establishment are not random.

A number of investigators have experimentally studied dispersal and ability for establishment of mangrove propagules, chiefly by recording the length of time they float in salt and fresh water. Some "mark-and-capture" experiments have been carried out. Rabinowitz (1978a), for example, established "dispersal" parameters for the common mangroves of Central America. These include longevity (how long the propagule floats while retaining viability), period of floating, period for establishment (permanent rooting), and period of "obligate dispersal" (how long a seedling floats before it can initiate the root system that will allow it to anchor). Clearly, the last is an absolute minimum value for time available for dispersal. She gives values from 35 days to more than a year for longevity, from 8 to 40 days for obligate dispersal, and a flotation time from as little as 1 day for *Pelliciera* in fresh water (but 6 days in salt water) to an "unlimited" period for *Avicennia*. It takes most mangroves 5 to 15 days to develop an anchoring root system. Differences between values for experiments carried out in seawater and fresh water are small and do not really affect performance. Some seedlings regularly regain buoyancy after they have sunk, increasing their dispersal ability. These observations are built later into the hypothesis that propagule weight affects ecological distribution, because larger units like those in *Avicennia* and *Pelliciera* are supposed to be more resistant to tidal buffeting and abound in open water. Hence, propagule weight determines zonation, according to Rabinowitz (1978b, c).

Establishment

An early expressed notion that *Rhizophora* propagules are pointed at the radicular end to promote self-planting as the seedling falls from the trees deserves little serious consideration, even though it is not difficult to watch it happen. Egler (1948) dismissed it with the disparaging appellation "plunk hypothesis." The establishment of propagules, once dispersed, is a more critical problem because the figures presented by Rabinowitz (1978a) and Chai (1982) show that establishment cannot occur during a single low tide. *Rhizophora* seedlings in still water seem to be able to take root while regularly submerged. The radicle (embryonic root) is not necessarily involved in the process; in *Rhizophora*, for example, it aborts and anchorage is the responsibility of the lateral (adventitious) root primordia developed subapically by the short root segment of the propagule. In *Bruguiera*, on the other hand, the radicle is developed. Another problem is how the elongated hypocotyl of the Rhizophoreae, which is usually stranded in a horizontal position, becomes erect. This is now understood and can be explained in some detail.

As an example of the problems facing mangrove seedling establishment, Krauss and Allen (2003) studied the factors that affect the regeneration of seedlings in *Bruguiera gymnorrhiza*. These factors include predation by insects and crabs, location within the intertidal zone, salinity, and degree of shade. The conclusion was that none of these was particularly a strong regulator of seedling success (Floreat *Bruguiera*!).

Hypotheses for Vivipary

The viviparous condition, which is pronounced in mangrove Rhizophoraceae and approached in the cryptoviviparous genera *Aegiceras*, *Avicennia*, *Pelliciera*, and *Nypa*, is so strongly associated with mangal that it is suggested to have adaptive significance in an intertidal environment. No fully acceptable explanation for the correlation has been offered, however. The mangal environment is unique, so that it has been difficult to assign a single adaptive factor for vivipary. Halophytes and tropical swamp forest species do not show vivipary, therefore the condition is neither a response to salinity nor to wet soil alone. On the other hand, mangrove propagules are widely distributed and must become established in slightly to extremely unstable substrate conditions. Rabinowitz (1978b) suggested that the length of the seedling promoted establishment in mangal and that the zonation could be a simple correlation of propagule sorting by wave action. A longer propagule is the product of hypocotyl extension before the embryo is detached; in Rhizophoraceae it is an exaggeration of normal development in terrestrial environments. On the other hand, the diversity of seed and fruit size, especially in IWP mangroves, shows little correlation between propagule size and a species position in mangal zonation.

In contrast to these mechanical explanations, other authors have turned to aspects of the physiology of viviparous seedlings and have investigated variations in salt concentration, respiration activity, and enzyme distribution (notably the series of articles by Pannier 1959, 1962; Pannier and Rodriques, 1967; Pannier and Pannier, 1975). It was established early (Walter and Steiner, 1936) that developing viviparous seedlings in Rhizophoraceae retain lower salt concentrations than the parent shoot and especially leaves; these observations were subsequently verified by Lotschert and Liemann (1967). There is a two-stage drop in salt concentration from the pedicel to the fruit and then from the testa to the embryo. The same condition is found in the cryptoviviparous *Aegiceras* (Bhosale and Shinde, 1983). Localization of a physiological barrier in the floral pedicel or testa (until its rupture) is uncertain, even though there is a specialized "endothelium" or endothelium-like layer, the cells of which have the properties of transfer cells, at the limit of the endosperm (Juncosa, 1982b). In a later study, Wise and Juncosa (1989) showed that transfer cells in *Rhizophora* are unique both in extending around the entire surface of the integument and in their prolonged activity, which lasts for a year. Even so, the authors suggest that the extended green hypocotyl may be fixing all its own carbon compounds and receives only water and mineral salts from the parent plant. No barrier in terms of ionic restriction has been identified. A physiological barrier is not itself necessary to account for different salt concentrations, as the differences could result from a diminished passive accumulation of salts in organs that do not transpire or that transpire slowly. Structural evidence for this is the absence of stomata from the hypocotyls of *Ceriops* and *Rhizophora* seedlings (Wilkinson, 1981). Stomata are present in the seedlings of *Bruguiera* (Rhizophoraceae; Kipp-Goller, 1940) but in limited numbers. On the other hand, seedlings develop numerous and often conspicuous lenticels (Roth, 1965).

This lowered salt concentration, whatever its origin, is interpreted as a mechanism to "protect" the embryo until maturity from the deleterious effects of high salt concentrations. Sudden submergence in seawater is presumably not detrimental, and because most propagules spend some time in seawater before they become established, there is a presumed gradual adjustment to saline conditions (Joshi et al., 1972). Comparable studies for other halophytes that germinate in fresh water but where vivipary does not occur could be informative.

The seedling is first nutritionally dependent on its parent shoot (Pannier and Pannier, 1975; Bhosale and Shinde, 1983), but makes a contribution to its own energy requirements via photosynthesis after the plumule is exposed. Studies have shown that individual species have different rates of seedling respiration (Chapman, 1962a, b). Brown et al. (1969) studied the capability of seedlings to respire anaerobically in the dark using thick slices of hypocotyl tissue, but the methods seem somewhat artificial in relation to normal seedling biology.

Establishment of the elongated seedlings of Rhizophoraceae has occasioned discussion and experimentation, as observers have had difficulty accepting that seemingly rigid axes can become erect after rooting (Bowman, 1916, 1917; Egler, 1948; Larue and

Muzik, 1951). Growth curvatures are possible; and may be a characteristic of established seedlings, as in *Bruguiera parviflora*, but this is a result of initial curvature of the attached hypocotyl, which is determined by its positive geotropic orientation on the shoot (other than when directed downward). A developed radicle is present only in *Bruguiera* (a presumed primitive feature); in other genera, rooting is initiated by lateral roots at the radicular end of the seedling, with the radicle itself aborted. Establishment in most species is rapid; Chai (1982), for example, gives values of as little as one to two days for the initiation of root development in mangrove Rhizophoraceae, with plants firmly rooted after 10 days. Root development is most rapid in seedlings that are planted erect.

Vivipary Explained

Despite suggestions as to why vivipary is so common in mangroves, no successful explanation had been forthcoming until the suggestion of Tomlinson and Cox (2000) that a simple solution could be found in the method of seedling erection in the Rhizophoraceae, a topic that had engendered some earlier discussion and centered on the method of seedling dispersal, prior to establishment. In principle, seedlings are most likely to become stranded in a horizontal position but must rapidly become erect in order to develop an erect axis and bring the shoot apex above the level of tidal influence, and so minimize its effect. The mechanism involves the development of tension wood fibers ("gelatinous fibers" of wood anatomists) in the newly formed secondary xylem, after the seedling has become anchored by adventitious roots. The background to this idea merits a discussion of some of the earlier ideas about the topic. It also makes relevant a reference to the earliest illustration of the seedling of *Rhizophora* provided by Rumphius (Fig. 1.2).

The "plunk" hypothesis. An early explanation, unsupported by any direct observation, was that the mature seedling shape provides a self-planting missile ("torpedo" in some descriptions), so that the released seedling, with or without the fruit, plunges directly into the underlying soft mud so that it remains straight and becomes anchored by the formation of adventitious roots. This idea, the "plunk" hypothesis as named by Egler (1948), was easily debunked by him and subsequent authors (Lawrence, 1949; LaRue and Muzik, 1951). First, the propogule is as likely to fall into tidal water as mud; second, it is an elementary observation that *Rhizophora* seedlings mostly float (in a horizontal or erect position), and thus are carried away from the parent tree by tidal influence, as Rumphius described; third, established *Rhizophora* seedlings do not survive in shade because they are light demanding and occur only in numbers in a closed community where there is a canopy gap. In this respect, seedlings are most likely to survive in open communities.

Yamashiro (1961), working with *Kandelia* in southern Japan (the northernmost mangrove "forest" in the world), demonstrated that initially attached marked seedlings, after they have fallen naturally, are transported within 30 days at least 50 m away from the mother tree. Out of 1854 marked seedlings, 1627 (87.76 percent) were "lost"; that is, were carried beyond the 50 m study plot. Only 31 (1.67 percent) were collected under the mother tree. Figures for trees in different tidal regimens were comparable. As stated earlier, the absence of an understorey in a mature *Rhizophora* forest is explained by the inability of seedlings to grow to maturity in shade.

Later studies thus concentrated on the dispersability of floating seedlings, as in mark-and-capture experiments in which (understandably!) few were recaptured but then at great distances (sometimes kilometers) from the point of release (Davis, 1940). Similar experiments chose to measure the long retention of viability in seedlings floating in either fresh or salt water.

The hook. Once the plunk hypothesis was abandoned, the stranding of elongated seedlings in a horizontal position drew forth further commentary. In terms of dispersal, Egler (1948) suggested that, at times, buoyancy of self-erecting seedlings would be provided by a high tide, but obviously this is not necessary for erection, as our dry land experiments show (Plate 15). The development of a basal hook, as first illustrated by Rumphius (Fig. 1.2), had been well described and further illustrated (Egler, 1948; Lawrence, 1949; LaRue and Muzik, 1954). An illustration by Ding Hou in his account in Flora Malesiana (1958) shows the hook in *Rhizophora mangle*, although this is not an Asian species. It was generally agreed that the hook is the result of the process that re-erects the seedling and allows it to become erect. Some writers were content to illustrate a series of disinterred seedlings to represent the process (Lawrence, 1949), but LaRue and Muzik (1954) observed, over an extended period, the process in *R. mangle* sown horizontally in shallow flats, hence measuring how long erection took. Tomlinson and Cox (2000) conducted similar experiments in Hawaii on all four genera of mangrove Rhizophoraceae (*Bruguiera, Ceriops, Kandelia,* and *Rhizophora*), which showed that all became erected by formation of a basal hook. The experiments were repeated in Florida on *R. mangle*; that is, in its native location (Fisher and Tomlinson, 2002), and were the substance of a refutation of previous mistaken observations (Fisher and Tomlinson, 2012).

None of the early observers seemed to have considered the mechanism by means of which the hook is developed and the seedling axis erected, other than the assumption that it was by a "push" from the lower side in contact with the substrate; that is, by unequal extension growth of the lower surface. There was no anatomical observation of the structure. In contrast, Tomlinson and Cox (2000) showed that the movement was caused by a "pull" via the development of tension fibers (gelatinous fibers) in the secondary xylem produced eccentrically within the hook; that is, on the upper surface and opposite to where the adventitious roots occur most commonly. These observations were confirmed by Fisher and Tomlinson (2002). This shows the efficacy of tension wood even in circumstances of a mechanical disadvantage – rather like pulling a flagpole erect by applying lift at the base rather than the apex. The mechanics would indeed be aided at high tide by flotation.

Once the seedling is erect, the work of tension fibers is done; further but still limited development of secondary xylem throughout the hypocotyl persists without their continued formation, but the hook endures because the hypocotyl as a whole undergoes little further secondary thickening, so that the hook is visible on older seedlings (e.g., Fig. 1.3). We must recall Rumphius' illustration of the later stages of trunk development (Fig. 1.1) and realize that secondary growth of the axis increases most rapidly in a distal direction and in association with the aerial roots formed in succession up the trunk. The hook persists unobscured by this later growth and as the vestige of the initial erection process. Later-formed roots in a distal direction become progressively wider by secondary thickening, in reverse order of their age, to form the massive flying buttresses of the mature tree (Plate 13A).

The function of the hook is therefore seen as a mechanism that facilitates the rapid erection of an axis that otherwise is regularly inundated by the tide. Consequently, the plumule is positioned as early as possible in a completely aerial environment, something like a swimmer's snorkel. This adaptation is anticipated in the necessarily viviparously extended hypocotyl. Gas exchange in plumular leaves is promoted as they are least inundated. The hypocotyl alone is less able to promote adequate gas exchange because it is distinctly devoid of extensive stomatal development, which in part explains how it can float for an extended period without becoming waterlogged, part of its earlier adaptive construction.

The development of reaction fibers in a viviparous seedling axis requires no unusual mechanism because they occur routinely and eccentrically in the tilted seedlings of terrestrial plants, where they have been used extensively in studies of reaction wood formation. They occur, for example, in tilted seedlings of other mangroves (e.g., in *Avicennia* and *Laguncularia*). The eccentric production of tension fibers is thus a normal condition of seedling growth, but in the elongated seedlings of the Rhizophoraceae, it is clearly restricted to the base of the axis where the adventitious roots are also eccentrically developed by contact of the lower surface with the substrate, and where secondary xylem is first produced. The association of secondary growth with root formation seems physiologically as well as functionally efficient.

Hence, vivipary is explained.

Distal aerial roots themselves may form reaction fibers in later development, a topic not yet explored fully. That reaction fibers can be very significant in roots and play a major role in tree architecture, is demonstrated in *Ficus*, where tension fibers are produced concentrically in aerial roots once they reach the ground, ultimately resulting in straight columns that support the massive crowns of many species (Zimmermann et al., 1968).

Hypocotyl Polarity

Polarity of Seedlings

A distinctive feature of rhizoporaceous seedlings is the marked polarity of the hypocotyl in terms of its ability to regenerate new organs. Plumular and radicular ends are contrasted markedly in this respect, as initially emphasized by LaRue and Muzak (1954). The difference is easily demonstrated experimentally (Plate 15).

The results show that when terminal portions of the hypocotyl are cut off, radicular cut surfaces generate roots (Plate 15A, B), and plumular portions generate shoots (Plate 15C, D). This makes sense adaptively, but the mechanism remains unexplored. In damaged seedlings, the regeneration of new shoots can be seen often in the wild. This polarity seems universal for other genera in Rhizophoreae. Ogita et al. (2004) show it in explant experiments on *Kandelia candel*, but add the novelty of showing how the initials of the new shoots originate from a phellogen-like meristem at the cut surface and in the region close to preexisting vascular bundles. Regeneration did not require added growth substances. An interesting model system is thus revealed.

Summary

Summarizing this information produces the hypothesis based on the following observations and logical extensions.

1. Vivipary is a mechanism generated by absence of a dormant phase in the development of the embryo in the fruit.
2. All subsequent features are normal features of a seedling with epigeal germination (the presumed ancestor).
3. The seedling develops an extended hypocotyl with marked polarity while still attached to the parent plant and capable of penetrating the pericarp of the fruit (i. e., it germinates).
4. The elongated seedling is detached (usually from the fruit) and dispersed, often widely, by flotation, its epidermal features preventing waterlogging.
5. It is stranded, most usually in a horizontal position, becoming anchored via adventitious roots formed on the surface in contact with the substrate.
6. Secondary xylem is rapidly developed but in an asymmetric condition, as an evident gravimorphic response.
7. Gelatinous fibers are produced on the upper side, normal fibers on the lower side.
8. Gelatinous fibers function as tension fibers and contract, creating a tension that shortens the axis on the upper side and erects the seedling axis relatively rapidly.
9. This raises the plumular end of the seedling, so that if inundated (as is periodically possible in a tidal environment), the shoot can function more rapidly in an aerial environment and in a gaseous environment, gas exchange facilitating photosynthesis.
10. The erect seedling can function at a competitive level in a high light environment, but once erect, it is no longer subject to an eccentric gravimorphic stimulus.

Taking all other factors into consideration, the unique feature of the mangal environment (because it shares other features with halophytic and heliophytic habitats) is the diurnal fluctuation in water level, which is most pronounced in the seaward direction. The viviparous condition under these circumstances is a unique method of adjusting to this habitat, and is a mechanism widely deployed by the "true" mangroves whatever their evolutionary origins, because it is a simple modification of developmental processes that characterize most flowering plants; that is, Occam's razor applies. Further studies should explore this mechanism.

11 Mangroves and People

The great impediments to this stage even worse than the tea-tree swamps, has been
the positively evil mangrove swamps. It is the red mangrove that rules hereabouts;
a greedy usurper, which occupies the no-man's-land between dry land and water,
a wide and flat space shunned by other plant life. This grasping tyrant seizes the
unwanted, and consolidates its position so as to become unassailable, surrounded by
salt water at high tides and mud at low tides, when it is irrigated by ooze from sluggish
backwaters. The mangrove spreads widely, its roots looping, branching, grappling,
entwining, revelling in the sludge; from its branches it drops aerial roots into the static
convulsions below. And over all is a dense roof of glossy leafage, which effectively
excludes sunlight from the stinking, impenetrable chaos beneath. One can do nothing
with mangroves but avoid them.

From *Kennedy of Cape York* by Edgar Beale (1970)

Introduction

Because of their low floristic diversity and absence of a specialized understorey,
mangroves may appear to have limited resource diversity, but this is misleading. Human
interaction with and use of mangroves must have gone back to unrecorded times, such is
the propensity of people in the tropics to occupy coastal regions. There is a record, dated
200 BC, of trade with poles harvested in the mangal in East Africa and exported to the
Middle East (Spalding et al., 2010). This represents utilization, although people have
rarely dwelt within mangroves (despite the existence of a self-flushing sewage [seaw-
age?] system) because of their general inhospitable and often mosquito-infested nature.
On the other hand, people in the adjacent hinterland have derived protection, as from
windstorms and tsunamis, by the ameliorating effects of an existing tidal forest,
although forced evacuation is a better solution. Storm surges can drive massive wave
formations many miles inland so that any means of protection is desirable. Documentation exists showing that a seawall of mangroves saves lives (Das and Vincent, 2009).
Mangroves otherwise provide a diversity of raw materials: timber for construction
(including boats and fish-traps), pulpwood (paper) or fuel wood (mostly as charcoal),
fodder for livestock (despite a high salt content), tannins, but rarely food. Mangroves
are a cheap source of medicines and pharmaceuticals (Patra and Thatoi, 2011). They can
be an important source of honey, in competition with nectarivorous animal pollinators.
However, the greatest benefit is indirect, as seafood, because mangroves are the
nurseries for juvenile stages of commercial crustaceans and shellfish, as well as fishes

harvested offshore. These contrasted uses are documented extensively in Spalding et al. (2010) and described in dramatic detail in Warne (2011).

Human use of mangroves can be abusive and with long-term effects, as when they are overexploited (as for charcoal, timber, and pulpwood) or totally destroyed in some form of conversion. The subject is described at great length by Warne (2011), with observations of many of the socioeconomic consequences. Destruction includes urban development along heavily occupied coasts, conversion to agriculture (rice paddies with salt tolerant rice varieties), or more extensively, mariculture (usually shrimp farming, although the economic benefit may be short-lived), or development of artificial salt pans. On a global basis, these processes are extensively commented upon in Spalding et al., 2010. This overexploitation can be followed by limiting legislation, which is intended to protect mangroves, but often (and ironically) it occurs where urbanization is most extensive, as in South Florida. Enforcement can be difficult or easily avoided. Degraded mangal is not necessarily easily regenerated because of substrate modification, often by erosion. Mangal occupies a high-energy environment. In addition, mangal is highly susceptible to pollution, most obviously by oil spills from offshore wells, and especially, adjacent to coastal refineries.

Toxic waste of various kinds may accumulate in the mangal, which can filter and retain flotsam, very evident where mangroves impinge on large cities. A cohort of volunteers regularly removes accumulated garbage from the mangal at the Kampong Garden of the National Tropical Botanic Garden in Coconut Grove, Florida. This Garden has also been the site of an accidental experiment that has demonstrated directly the sensitivity of mangroves to toxic chemicals. This involved the eradication of a small population of the introduced mangrove *Bruguiera gymnorrhiza* designated as "invasive," although in its 70 years since its planting, it had extended its seedlings only about 30 m and nowhere beyond that (Fourqurean et al., 2009). The population had been used extensively for teaching and research. Plantings of this species in gardens elsewhere had not survived in South Florida in competition with adjacent native plants. As part of the eradication process, a herbicidal drench was applied. This immediately killed 15 large trees of *Rhizophora* and *Avicennia*, even though tidal flush would have diluted the toxin. The delicacy of such robust trees was thus demonstrated.

Economic Complexity

Mangroves, in a sense, are an entrepreneur's dream: they produce raw material (lignocellulose) from seawater by using renewable energy sources (Clough, 1992). Most energy is supplied by sunlight. Tidal energy is also used, because nutrients are replenished by tidal action and estuarine runoff and detritus are conveniently flushed away (Fig. 2.1). Furthermore, if this energy is to be harvested, immediate transport by water in and out of mangal can be facilitated by the proximity of river mouths and the sea. If the products are to be exported, they are already on the coast. Despite these benefits and extensive exploitation, the question has been asked: "Mangroves – what are they worth"? (Christensen, 1983). The answer depends on whether one sees mangal as a resource, either to be harvested once as a realizable asset or to be sustained on a renewable basis as a dividend from a fixed investment, or as land to be converted into a more profitable resource by a conversion of capital. The question is also phrased in terms of a direct economic assessment of mangroves per se, so that some monetary value can be put on both capital and its yield. The question is not exactly an academic one and litigation has, in one instance, established a value of $751 368.30 per hectare for the restitution of mangrove in Puerto Rico damaged by an oil spill (Lewis, 1983).

In contrast, one may express a concern for the long-term socioeconomic implications of mangrove exploitation at various levels: from the extremes of a national economy to a local resource important only in a peasant economy. It may be possible to give reasonably accurate dollar values at the regional level (Warne, 2011), but it is much more difficult to do so at the local level, even though the greatest total utilization occurs locally and probably benefits most individuals. The total number of people who derive a livelihood from mangrove plants (many millions) may be more significant in social terms than total profits to some industrial complex. The covert uncertainty of who benefits most underlies any discussion of mangrove utilization, no matter how precisely assessments have been made. Decisions about appropriate policies for mangrove use are thus made difficult.

A major problem in economic analysis is the difficulty of recognizing or distinguishing indirect and direct benefits, apart from problems of establishing economic benefits where either there is no cash flow or it cannot be measured very accurately (Silva et al., 1990). Mangal can also represent an interphase community (which makes it such an attractive ecosystem for study by biologists), and therefore creates administrative complexity, given that competing economic interests are at work. Major industries, such as forestry, fisheries, and agriculture, may be in conflict because each can claim mangroves as their administrative domain, and the policy that is best for one industry may be detrimental to another (Christensen, 1983). A forestry department will emphasize utilization that may degrade the resource; a fisheries department will emphasize conservation with a minimum of disturbance; and an agricultural department may advocate conversion and replacement by some putatively more valuable resource. Finally, coastlines are valuable real estate. These conflicts are the background to mangrove management as well as much of the economic justification for extended research on mangrove communities. In addition to summaries by Christensen (1983) and Walsh (1977), there are extensive older accounts, such as by Foxworthy (1910), Brown and Fischer (1920), Brown and Merrill (1920), and Watson (1928). The literature on these topics is voluminous, but the overview in Spalding et al. (2010) asks many penetrating questions and is a lead source.

Forestry

Mangroves constitute a minor part of the forest resources of most tropical countries, as they can occupy only a limited area, generalized as about one percent of forest area in most tropical countries, but two percent of all forest types in Papua New Guinea

(Percival and Womersley, 1975). The largest *per capita* values occur on small tropical islands, with the Turks and Caicos possible leaders; Isle de Bassin in the Indian Ocean has about 50 percent mangal, but is uninhabited.

Nevertheless, some forestry departments have emphasized mangrove management and utilization, notably in Malaysia. The classic study by Watson (1928) was an attempt to produce information from which a management program was to be developed. The main objective in the program was to produce biomass on a sustained yield basis. Changing technologies tend to move faster than recommended rotation cycles, so that a growth rate designed to support a sawn timber industry might be ineffective if directed toward a pulpwood or a pole timber resource. A short-term rotation might be suitable for pulpwood but allows no thinning for firewood. In terms of actual harvesting, it is usually agreed that clear felling is the preferred procedure; discussion then centers on whether subsequent regeneration should be either natural or artificial and to what extent selective culling is to be allowed during the regeneration cycle.

Sawn timber. A number of major mangal constituents produce timbers with desirable qualities, such as high density (e.g., Plate 10G) and termite- and marine-borer resistance, so that woods may have special uses, such as in boatbuilding and fishtraps. Low floristic diversity is an asset when extraction is necessary and can be used as an argument for clear felling. Large-scale exploitation using heavy equipment is very disruptive of the community. Mangroves are a decreasing source of merchantable timber, however, because of their decreasing average maturity as they are exploited. A few species of the back-mangrove community; for example, *Heritiera* and *Xylocarpus*, produce high-quality timber but their scarcity and difficulty of access have made them unattractive for harvesting. *Heritiera fomes* in the Sundarbans has been the most actively exploited species.

Poles. Unsawn poles are now the most common extraction product, especially at the local level, because they can be harvested readily by simple manual methods and because of short rotations. Such use is known to have existed for over two millennia. An example from Thailand is illustrated in Plate 16D, but is unusual as it is a silvicultural activity, with *Rhizophora* planted in straight rows and the aerial root system suppressed in some way. Poles are also suited to the short rotation often practiced in managed forests (frequently by necessity where population pressure is great), and are relatively easily harvested and transported (Plate 16B). These practices, although they do not allow the regeneration of tall forest, are probably minimally disruptive of the community. Regeneration from seed is not inhibited because most mangroves flower as saplings, but there could be selection for slow-growing species.

Fuel wood. This may be used directly or after conversion to charcoal, and is probably the main biomass utilization of mangroves at the local level. Mangrove management as a fuel resource is also suited to minimally disruptive short-term rotations, as only small trees are needed and these are most easily harvested by manual woodcutting, which minimally disrupts the substrate. Charcoal manufacture can be either on an industrial scale (Plate 16E), or as a subsistence activity (Plate 16F).

Tannins and dyes. Use of mangroves for tan bark has, at times, been extensive; the entries under "tannin" in the index by Rollet (1981) amount to eight columns. Industrial development of synthetic or more accessible alternatives has tended to shut down most commercial operations (e.g., in Papua New Guinea, as described by Percival and Womersley, 1975). The high tannin content of most mangroves, especially in the mangrove Rhizophoraceae, presumably increases their resistance to herbivores, so that this biological attribute has had a direct industrial spinoff. Local use of mangrove tannins may still be common, as in the curing of fishing nets. Mangrove sap is the source of a black dye used in the manufacture of Polynesian tapa cloth.

Pharmaceuticals. Although medicinal use of mangroves and especially mangrove bark is commented on in the literature, the review by Thatui et al. (2014) explores potential use on the basis that such plants, living in highly stressful environments, are likely to be rich in antioxidants. Much of the recent literature on mangroves consists of identification of their biochemical constituents. Mangroves have had limited local use in medicine (Burkill, 1935), but many of their properties are assumed to be, if not, magical. Fortunately, there are few poisonous plants in the mangroves. *Gluta* (Anacardiaceae) has an irritant sap that causes skin rashes; *Excoecaria* and *Hippomane* (Euphorbiaceae) have an irritant sap that causes skin ulcers. The seeds of *Cerbera* are said to be poisonous. Where plants contain saponins, they have been used as fish poisons (e.g., *Barringtonia* and *Derris*).

The Nypa palm. The mangrove (nipa) palm, which often dominates quieter estuarine and lagoon areas in southeast Asia, is exceptional in that it provides a diversity of minor products (Fong, 1987a, 1989). Industrial ethanol is made by distilling the fermented sugary phloem sap, which is obtained by tapping the lateral inflorescence (Päivökee, 1984), and at times has been produced on a large scale. It is a suggested source of wine (Fong, 1987b). An early account of the industry in the Philippines is given by Brown and Merrill (1920), who comment that the source at that time was the cheapest known form of industrial ethanol; a statement hardly likely to apply in modern times. The physiology of sap production is not understood; it is presumably similar to the process in other palms tapped for sugar and wine (notably *Arenga, Borassus, Cocos, Phoenix,* and *Raphia*). Even though *Nypa* ethanol represents a renewable fuel resource, it seems not to be suited to large-scale, successful commercial production. *Nypa* provides other minor commodities: attap or thatch from its leaves, which are rated superior to any other palm thatch (Plate 16C), cigarette papers from the stripped surface layers of the leaflets, and a sort of salt from the ashed leaflets (Burkill, 1935). The edible jelly from the unripe endosperm is a sweetmeat and one component of "gulah malacca" (a three-palm product, with coconut and sago), a traditional Malaysian dessert valuable in quenching the fires of hot curries.

Fisheries

The indirect relationship between mangroves and some inshore commercial fishing and shrimping is well recognized. First, mangroves supply nutrition (mainly via a detritus chain) to marine communities. Second, they may provide a habitat for some commercially exploited marine organisms at critical phases of their life cycle; they function as nurseries. Third, they may directly support some shellfish, notably oysters, either growing on aerial roots (Plate 2B, C), cultivated, or growing naturally in mud. Finally, mangroves stabilize shorelines and protect inshore fish habitats from sediment pollution. Figures for the cash value of commercial fish catches are relatively easily assembled and can be used to demonstrate the indirect commercial value of mangroves (Christensen, 1983; Warne, 2011).

Plant–Animal Interactions

Mangrove plants are demonstrably unpalatable and provide little direct human food. This is a reflection of their high tannin content and leathery texture, which may even involve extensive development of ideoblastic sclereids (notably *Rhizophora*, *Sonneratia*, and *Aegialitis*). There are reports of young viviparous seedlings of *Bruguiera* or of *Avicennia* seeds being boiled and eaten, but largely as famine food. In general, mangroves are designed to resist herbivores. This may account for the lack of close plant–animal interactions within the community. Despite this, there are records of ungulates seen browsing in mangroves, and attempts to cultivate them in the Middle East as goat and camel fodder are known (Spalding et al., 2010). Planting mangroves for fodder has been attempted in Eritrea (Warne, 2011), but somewhat controversially, as they require added fertilizer.

Although pollination by animals is common, these pollinators are very generalized and probably not residents. Parakeets in Australia can be a destructive element in their search for nectar in *Sonneratia* flowers. Mangrove fruits, seeds, and seedlings are all dispersed by water. Fleshy, edible fruits are not produced; the most conspicuous exceptions are the mistletoes (e.g., *Amyemena*) and epiphytes (e.g., *Hydnophytum*), where bird dispersal of fruits is essential to their establishment. On the other hand, seeds of a few mangrove associates are edible (*Barringtonia edulis* and *Inocarpus fagifer*). Animal-pollinated flowers in most instances have nectar as a pollinator reward, and in some areas, mangroves are an important commercial source of honey. Ninety percent of all natural honey in India is produced from mangroves. More details of these interactions are given in the accounts of individual species.

Ecotourism. Perhaps a very modern form of plant–animal interaction is seen in the development in recent decades of tourism within mangroves and certainly one that justifies the preservation of the mangal as a direct source of local income. Visitors may enjoy bird watching or the sight of proboscis monkeys. Tours may be conducted along boardwalks or in shallow draft boats, producing a sense of adventure, especially if the presence of crocodiles and tigers is mentioned. Perhaps the most romantic are tours conducted at night to view the abundance of fireflies, as in Malaysia (Wah Juliana et al., 2012). Ironically, tours can best be conducted in areas adjacent to modern hotels built where mangroves have been cleared!

Wildlife management. Christensen (1983) and MacNae (1968) provide information about the relationship between faunal communities in mangals and the need for preservation measures, listing species of mammals, birds, and reptiles. There is not a high degree of interdependence, as we have seen, because few animals are exclusive mangrove dwellers and occur in small numbers (Luther and Greenberg, 2009). The mangrove forest does not make "hunting" on foot an attractive proposition. It is of interest, however, that the few "dangerous animals" in mangroves are largely themselves "endangered"; for example, crocodiles (world-wide) (Plate 3D) and tigers (Bangladesh) (Plate 3A, H). Mangroves have been a refuge for the "Key deer" in South Florida. Spalding et al. (2010) include an enumeration of large animals that have been recorded. It should be emphasized that most animals in the mangal are visitors, as with migrant birds or animals that naturally inhabit wet habitats, like hippopotami (Plate 3B), elephants, water buffalo, but where browsing is limited because of high salt content – or even favored for the same reason. Any marine mammal may be encountered in the mangal; manatees in South Florida and seals elsewhere (Plate 3C). Efforts to expand the mangal artificially in Eritrea have had the objective of providing browse for goats, sheep, and camels.

Agriculture and Aquaculture

Agricultural use of mangal is directed mostly toward its conversion, and salt-resistant rice varieties have been successfully cultivated, notably in Sierra Leone in converted, impounded mangal (Walsh, 1977). Otherwise, conversion is usually difficult either because the texture of the mangrove substrate is unsuited to the subsequent use of standard agricultural equipment or because the previously anaerobic soils, when oxygenated, become highly acidic. Conversion of mangal to rice fields in Costa Rica has been legislated against as a conservation measure.

More successful conversion for mariculture and aquaculture has been achieved, yielding fish, shrimp, and shellfish. Oysters have been successfully cultivated with little or no mangrove conversion. This simply uses artificial supports for the crop, such as stakes at the edge of estuarine creeks. In aquaculture of this type, the mangal simply coincides with a suitable substrate and a salt-water medium for fish culture (Plate 16A). Nutrient resources are made available via the mangal so that enterprises of this kind again depend on a minimally disturbed plant community. On the other hand, extensive conversion to developed shrimp farms has been the most destructive of tidal forests and on a global scale. Hence, the title of the book by Warne (2011) – *Let them eat shrimp* with the subtitle *The tragic disappearance of the rainforests of the sea*. His theme is that such activity destroys the livelihood of many millions of coastal dwellers in the tropics.

Other Industrial and Quasi-industrial Uses

Raw materials. Mention has been made of industries that use the mangrove biomass with little or no subsequent industrial conversion. In contrast, modern technology may demand little more than a supply of lignocellulose as a manufacturing substrate. Recent development toward the total chipping of mangroves for pulpwood and cheap synthetics (notably rayon) has been extensive in southeast Asia, especially Malaysia, the Philippines, and Papua New Guinea. The material is exported for conversion to industrially developed countries, such as Japan. This is another kind of destructive exploitation. The disturbance to the community is maximal because all standing bio-mass is removed. Soil textures are disrupted and there may be little subsequent regeneration. Furthermore, there may be little direct benefit to local communities, because the foreign currencies that are generated do not percolate to the local level. The employment benefit may be short term. An ironic situation on Bougainville, Papua New Guinea, saw the development of the massive copper mine at Panguna in the 1970s, destroying lowland forest and disrupting local communities. The massive amounts of coastal sediments generated from inland erosion were suggested to be an ideal substrate for the development of mangal!

Salt conversion. In drier areas, mangroves can be converted to salt "pawas"; there may even be an alternation between salt and shrimp production within the same enclosure in dry and wet seasons, respectively. The product may not compete commercially with the large-scale production in the drier subtropics (e.g., the Bahamas and Western Australia), where tides, topography, and climate are optimally favorable. Ashing of mangroves is a local alternative. Some mangrove plants secrete salt, presumably very pure (e.g., *Avicennia*), but no attempt seems to have been made to capitalize on this natural process.

Sewage treatment. Mangrove swamps close to major centers of populations are familiar to travelers as unofficial garbage dumps and local sewage-disposal units. The detritus of the consumer economy is often not readily dispersed because of the restricting aerial root systems of mangroves. A common (but discrete) sight around Asian coastal villages is the "thunder box" accessible by a boardwalk and built into the mangal. Tidal scour makes the unit self-flushing. A more serious consideration for the use of mangroves as natural sewage-treatment plants has been presented by Australian ecologists, usually with scrubby mangal in mind (e.g., at Westernport Bay, Victoria). The effectiveness of this proposal depends on the knowledge of nutrient cycling in these communities. An extended evaluation has been provided by Clough et al. (1983).

Coastal protection and accretion. There is much discussion in the ecological literature on the extent to which mangroves trap sediment and so contribute to land building. The general conclusion seems to be that they do so only in geomorphologically prograding regions; that is, areas that would undergo sedimentation even in the absence of shoreline vegetation. Mangroves nevertheless stabilize these prograding shores and prevent excessive shifting of coastlines. Consequently, they are particularly important on coasts that are subject to major tropical storms, and they buffer the destructiveness of wind and storm tides during hurricanes and typhoons. In North Queensland, small sea-going vessels are routinely moored in creeks within mangroves as an impending cyclone advances, as a proven protection from damage, a "cradle of safety." Elsewhere, "shelter forests" of *Sonneratia* species have been constructed. A major part of this natural barrier to storm destruction is the aerial root system of *Rhizophora*, as pointed out by Méndez-Alonzo et al. (2015). Hydraulic drag of these roots has been estimated by computer simulation (Ohira et al., 2013). Deforestation of deltaic regions in Bangladesh (the Sundarbans) has been widely accepted as a cause of increased property damage and loss of human life during tropical storms, as noted earlier (Das and Vincent, 2009).

Direct economic measurement of the benefit of the protection of undisturbed mangal to coastal regions is again difficult, except where one can estimate property damage in adjacent unprotected regions. The buffering effects of mangal are comparable to the benefits of terrestrial watershed forests in preserving water quality and preventing erosion. In some areas, real estate development of mangroves may be economically attractive in temporarily increasing land value (or by dredging operations, actually creating saleable building sites.) The "developer" may adopt a Janus-like attitude, expressing concern for the quality of the environment and even seeking limited re-establishment of natural communities, such as seagrass meadows and mangal, on sites that would have retained their economic and aesthetic status in the absence of "development." This ethical dilemma has been acute in areas like South Florida and Singapore, where affluence and population density demand the ready conversion of a resource. At the same time and somewhat paradoxically, Singapore provides a good example of attempts at regeneration of mangal in the study by Lee et al. (1996), leading to a discussion of the effect of substrate on such efforts.

Rising sea levels. A hot and fundable topic in the field of ecology is the impact of climate change in raising sea level, and mangroves are at the literal frontier of this effect, as discussed by Krauss et al. (2014) in a paper that was highlighted on the journal cover. The topic has been studied experimentally by Ellison and Farnsworth (1997), albeit with a simulated sea-level rise. Historically, of course, mangroves are adapted to such changes and not necessarily over the long term (Roth, 1992).

Productivity

The economic utilization of mangroves depends ultimately on biological productivity, usually measured in the amount of biomass produced per unit weight of standing biomass. I make no attempt here to summarize or review the extensive literature on the subject, because it is not entirely relevant to the scope of the book, and it is difficult to make generalizations based on studies that have been carried out in different parts of the world and in communities of different ages that may have very different canopy heights and very diversified nutrient patterns. Early reviews are provided by Christensen (1978, 1983), Lugo and Snedaker (1974), and Chai (1982). One notes in this literature the wide range of values (from 2 to 16 $m^3.ha^{-1}.year^{-1}$ of wood mean annual increment). The higher values are comparable to those that can be produced for both temperate and tropical forests, or even exceed them. Bearing in mind that wood is not necessarily the only commercially utilizable product of mangroves, this suggests that carefully managed mangal is indeed an excellent economic resource. A modern, comparative assessment of mangal productivity at the hands of skilled production ecologists is urgently needed.

Conservation

A recurrent theme in recent literature has been the concern about the global loss of mangal (Spalding et al., 2010; Warne, 2011). To some extent, this increased destruction in many tropical countries is a post-colonial condition occasioned by an increase in population and a global economy. The spectacular rise of interest in mangroves is the result of two interlinked philosophies. First is the concern of conservation biologists, but directed toward a community rather than individual species (because no species of true mangrove is listed as endangered, such is their wide distribution). Nevertheless, that a species may locally be exterminated is well documented by the disappearance of *Bruguiera* from Taiwan. In India, there has been concern for the disappearance of some mangrove species over large areas. Second is the effects that global warming might have on rising sea levels, where the mangal obviously will be directly affected.

The extent of disappearance has many different causes and cannot always be laid necessarily at the door of local populations. An example is seen in the Sundarbans, the deltaic mangal of the River Ganges. Here, increased sedimentation is a primary cause of mangrove loss because of the deforestation around headwaters far inland. Here, positive measures have been the artificial planting of mangroves, as extensively documented in Spalding et al. (2010).

Conclusion

Preservation and utilization of mangal directly or indirectly for agriculture, fisheries, and forestry at both the local and industrial level are seen as competing influences for the same limited and often sensitive resource. The mangrove community is almost a microcosm of the socioeconomic complications attendant on the human use of a natural resource. Undoubtedly, mangal is useful, although "what is it worth" is not a very specific question until it is qualified by "to whom" and "in what terms." Mangal may be a simple ecological community in terms of species diversity, but it is an exceedingly complex ecosystem to evaluate. The biological basis for sustained use and management is still deficient; the safest policy minimizes direct utilization and especially disturbance. The woodcutter puttering his way through mangrove creeks with a load of firewood in a shallow-draft boat (Plate 16B), or the family fisherman returning with his catch to feed his family (Plate 16A), or women (*concheras*) who have some income from collecting shellfish, or the crab collectors (*catadores de carangueyo*) who supply local restaurants, may not be very dramatic figures in entrepreneurial terms, but this is probably the best image that can be associated with a mangrove forest treated as a resource to be sustained in perpetuity.

Part II

Detailed Description of Families

Family: Acanthaceae

A family of mainly tropical herbs, shrubs, or even small trees with zygomorphic, sympetalous, usually conspicuous flowers with (two) four (five) stamens; ovary bilocular; fruits usually capsular with the two or more seeds compressed and surrounded by the hardened funiculus, which functions as a jaculator when the capsule itself bursts. *Acanthus* is the only genus of the family that has representatives in mangrove communities. Modern systematic analysis suggests that *Avicennnnia* should be included in Acanthaceae; here it is retained in the traditional Avicenniaceae. One species appears in Rumphius (1741–1755) as "*Aquifolium indicum.*"

Acanthus L. 1753 [*Sp. Pl.*: 639]

A genus of some 30 species in tropical Asia and Africa but with a center of diversity in the Mediterranean region (including southern Europe). It is distinguished from related genera that also have one-celled anthers and only four calyx segments (e.g., *Blepharis* and *Crossandra*) by its usually spiny leaves, spicate terminal inflorescences, two bracteoles, and uniform anthers. It is distinctive within the family, which lacks stipules, in often having a pair of spines on each side of the stem in the stipular position at most nodes (Fig. B.1d). The spines persist, and the stem itself is commonly described as spiny. Sometimes the spines are double.

The species described here are characteristic associates of mangroves with a total range from India to the Western Pacific (New Caledonia), tropical Australia, the Philippines, and as far north as China. Three species are circumscribed (van Steenis, 1937) as indicated in the accompanying key, but there has been a tendency to treat them as one single variable species. They do not seem to differ in any consistent vegetative feature. They penetrate most extensively into mangal as a light-demanding understorey below canopy gaps.

1A.   Open flower 3.5–4 cm long; corolla in part light blue or violet (rarely white). Bracteoles persistent, in fruit up to 1 cm long. Ripe fruit 2.5–3 cm long or longer; seed approximately 10 mm in diameter. Inflorescence usually longer than 10 cm. Plants typically robust with spiny to very spiny leaves. Common in the back mangal throughout the Asian tropics from India to Polynesia and Australia; rarely inland. . . . .*A. ilicifolius*

1B.   Open flower 2–2.5 cm; all parts smaller than in previous species. Corolla mostly white, less commonly deep purple. Bracteoles small, inconspicuous, early deciduous, or absent. Ripe fruit shorter than 2.0 cm; seed 5–7 mm in diameter. Inflorescence variable. Plants sometimes delicate. 2

2A.   Plant usually spiny with thick stems. Leaves usually widest below middle. Bract distinctly shorter than the calyx; deciduous before flowering. Bracteoles usually present but early deciduous. Common, often sympatric with the previous species, but with a more limited geographic range, mostly south and west. . . . .*A. ebracteatus*

2B.   Plant usually unarmed, with slender delicate sprawling stems. Leaves usually widest above the middle. Bract always longer than the calyx; deciduous during flowering. Bracteoles not present. Uncommon, locally present in Indonesia and New Guinea. . . . .*A. volubilis*

Acanthus ilicifolius L. 1753 [*Sp. Pl.*: 639] (Figs. B.1, B.2)
Acanthus neo-guineensis Engl. 1886 [*Bot. Jahrb.* 7: 474]

A low sprawling or somewhat viny herb, scarcely woody, to a height of 2 m. Hermaphroditic. Axes initially erect but reclining with age; branching infrequent and commonly from older parts. Aerial roots from lower surface of reclining stems (Fig. B.1b). Leaves decussate, usually with a pair of spines at the insertion of each leaf. Leaves glabrous; petiole short (1–1.5 cm); blade up to 20 cm long, gradually tapered below, either broadly lanceolate with an entire margin (Fig. B.1a), the apex rounded or mucronate, or more usually with a sinuous, spiny margin (Fig. B.1c), the apex broadly tridentate including an apical spine; spines sometimes also present above and below on major veins. Major spines at end of leaf lobes horizontal or erect; minor spines between major lobes erect. Inflorescences terminal, forming bracteate spikes 10–20 cm long, the spike extending with age. Flowers in four ranks (Fig. B.2a), up to 20 pairs, the bract below each flower 5 mm or shorter, often caducous; lateral bracteoles two, conspicuous and persistent. Flowers perfect; calyx four-lobed, the upper lobe conspicuous and enclosing the flower in bud, the lower lobe somewhat smaller; lateral calyx lobes narrow, wholly enclosed by upper and lower sepal (Fig. B.2f). Corolla zygomorphic, sometimes described as "mauve," at least 3 cm long with a short tube closed by basal hairs; abaxial lip broadly three-lobed to entire, adaxial lobes absent. Stamens four, subequal, with thick hairy connectives (Fig. B.2g), anthers medifixed, each with two cells, aggregated around the style (Fig. B.2h). Ovary bilocular, with two superposed ovules in each loculus, style enclosed by stamens, the capitate to pointed stigma exposed. Fruit a capsule 2–3 cm long and 1 cm wide, usually with four rugose angular seeds about 1 cm long; testa delicate, wrinkled whitish green. Jaculator conspicuous. Germination hypogeal.

Growth

Acanthus ilicifolius is typically a low woody herb that owes its ability for vegetative spread to its reclining stems so that it forms large patches by vegetative means. Growth

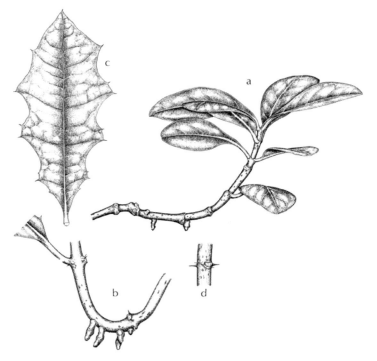

Figure B.1. *Acanthus ilicifolius* (Acanthaceae). Vegetative morphology. (a) Distal end of reclining stem (x1/2), leaves not spiny. (b) Adventitious roots on lower surface of bent axis (x1/2). (c) Distal leaf with spines (x1/2). (d) Node with spine pairs in stipular position (x1/2). (Material cultivated at Fairchild Tropical Garden, Miami, FL.)

is continuous in the sense that there are neither resting terminal buds nor obvious articulations; in unfavorable seasons, shoot growth may be inactive.

The distal, leafy shoots are at first erect, to a height of as much as 2 m. They are determinate by flowering, and the old terminal spikes are substituted by one branch (sometimes two) that originates from buds in the axils of leaves just below the inflorescence. Axes eventually recline under their own weight, apparently for mechanical reasons. This reclining growth in itself is not proliferative and is partly a consequence of the limited production of secondary tissue. In addition, proleptic branches develop from the older horizontal axes at infrequent and seemingly irregular intervals. This lack of self-support coupled with infrequent branching accounts for the common description of the plants as viny, as they may scramble over any adjacent vegetation that will provide support. The development of adventitious aerial roots is important in the sprawling habit because it anticipates the need for secondary rooting of shoots after they become horizontal. The aerial roots are initially thick, somewhat fleshy, and with a well-developed periderm.

Leaf morphology and variation. There is considerable variation in leaf form, mainly in relation to the degree of spinyness. Leaves are sometimes entirely spineless,

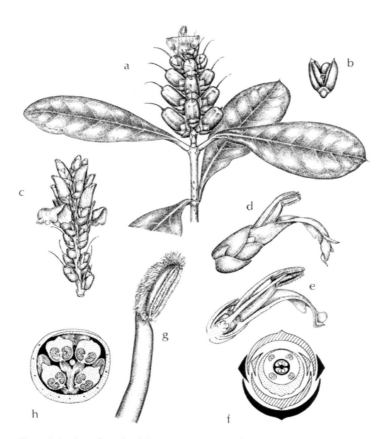

Figure B.2. *Acanthus ilicifolius* (Acanthaceae) flowers and fruit. (a) Old flowering head with maturing capsules (x1/2). (b) Dehisced but unexploded capsule (x1/2). (c) Flowering head (x1/2). (d) Flower in side view (x3/2). (e) Flower in L.S. (x1/2). (f) Floral diagram. (g) Single abaxial stamen (x3). (h) Flower bud in T.S. (x3) to show interlocking stamen hairs. (Material cultivated at Fairchild Tropical Garden, Miami, FL.)

ovate-lanceolate with a rounded apex and entire margins, but more commonly they have a sinuous dentate margin with spines on the margin and both surfaces, as is common in the terrestrial (non-mangrove) species of the genus. The variation is part genotypic and part phenotypic. Lack of spines seems to be a juvenile character, but one that may recur on the leaves immediately below the inflorescence or on proleptic branches. Leaf spinyness also seems to be accentuated by the degree of water stress, which may be related to salinity, seasonality, and light intensity. Some populations characteristically have a high proportion of spineless leaves, however, and have been considered of potential horticultural value because of their salt tolerance.

Anatomy. The epidermal glands of *A. ilicifolius* (Fig. 8.4E, F) have been described (Areschoug, 1902; van Steenis, 1937). These apparently are the source of secreted salt, which gives the upper leaf surface a greasy feel on occasions.

Reproductive Biology

Flowers are probably pollinated by both birds and insects. As the corolla expands, the lip is formed by marginal recurving, which physically tears the corolla on each side at its base. This allows the distal part of the corolla to separate. The four stamens are in two slightly unequal pairs with their anthers pressed together under tension facing the lip of the corolla. The anthers dehisce longitudinally by a single slit, with a thick line of hairs bordering the split so that pollen is presented on the lower side of the stamens. The stigma rests on top of the anthers but is prevented by their hairs from coming near the pollen (Fig. B.2d, g). There is an additional ring of dense hairs at the base of the stamens where the floral tube narrows. All these hairs point upward and outward, preventing small insects from crawling into the floral tube.

A large animal visiting the flower probes into the channel, about 12 mm long, formed between the stamens and the large corolla lobe below. When the bases of the stamens are forced apart by the visitor, they diverge in two lateral pairs and the style and stigma descend. When pressure on the stamen bases is released, the stigma lifts up and the stamens return to their original appressed position. The result is that the two divergent lobes of the stigma touch the back of the flower visitor first and pick up any pollen it may be carrying, whereas pollen is deposited by the flower as the stamens come together and the visitor withdraws. As a result of this functional morphology, the stigma and the dehisced anthers rarely contact each other. Self-pollination can occur if the stigma does not return to its original position sufficiently rapidly.

The stigma does not become receptive until the second day of anthesis. Flowers usually last two days (exceptionally, three or four). This weak protandry is likely to restrict self-pollination.

Observations by Primack et al. (1981) in Queensland show that flowers are visited by sunbirds (e.g., *Nectarina jugularis*); large bees (*Xylocopa*) were suspected of being pollinators, although no visits were actually observed. The size of the flower and its mechanics indicate that a relatively large pollinator is needed for effective pollen transfer. From the abundance of fruits in the population studied by these authors, it seems that pollination is normally very effective.

The seeds, with a wrinkled whitish testa that tears easily as the seed matures, consist of the two flattened green cotyledons enclosing the minute seedling axis. Usually all four ovules form seeds. Release is explosive, with the capsule splitting violently in the dorsiventral plane and the seeds propelled away with a spinning action like a discus. They are dispersed up to about 2 m.

Acanthus ebracteatus Vahl. 1791 [Symb. Bot. 2, 75, t. 40] (Fig. B.3)

Where this species grows with *A. ilicifolius*, the two seem distinct in the characters used in the key, but they are often confused. Duke (2006) distinguishes two subspecies as follows:

   A. ebracteatus ssp. *ebracteatus* Vahl 1791. Flowers mostly white, anthers also hairy on upper suture; leaves both entire and spiny.

Figure B.3. *Acanthus ebracteatus* (Acanthaceae), the mangrove "thistle"; Semetan, Sarawak.

> **A. ebracteatus** ssp. **ebarbatus** R.M. Barker 1986. Flowers deep purple, anther not
> hairy on upper suture; leaves mostly entire, rarely spiny.

> **Acanthus volubilis** Wall. 1831 [*Pl. As. Rar.* 2, 56, t. 172]

This species is much less robust in its vegetative parts than the other two species and is
described as unarmed. The inflorescence usually has fewer flowers and is shorter. Its
flower color is not described.

Taxonomy and Nomenclature

That *A. ilicifolius* and *A. ebracteatus* are conspecific has been suggested by earlier
authors and accepted by some recent ones (Percival and Womersley, 1975). The
situation is confusing because Vahl, in his original description of *A. ebracteatus*,
emphasized differences in leaf shape (which are not consistent) as diagnostic, and as
pointed out by C. B. Clark (1884, *Hook. Fl. Br.* 4:481), his description of a four-fold
difference in flower size between the two species is certainly incorrect. His diagnosis
also emphasizes *A. ebracteatus* as having the largest flowers, a difference reversed in
modern treatments.

Family: Anacardiaceae

A major tropical family of some 80 genera and 600 species. It includes some commercially important timber species as well as the mango and cashew nut. Many species have active alkenyl phenols, which produce a contact dermatitis (e.g., poison ivy, *Toxicodendron radicans* (L.) O. Ktze [syn. *Rhus radicans* L.]) The following genus represents the occurrence of the family as a mangrove associate.

Gluta L. 1771 [*Mant. Pl.* 2: 293]

A genus of about 30 species with a range from Madagascar (one species) to New Guinea, but with a concentration in Indo-Malaya. The following species, with a wide distribution from Myanmar and Thailand, the Malay Peninsula, and Malesia to Sumatra, northern Borneo, and West Java, is common along the edge of tidal rivers and is associated with mangroves, most characteristically behind *Nypa* swamps. It is said to be an important component of Malaysian mangroves because it supports firefly populations (Wah Juliana et al., 2012). It is included in this account because it has an irritant sap.

Gluta velutina Blume 1850 [*Mus. Bot. Lugd. Bot.* 1: 183]
There is a full account in Ding Hou (1977).

A large shrub or small tree growing to 7 m high, with numerous branched aerial stilt roots, developing an extended fluted stem base. Leaves simple, spirally arranged, elliptic, 12–30 cm by 5–8 cm, cuneate at the base, blunt apically, somewhat resembling those of mango. Inflorescence a terminal panicle up to 12 cm long with numerous small white or pink flowers with velutinous calyx and pedicels. Flowers each with five stamens surrounding the ovary raised on a central stalk (torus). Fruit usually solitary, a one-seeded drupe up to 7 cm in diameter but with a distinctive scurfy brown tuberculate surface and ridged toward the base. Germination hypogeal.

The best field characters are the black-spotted older leaves and the clear sap, which exudes readily and rapidly turns black so that the trunk becomes black stained. Any tree with this character should be avoided because it usually indicates a member of the family and the sap is often slightly to extremely irritant and toxic.

Family: Apocynaceae

A mainly tropical family of herbs, shrubs, lianes, and trees, typically exuding a white latex from cut surfaces; well-studied chemically because of the abundant presence of glycosides. Two genera are treated here; *Cerbera* occurs as trees in the Old World, and *Rhabdadenia* as vines in the New World.

Cerbera L. 1753 [*Sp. Pl.*: 208]

A genus of six species, of which at least three have a coastal distribution and may be constituents of back-mangal communities. They are readily recognized by their stout twigs with broadly oblanceolate leaves in dense terminal spirals, copious latex, and terminal cymes of conspicuous white, fragrant flowers. The genus superficially resembles the cultivated species of *Plumeria* (frangipani) but is distinguished by the large calyx segments, the stamens inserted toward or at the top of the corolla tube, the well-developed style with a thick disciform stigma, and particularly the fleshy, drupaceous (not capsular) fruit, which is dispersed by water. One species, *C. odollam*, is sometimes cultivated, and its range may have been extended artificially. It is, for example, a common street tree in Singapore. The fruits are poisonous, causing cardiac arrest, and used as a means of committing suicide by the Marquesans (Brown, 1935); leading to the unfortunate appellation "suicide tree."

Two closely related species (*C. odollam* and *C. manghas*) are occasional elements of coastal forest and back-mangal communities but with only limited salt tolerance. They have been much confused, even though they were clearly differentiated by Valeton (1895). (See also Backer and Bakhuizen van den Brink, 1964; 2:233.) Commenting on the confusion, Corner (1939) says: "How backward is botany in Tropical Asia, this genus will illustrate." They occur most typically in sandy soils, not regularly inundated tidally. A third species, *C. floribunda*, restricted to New Guinea, may also be encountered. The three are very similar vegetatively but may be distinguished as follows:

1A. Stamens inserted below middle of corolla tube; flowers numerous (40 or more) per cyme, each with a short inflated corolla tube, scarcely 1 cm long. . . . . *C. floribunda*
1B. Stamens inserted at or above middle of corolla tube; flowers relatively few (less than 30) per cyme, corolla tube slender (at least at base), longer than 1 cm. 2

2A.   Corolla with a yellow eye; tube short (1.5–2 cm long), but free corolla segments long (2–3.5 cm); stamens inserted at or shortly above middle of locally widened tube; throat of corolla not closed by sparsely hairy connective ridges. Leaves apiculate, with a fine point, primary veins perpendicular to midrib, terminating in a fine marginal commissure. . . . . *C. odollam*

2B.   Corolla with a red eye; tube long (2.5–4 cm), but free segments short (2–2.5 cm); stamens inserted in mouth of tube, which is closed by densely hairy connective ridges. Leaves bluntly pointed, primary veins arcuate, terminating in a well-developed commissure. . . . . *C. manghas*

The status of *C. dilatata* S.T. Blake 1948 (*Proc. R. Soc. Queensl.* 59(8): 161), is uncertain.

> **Cerbera manghas** L. 1753 [*Sp. Pl.*: 208]
> *C. lactaria* Hamill. ex DC 1844 [*Prodromus* 8: 353]
> *C. odallam* Bl. 1826 [*Bidjr. Fl. Ned. Ind.*: 1032]

Tree to 15 m, with grayish smooth bark; twigs stout (pachycaulous), branching consistently below terminal inflorescences (Leeuwenberg's model) but with frequent reiteration. Leaves spirally arranged, often aggregated in terminal rosettes, ovate-oblong to elongate-obovate up to 30 cm long, 7 cm wide, apex bluntly pointed, base gradually narrowed to a short petiole 1–2 cm long. Inflorescence an irregular, often one-sided cyme, with broad ephemeral bracteoles 1–2 cm long. Flowers perfect, pentamerous; calyx five-lobed, greenish white, the lobes 1.5–2 cm long by 5–6 mm wide, narrowly lanceolate and often persistent. Corolla tube white, up to 4 cm long, scarcely 3 mm wide at the base, gradually widened upward to the expanded throat 5– 6 mm wide; corolla lobes five, shorter than the tube (to 2 cm), broadest about the middle and abruptly narrowed toward insertion. Stamens five, sessile, inserted in the mouth of the tube with a hairy spur-like extension of each stamen more or less closing the throat, together with a glandular extension below each stamen. Ovary superior, with two free carpels, each with several ovules inserted on a thick ventral placenta. Style slender, expanded abruptly at level of stamens into a thick discoid, shortly bifid stigma. Fruit (often twinned) an ovoid or ellipsoid one- or two-seeded drupe 5–6 cm long and 3–4 cm wide, with green flesh, ripening purple, the endocarp somewhat transitional to fibers of the inner mesocarp. Germination hypogeal; seedling leaves narrowly lanceolate.

Growth

Leaves are arranged in a dense 5/13 phyllotaxis. There is no development of bud scales, although inactive terminal buds are protected by a resinous varnish. Each leaf subtends an inconspicuous lateral bud also protected by the resin, which appears to originate from a series of glands adaxially on the petiole at its insertion. These glands shrivel early and may persist as a black fringe along the top of the leaf scar. The young leaves themselves may be coated similarly with the resin, which flakes off as they expand.

As the terminal inflorescence develops, from one to five renewal shoots expand by syllepsis, with the base of the branch and its subtending leaf commonly fused together for as much as 3 cm.

Floral Biology

Inflorescence construction is irregularly cymose, the nodes developing a series of branches each with a single, somewhat petaloid and ephemeral bract that almost encloses the axis at its insertion. The inflorescence seems to develop with progressive overtopping of older by younger axes so that there is a rough acropetal sequence of flowering over an extended period.

The floral mechanism has been well described by Valeton (1895). The pollen is presented early, almost before the flower opens, with each half-anther depositing its pollen as a sticky mass on the stigma, which secretes mucilage over its upper surface. The pollen is further prevented from falling into the tube by a shallow rim at the base of the stigmatic disk. The mouth of the tube is partly occluded by the hairs of the five ridges, which protrude both above and below but leave five narrow channels with access to the lower, narrower part of the tube. Details of pollen transfer are not known, but clearly the flowers are attractive to animals; crossing is promoted by the flower visitor, which is likely to encounter the stigmatic disk before it touches the pollen mass.

The pollination mechanism of *C. odollam*, which is likely to be somewhat different in view of the different floral construction, has still to be clarified.

Nomenclature

Linnaeus, in his *Species Plantarum* (1753), described three species. Confusion arose because the name he applied to his second species, *C. manghas*, was based on *Manghas lactescens* in Burmann's *Flora Zeylanica* (150, t. 70, f. 1) and also on *Odollam* of Rheede's *Hortus Malabaricus* (1678–1703;1, p. 79, t. 39). The other two Linnaean species refer to elements now included in *Thevetia. Cerbera odollam* was correctly circumscribed by Gaertner in his *Fructus et seminus plantarum* (2: 193, t. 124) as a species distinct from the *C. manghas* of Linnaeus. Nevertheless, subsequent authors have used the names *manghas* and *odollam* for both entities, despite the clear distinction between them, so there is a confusing pattern of mutual synonymy.

Cerbera odollam Gaertner 1791 [*Fruct. Sem. Pl.* 2: 193, t. 124] (Fig. B.4; Plate 5A)

This species differs from the previous one in the following important features: flowers with lanceolate green or white sepals (15 mm by 2–5 mm), ephemeral, falling at or before anthesis; corolla trumpet-shaped, equal to or scarcely exceeding calyx, inflated distally in region of attachment of stamens but not at the mouth, filaments short,

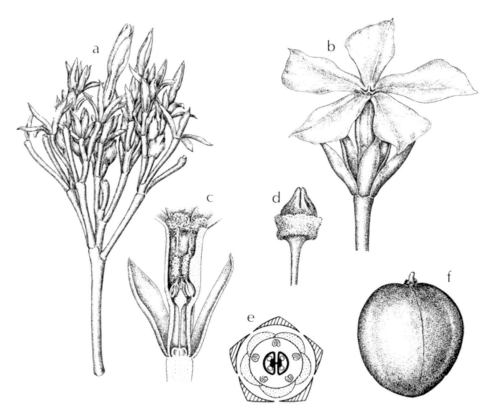

Figure B.4. *Cerbera odollam* (Apocynaceae) flower and fruit. (a) Inflorescence (x1/2). (b) Single flower (x1). (c) Corolla tube in L.S. (x3/2). (d) Detail of stigma (x4). (e) Floral diagram. (f) Fruit (x3/8). (Material cultivated at Forestry Department Headquarters, Kuching, Sarawak.)

scarcely hairy; mouth of corolla tube open with a yellow eye, corolla lobes asymmetrical and broadest above the middle, abruptly narrower basally with a narrow insertion. Fruit always solitary, without a clear distinction between mesocarp and endocarp. Germination hypogeal.

According to Valeton (1895) there are also differences in leaf shape and venation as indicated in the key, but these have not been elaborated by subsequent authors. Corner (1939) says the species cannot be so distinguished.

> **Cerbera floribunda** K. Schumann 1889 [*Schumann-Hollrung, Fl. Kais.-Wilhelm Land*: 111] *C. micrantha* Kanehira 1931 [*Bot. Mag. Tokyo* 45: 343]

This species was originally described as endemic to New Guinea and is distinguished by its numerous flowers (up to 100) in each compact cyme, the short corolla tube with the stamens inserted below the middle, and the elongated stigma. It is recorded from several coastal and estuarine stations, including an association with *Rhizophora*, but otherwise chiefly inland (Markgraf, 1927). Material from Palau identified by Markgraf as *C. floribunda* suggests an interesting range extension.

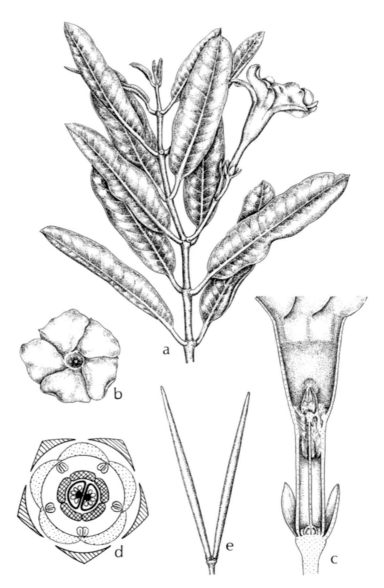

Figure B.5. *Rhabdadenia biflora* (Apocynaceae) flower and fruit. (a) Flowering shoot (x1/2), flowering axis is terminal. (b) Flower from above (x1/2). (c) Base of flower in L.S. (x3/2). (d) Floral diagram. (e) Fruits (x1/2). (From Correll and Correll, 1982.)

Rhabdadenia Mueii.-Arg. 1860 [*Mart. Fl. Bras.* 6(1): 173]

A viny genus of three species, in subtropical and tropical America. The following species is widely distributed in the Caribbean region, and occurs characteristically in fresh- or salt-water swamps. Typically, it is found in back-mangrove communities.

Rhabdadenia biflora (Jacq.) Muell.-Arg. 1860 [*Mart. Fl. Bras.* 6(1): 175] *Rhabdadenia paludosa* (Vahl.) Miers 1878 [*Apoc. S. Am.*: 119]
(Fig. B.5)

A slender vine with pliable stems several meters long. Leaves simple, opposite, often in distant pairs. Growth of flowering shoots is sympodial below terminal cymes of flowers, with the renewal shoot displacing the inflorescence into an apparently lateral position. Flowers are conspicuous, white or pink, up to 7.5 cm long and 4 cm in diameter. Fruits are slender paired follicles up to 14 cm long, dehiscing to release the numerous hairy seeds. The plant is unusual in mangrove-associated communities in having primarily wind-dispersed seeds.

Family: Arecaceae (Palmae)

One of the larger, most natural, and economically important monocotyledonous families with about 180 genera and 2600 species; typically plants with aerially unbranched, woody trunks and large compound leaves (palmately or pinnately dissected) in a terminal crown, and forming extensive pure colonies (Plate 17D). The family is divided into five subfamilies and much further subdivided (Dransfield et al., 2008). *Nypa* is a distinctive rhizomatous mangrove element in the Old World. It forms the monotypic subfamily Nypoideae, an early diverging taxon, as supported by its unique morphology and distinctive anatomy (Tomlinson et al., 2011) and extensive early fossil record, beginning in the early Cretaceous (Dransfield et al., 2008, p. 76). A number of other palm species may occur as mangrove associates.

The taxa treated here may be distinguished as follows:

1A. Unarmed, rhizomatous palm, massive axis buried in estuarine mud; leaves pinnately erect to a height of 7 m; fruiting head globose with numerous fibrous fruits (Plate 18A, B). . . . . *Nypa fruticans*

1B. Spiny, trees or lianes, leaves 1–3 m at most; fruiting heads paniculate. 2

2A. Lianes, supported by extended grapnel-like structures; stem diameter 3–4 cm; fruits with numerous reflexed, regularly overlapping scales. . . . . *Calamus erinaceus*

2B. Stems self-supported (Fig. B.9A), clustered, axes to about 10 cm in diameter, armed with spines that become erected after leaf fall (Fig. B.9B); fruits not scaly. . . . . *Oncosperma tigillarium*

Additional palms that may be encountered, usually as outliers of swamp communities, include *Bactris major* Jacq. (tropical America), *Phoenix reclinata* Jacq. (tropical Africa), species of *Euterpe*, *Manicaria*, and *Mauritia* (tropical America), and species of *Raphia* (Central America and tropical Africa). Such palms may have an appreciable tolerance of salt water (e.g., *Phoenix paludosa* Roxb. in Asia). The only other arborescent monocotyledon that might be mentioned in this context is *Pandanus candelabrum* as a constituent of the back-mangrove in West Africa.

Nypa Steck 1757 [*De Sagu* 15; see Moore 1962 *Taxon* 11: 164–166]
Nipa Thunberg 1782 [*Kongl. Vet. Acad. Nya Hanal.* 3: 231]

Nypa fruticans (Thunb.) Wurmb. 1781 [*Verh. Batav. Genootsch*]
Nipa fruticans Thunberg 1782 [loc. cit. 3: 234]
(Figs 5.4, 5.5, 5.6, 10.1; B.6, B.7, B.8; Plates 17, 18)

A monoecious palm with a thick, subterranean, somewhat fleshy, dichotomously branched, rhizomatous axis (Fig. 5.6) up to 70 cm in diameter, with oblique raised leaf scars; axis and leaf bases usually submerged in mud and inundated at high tide (Fig. 5.5), rarely exposed by erosion (Fig. 5.7). Each terminal shoot supporting a cluster of erect paripinnate leaves to 7 m long. Leaf with a terete petiole 1–2 m long and a somewhat bulbous leaf base, closed at the insertion. Leaf base with a pronounced groove accommodating the next youngest leaf, but two grooves at the point of dichotomy (Plates 17E, 18C). Leaflets reduplicately folded, numerous (30–40), regularly arranged in the grooved rachis, and each lanceolate, to 70 cm long, with a single prominent adaxial midrib. Indumentum absent except for frequent scattered abaxial scales, up to 1.5 cm long, below the midrib. Inflorescence (Uhl, 1972) on a long peduncle, axillary, arising above the water surface at high tide; bicarinate prophyll (Plate 17F) and initial series of spirally arranged bracts, sterile and enclosed by the leaf base; distally 7–9 similar bracts, fertile and subtending lateral branches, the branches themselves branched to three or even four further orders. Branches usually somewhat adnate to parent axis but without a conspicuous prophyll. Main axis of inflorescence ending in a globose cluster of congested female flowers, at first enclosed by a series of short sterile bracts; all lateral axes ending in a club-shaped spike of densely arranged male flowers (Plate 18F). Flowers (Fig. B.6) dimorphic and sessile, each subtended by an inconspicuous, even filiform bract. Male flower (Fig. B.6i, j, k) with six short tepals 4–5 mm long, in two whorls, the outer tepals somewhat widened above; stamens three, united to form a central column, the anthers dehiscent extrorsely. Pollen spinous (Fig. 9.1). Female flowers (Fig. B.6c–f) with six tepals much as in the male flower, but exceeded by the three (or four) separate angular carpels about 1 cm high, the stigmas sessile, funnel-shaped. Carpel unilocular with a single basal anatropous ovule. Fruiting head developing into an aggregate of fertile and partly developed but sterile carpels (Plate 18B, G), the individual fruit an angular drupe with a smooth, dark brown epicarp, mesocarp fibrous, endocarp thick, the seed grooved adaxially. Endosperm usually with a central hollow; embryo small, basal (Fig. B.7a, b). Germination incipiently viviparous (Plate 18D, E; Fig. B.7c, d), essentially hypogeal, but initiated on the fruiting head with the plumule protruding before the fruit is released (Fig. B.8b, c).

Embryo Morphology

The configuration of the embryo in the germination process is unusual because it is seemingly the reverse of that in all other palms (Fig. B.8; Plate 18D). The plumule is the first structure to appear, which implies that it is the root that functions as a haustorium, not the cotyledon, as in all other palms. The fruit germinates in a horizontal position so that the first (scale) and subsequent juvenile (foliage) leaves all become erect by rotation of the leaf base (Fig. B8e, f). To date, I have not succeeded in unraveling this morphological conundrum.

   In virtually all respects, *Nypa* contrasts remarkably with other palms, reflecting its highly adaptive morphology and systematic isolation.

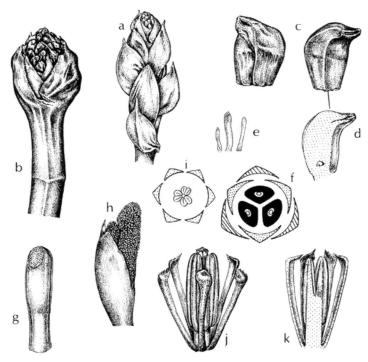

Figure B.6. *Nypa fruticans* (Arecaceae-Nypoideae) inflorescence and flower. (a) Distal part of inflorescence at female stage (x1/5; see Plate 18F). (b) Female flower head with lower bracts removed (x1/2). (c) Single carpels from different aspects (x3/4). (d) Carpel in L.S. (x3/2). (e) Tepals, one outer, two inner, from a single female flower (x3). (f) Floral diagram of female flower. (g) Young male spike still enclosed by bract (x1/2). (h) Male spike at anthesis (x1/2); apex of next higher order spike protruding on left. (i) Floral diagram of male flower. (j) Male flower just before anthesis (x9). (k) Male flower in L.S. (x9); staminal column unextended. Note: the perianth has been described as two whorls (Uhl, 1972); this is most obvious in the female flower (e) and least so in the male (j). Staminal column (k) extends about two-fold at anthesis, with the stamens standing above the tepals. (Material from Kuching River, Sarawak.)

Nomenclature, Taxonomy, and Phylogeny

The monogeneric subfamily Nypoideae. It is considered to represent an independent line of specialization, and because of its numerous distinctive features, has no obvious close relatives. Relationships with the Pandanaceae have been suggested on the basis of superficial similarities of the fruiting head, but the basic morphology of the two forms is quite different. In its vegetative anatomy, it shows all the diagnostic features of the Arecaceae (Tomlinson et al., 2011); any attempt to segregate it as a distinct family Nypaceae should be discounted.

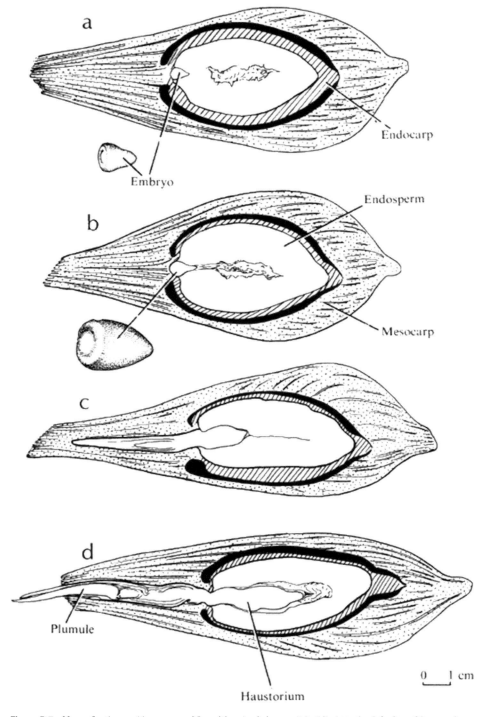

Figure B.7. *Nypa fruticans* (Arecaceae-Nypoideae) vivipary. (a)–(d) Attached fruits of increasing age to show development of viviparous embryo. (d) Corresponds to stage when fruit is released. (After Tomlinson, 1971.)

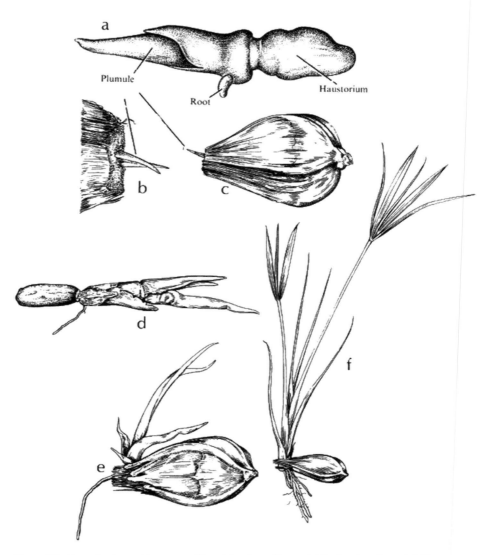

Figure B.8. *Nypa fruticans* (Arecaceae-Nypoideae) seedlings. (a) Isolated embryo (x3/2), corresponding approximately to stage of development in Fig. B.7c. (b) Detail of protruding plumule (x1). (c) Fruit at time of release (x1/2). (d) Isolated embryo (x1/2) from detached, floating seed. (e) Established seedling (x1/2). (f) Older seedling (x1/8). (After Tomlinson, 1971.)

Distribution and Ecology

Widely distributed from Sri Lanka, the Ganges Delta, Myanmar, and the Malay Peninsula, through Indonesia to New Guinea and the Solomon Islands, northward to the Philippines and Ryukyu Islands, and southward to north Queensland. *Nypa* has had a much wider distribution outside its present range because fossil fruits are known from the Paleocene in Brazil and the Eocene in Europe, Africa, and India. Its pollen (or at least pollen closely similar) is one of the oldest known identifiable angiospermous pollens referable to a modern species, being recorded in sediments as early as the Upper Cretaceous.

In its present range, it is a palm of quiet estuaries or shallow lagoons (Fig. 5.5) into which fresh water may run. It makes extensive pure stands by virtue of its rhizomatous dichotomizing habit (Fig. 5.6). It seems to favor brackish waters, often forming a wide border beyond the fringe of adjacent mangroves or swamp forest (Plate 17D). It does not occur on shores exposed to high wave action and not in hypersaline conditions. This plant is an unusual mangrove because of its colonial habit. The axis is normally buried, but when occasionally exposed, resembles a series of overlapping cowpats. The stem dichotomy is then evident (Fig. 5.7). Isolated stems that are washed out cannot re-root, but Chai (1982) suggests that large clumps that retain much of their original root system may become re-established; the only example of vegetative dispersal in mangroves.

Nypa has been introduced outside its natural range, notably in West Africa where it is usually regarded as an invasive species despite its economic potential. In Panama, Duke (1991b) indicates that the small population there is introduced. It sets viable seedlings and is apparently pollinated by non-stinging bees (*Trigona*).

Pneumatophores

It is interesting to reflect on the ability of *Nypa* to provide an oxygen supply to the submerged rhizome growing within an anoxic environment. This is mediated by an unusual mechanism involving mature leaves that function in a way comparable to the modified root system of other mangroves. This has been investigated in detail by Chomicki et al. (2014) who provide compelling structural and developmental evidence that the persistent basal part of the leaf axis in *Nypa* functions in the same manner as a pneumatophore. The distal part of the leaf abscises at a precise level so that the stump of the leaf base persists as a vertical projection of uniform height, becoming sealed distally by tannin-rich tissues. The surface of the leaf base develops lenticels (possibly modified stomata) as ports of entry to the internal air space system, which is continuous with that of the stem. Measurements suggest that this function can persist for at least four years. Hence, part of the success of *Nypa* in the mangal comes from a unique set of developmental and functional attributes that are modifications of a standard palm morphology.

Growth

Growth of the axis seems to be continuous, and disinterred stems show no morphological discontinuities except for the branch forks. The leaves are spirally arranged on horizontal axes but are brought into an erect position by unequal growth of the leaf base (Fig. 5.6). The leaf sheath is a closed tube, as in all palms, but the tube is short and extends as a short groove into the base of the terete petiole. Branching as described by Tomlinson (1971) occurs at regular intervals by an equal dichotomy, which is evident in the leaf immediately behind the fork, and which has a characteristic double groove and always occupies the same position relative to the fork (Fig. 5.6a; Plates 17E, 18C). Architecturally, the palm may be referred to Schoute's model (Hallé et al., 1978), but with plagiotropic rather than orthotropic axes (cf. *Hyphaene*). The result is an axis that spreads by regular forking of monomorphic axes (Fig. 5.6c,d). No vegetative axillary meristems are produced. Because the leaves are spirally arranged, but must become vertical at maturity, there is a peculiar growth reorientation as the leaf matures, differing for each leaf in the phyllotactic spiral.

Reproductive Biology

Clearly, the reproductive mechanisms and genetic consequences of this palm are unusual among mangroves because of its extensive clonal spread. Most leaves are capable of subtending an axillary inflorescence, but not all develop completely. Each inflorescence is protogynous; the stigmas of the carpellate clusters become receptive as they are exposed by the separation of the enveloping bracts. Receptivity is indicated by the stickiness of the stigmatic surface. The pollen is released subsequently, as the individual male spikes are exposed in turn, elongating beyond the enveloping bracts (Plate 18F). In each male flower, the staminal column is extended, carrying the pollen as a mass onto the surface of the spike; flowers are more or less synchronous in anthesis.

Essig (1973) records two kinds of insects visiting *Nypa* inflorescences "in significant numbers." First, there are small bees of the genus *Trigona*; second, small flies of the family Drosophilidae. I can confirm these observations. Apparently, the flies complete their life cycle in the somewhat fleshy axes of the branches of the male inflorescence, because these become extensively damaged by larvae that are presumed to be mainly of the same flies. Essig suggests that the bees are not important pollinators because they visit male flowers almost exclusively. The flies visit both types of flower and pollen becomes attached to them. In cultivation, it is considered necessary to hand-pollinate flowers to achieve successful fruit set; the ease of this and the clonal habit indicate that plants are self-compatible.

Fruits may fill a single head completely, but isolated fruits commonly occur in a mass of partly developed but sterile carpels. The mature fruits are dispersed by water; their detachment is assisted by the plumule, which extends precociously before the seed is loosened, often penetrating the fibrous tissue of the inflorescence axis. The peduncle typically becomes pendulous under the weight of the fruiting head, but it is supported by water at high tide (Plate 18A). Tidal movement aids in the detachment of mature fruits. The stranded fruit continues to develop producing two or three scale leaves before the first bladed leaf (eophyll), which is pinnate (Fig. B.8f). The seedling axis is horizontal from the start.

Economic Uses

Nypa is one of the more commercially valuable plants of the mangroves on a sustained yield basis (Brown and Merrill, 1920). The leaves make a high quality thatch (attap; Plate 16C); the immature endosperm provides a jelly-like sweetmeat in Malaya; and in the Philippines, industrial alcohol is extracted by distillation after fermenting the sugar tapped from the inflorescence, in the manner of toddy (Fong, 1987a, 1989). It is also suggested as a fuel crop (Fong, 1987b).

> *Calamus* L. [*Sp. Pl.*: 325]
> See Beccari 1908 [*Ann. R. Bot. Gard. Calcutta 11 and Appendix 1913*]

A genus of over 370 species, mainly dioecious, spiny, climbing palms with whip-like grapnels that are either extensions of the leaf tip (cirrus) or sterile adnate inflorescences (flagellum). Leaves pinnate. Fruit covered with numerous reflexed scales. It is exclusively Indo-Malayan except for *C. deeratus* in West and Central Africa. In a modern analysis, several segregate genera have been sunk within an expanded *Calamus*; if accepted, this would include over 500 species.

> *Calamus erinaceus* (Becc.) Dransfield 1978 [*Kew Bull.* 32: 484; *see also* 1979 Malay. For. Rec. 29:130]

A multiple-stemmed rattan forming thickets on the landward edge of mangal or behind coastal sand bars. Stems climbing to 15 m, 2–3.5 cm in diameter but up to 6 cm wide including enclosed sheaths. Sheaths orange to yellowish green, very densely armed with horizontal or oblique whorls of spines. Petiole at insertion on sheath with a stout knee. Leaf about 4.5 m long with numerous leaflets up to 40 cm by 2 cm, leaf axis extended into a cirrus about 2 m long. Inflorescences as lateral panicles from axils of uppermost leaves, with numerous tubular overlapping bracts, ultimate axes (rachillae) with distichous series of flowers, the male much more dense than the female rachillae. Fruit globular, about 1 cm in diameter, covered with about 12 vertical series of triangular scales; embryo minute, endosperm homogenous.

The canes of this species have little commercial use. This species is probably the one referred to as "*Daemonorops leptopus*" in Watson (1928).

> *Oncosperma* Bl. 1838 [*Bull. Sci. Phys. Nat. Neerl.* 1: 64]
> (Fig. B.9A, B)

A genus within the subtribe Oncospermatinae (Dransfield et al., 2008), occurring from Sri Lanka widely in southeast Asia.

> *Oncosperma tigillarium* (Jack) Ridl. 1900
> [*J. R. Asiat. Soc. Straits Branch* 33: 173]
> *O. filamentosa* Bl. 1838–1839 [*Rumphia* 2: 97]

A tall, many-stemmed palm up to 25 m, but the stems are rather slender (scarcely 10 cm in diameter) with a crown of pinnate leaves up to 4 m long, each leaf with characteristic drooping leaflets. Stems armed with flat, sharp, black spines. Both Watson (1928)

a   b

Figure B.9. *Oncosperma tigillarium* (Arecaceae-Arecoideae). (a) A characteristic back-mangal species in Bako National Park, Malaysia. (b) Spiny trunk; stems are used in construction after the spines are shaved off.

and Chai (1982) record this palm as marking the inner limit of the mangrove that is found in regions transitional to non-littoral forest. Watson indicates that it grows inland and suggests it is likely to be killed by salt water. It is successfully cultivated at the Montgomery Botanical Center in South Florida adjacent to tidally influenced open water.

Family: Asteraceae (Compositae)

The largest family of dicotyledons, cosmopolitan and with more than 20 000 species, mainly herbs and often weedy. Mentioned here are two genera among the many members of this family that are likely to be encountered because of their weedy tendencies.

> Annual, with disk florets in compact heads; achenes with a well-developed pappus. . . . . *Pluchea*
> Perennial. . . . . *Tuberostylis*

Pluchea Cass. 1817 [*Bull. Soc. Philomel* 31]

A genus of about 40 species common in wet habitats. Species of *Pluchea* that occupy coastal areas may be encountered, like *Pluchea indica*. Less in the Asian tropics and *Pluchea odorata* in the Americas. This species is included and illustrated here as a plant likely to be seen in the back mangrove in Tropical America. Its albeit artificial inclusion is justified as representing the annual habit, not otherwise expected in the back mangrove, and also illustrating the large family to which it belongs.

Pluchea odorata (L.) Cass 1826 [*Dict. Sci. Nat.* 42: 3]
[Nesom, G. Sida. 21; 59] (Fig. B. 10)

Described as a fibrous rooted annual up to 120 cm tall, with aromatic leaves. Outer flowers of each head are male, inner flowers female. It is naturalized in Hawaii.

Tuberostylis Steetz 1953 [Seeman, *Bot. Voy. Herald* 142, t. 29]

The two species of this genus have been recorded as mangrove associates by Gentry (1982).

The spathulate leaves are opposite, about 2–4 mm long, with a flattened petiole, and an expanded, ovate blade with a crenate margin.

Tuberostylis axillaris Blake 1943 [*J. Wash. Acad. Sci.* 33: 265]

Recorded as a semi-scandent shrub in dense tidal forest in Colombia.

Tuberostylis rhizophorae Steetz 1853 [Seeman, *Bot. Voy. Herald* 142, t. 29]

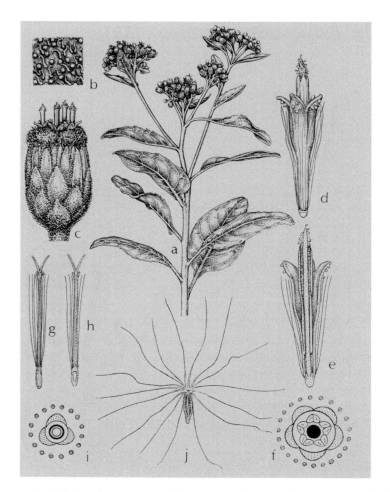

Figure B.10. *Pluchea odorata* (Asteraceae). (a) Summit of flowering stem (x2/3). (b) Section of leaf with glandular pubescence (x30). (c) Flower head, side view (x6). (d) Perfect disk flower, side view (x12). (e) Perfect disk flower, longitudinally dissected (x12). (f) Floral diagram of perfect disk flower. (g) Pistillate disk flower, side view (x12). (h) Pistillate disk flower, longitudinally dissected (x12). (i) Floral diagram of pistillate disk flower. (j) Achene with pappus (x12). (From Correll and Correll, 1982.)

The original description refers to this species as growing "epiphytically on the roots of Mangrove Trees, Southern Darien" (i.e., Colombia). The "plant affixes itself to the aerial roots of *Rhizophora* by means of adventitious roots which arise in pairs immediately below the node." The species has a wider distribution in Central America and has been recorded on the base of trunks of *Mora oleifera*.

Family: Avicenniaceae (Acanthaceae)

The genus *Avicennia* L. was initially included in the family Verbenaceae, from which it differs in important features such as its anomalous secondary thickening, leaf anatomy, placentation, incipient vivipary, and seedling morphology, which have been long recognized (van Tieghem, 1898); its segregation as a distinct and monotypic family was subsequently generally accepted. Its pollen also distinguishes it from Verbenaceae (Mukherjee and Chanda, 1973; McDade et al., 2008). Avicenniaceae is retained here, but *Avicennia* is most appropriately placed in Acanthaceae, but still somewhat isolated as a basal group (Schwarzbach and McDade, 2002). Borg et al. (2008) suggest that it is the closest relative of the Thunbergioideae, agreed upon by McDade et al. (2008). *Avicennia* is pantropical and the most widely distributed of all mangrove genera in terms of its latitudinal distribution, extending to 38 45' S in Victoria, a little further south than its limit in New Zealand.

A thorough early review and extended discussion of *Avicennia* were provided by Moldenke (1960, 1967), based largely on a study of herbarium material. *Avicennia* in Australia has been studied extensively by Duke, based on detailed field study, leading to the recognition of *A. integra* endemic to northern Australia, and *A. marina* with three varieties (Duke, 1988, 1990, 1991a, 2006). A modern and worldwide revision of the genus based on comparable field observations is much needed because it is one of the most consistently present constituents of mangrove vegetation. Some species and varieties have wide ranges, and there is additional geographical variation within individual taxa. A beginning to a detailed account of genetic diversity has been started by Cerón-Souza et al. (2005) and Nettel et al. (2005) in their study of *A. germinans*.

Clerodendron inerme (originally Verbenaceae, now Lamiaceae), a diffuse shrub of the back mangrove is widely distributed in Australasia. Its floral biology has been described by Primack et al. (1981).

> *Avicennia* L.1753 [*Sp. Pl.*: 110; *Gen. Pl. Ed.* 5: 119, 1754]
> For a full synonymy, see Moldenke (1960, 1967) and Bakhuizen van den Brink (1921).
> (Figs. 6.1A, 6.2D, 7.1, 7.4 lower, 8.4G, H; B.11–16; Plates 6, 8, 9, 13)

A pantropical genus of about eight species, occupying diverse mangrove habitats, either within the normal tidal range or in back mangal, with a high tolerance of hypersaline conditions. The genus is uniform in its gross morphology and anatomy, and although there are often useful diagnostic characters seen in the field (e.g., bark color and texture, and crown color), these are rarely transmitted on the labels of herbarium specimens. Consequently, available keys (notably that of Moldenke, 1960) are not easy to use and identification of herbarium specimens may be difficult. The following description refers largely to the rather consistent and uniform generic characters; specific diagnoses are outlined in the later key and descriptions.

Generic Description

Trees to 30 m, or when depauperate, low, much-branched, dense-crowned shrubs. Aerial stilt roots sometimes developed (e.g., *A. officinalis*; Fig. 7.1; Plate 13E). Underground horizontal roots extended, cable-like and supporting, as lateral branches; both erect exposed pneumatophores and descending absorbing roots (Fig.7.4 lower). In open situations, the extent of the root system, many meters from the trunk, can be mapped by the exposed pneumatophores (Plate 11E). Bark variously rough, black, hard, and fissured or checkered (e.g., *A. germinans*); gray and lenticellate (e.g., *A. marina*); to smooth, yellowish green, and flaking (e.g., *A. marina* var. *eucalyptifolia*, Plate 4G) or smooth (*A. officinalis*, Plate 4H). Secondary thickening anomalous; cylinders of xylem tissue alternating with cylinders including phloem, each cylinder produced by a new cambium. (Fig. B.11). Shoots glabrous or with a fine indumentum superficially, not readily recognized as made of separate trichomes; distinctly but finely woolly in *A. lanata*. Twigs somewhat jointed with swollen nodes, terete or more or less four-angled; internode variable, long on vigorous shoots, for example saplings, but commonly short on distal laterals. Sequential branching by syllepsis, continuous or diffuse.

Leaves decussately arranged, the youngest leaf pair enclosing the apex within the petiolar groove; bud scales not developed on terminal buds. Leaves shortly petiolate (1–3 cm), the petiole with a deep basal groove, the groove with black marginal hairs, the hairs continuous in a line across the node. Blade leathery or somewhat fleshy, with inconspicuous veins, the midrib prominent below; glabrous and light to dark green and shiny above, with numerous close-set microscopic depressions; indumentum of a uniform dense palisade of microscopic club-shaped hairs below (Plate 9E), imparting a characteristic gray, white (in *A. alba*, Plate 8C, below), yellow, or olive green (Plate 8C, above) color. Blade 3–15 by 1–6 cm, varying from ovate to narrowly elliptic to lanceolate; apex bluntly rounded and even slightly emarginate to acute or even attenuate; base acute. Margin entire, slightly thickened. Leaf shape generally useful in specific diagnosis (Figs. B.14, B.15).

Flowers (Figs. B12, B.13) protandrous, perfect, tetramerous to modified pentamerous, usually with 4 to 10 decussately arranged flower pairs in dense spicate to capitate units, the units themselves aggregated into terminal or axillary, somewhat paniculate assemblages on distal shoots (Fig. B.12b; Plate 6G–I), the lower units subtended by foliage leaves, the upper by reduced, often inrolled leaves. Each flower subtended by a convex, triangular bract and two lateral bracteoles, the three structures imbricate and enclosing the base of the flower. Sepals five, free, ovate, and extensively imbricate, largely enveloped by bracteoles; the calyx and bracteoles persisting in fruit. Corolla

Figure B.11. Avicennia sp. T.S. young woody stem showing successive rings of xylem and phloem produced from successive cambia.

tubular at the base, becoming equally or unequally lobed at the apex, the lobes either erect or reflexed; three main types of corolla may be distinguished, although this does not cover the total variation:

1. *Germinans* type (Fig. 12.8e, f, h; Plate 6F). Distinctly zygomorphic, white, 10–13 mm wide, corolla lobes at first slightly imbricate, later spreading, densely hairy on both surfaces, two lateral and abaxial (ventral) lobes rounded to at most slightly emarginate and somewhat narrowed toward insertion, adaxial (dorsal) lobe broad toward insertion and distinctly to conspicuously two-lobed. *Avicennia bicolor* has smaller flowers (6–8 mm wide) with a distinct yellow throat.
2. *Officinalis* type (Fig. B.13f,g; Plate 6H). Somewhat zygomorphic, yellow, 6–10 mm wide, corolla lobes at first slightly imbricate, later spreading and rather irregular, glabrous within, petal lobes as in type 1.
3. *Marina* type (Fig. B.13a–d; Plate 6G). Strictly actinomorphic, yellow or orange, 4–8 mm wide, corolla lobes at first valvate, later spreading, glabrous within except for minute capitate glandular hairs toward base. Five-lobed flowers occasionally present in some populations.

Stamens four, anthers about 0.5 mm, alternating with corolla lobes, dehiscing introrsely by longitudinal slits but retaining adherent pollen. Stamens in *Germinans* and *Officinalis* types inserted basally on corolla, slender, equal or slightly unequal filaments

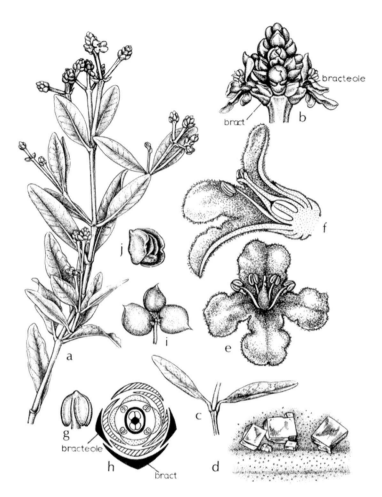

Figure B.12. *Avicennia germinans* (Avicenniaceae) shoot and flowers. (a) Distal shoot with flowers (x1/2). (b) Terminal flower cluster (x3/2). (c) Node (x1/2). (d) Leaf surface with salt crystals (x10). (e) Flower from front (3). (f) Flower in L.S. (x3). (g) Isolated placenta with four pendulous ovules. (h) Floral diagram. (i) Fruit cluster (x1/2). (j) Seed (seedling) at time of release with folded cotyledons (x1/2). (From Tomlinson, 2001.)

2–3 mm long, bent to orient anthers adaxially; in *Marina* type, inserted regularly and equally on the corolla tube at its mouth, anthers sessile or at most with a short filament scarcely exceeding the anther. Anthers are markedly unequal in *A. bicolor*, with an inner pair enclosed by an outer pair.

Ovary either completely hairy or hairy only at and above the middle, superior, unilocular, flask-shaped, extending into a prominent longer or shorter style with a bilobed stigma, the lobes at first appressed, later diverging in dorsiventral plane. Placentation essentially axile, the four ovules pendulous from a central stalk that has a terminal umbo projecting into the base of the stylar canal without closing it.

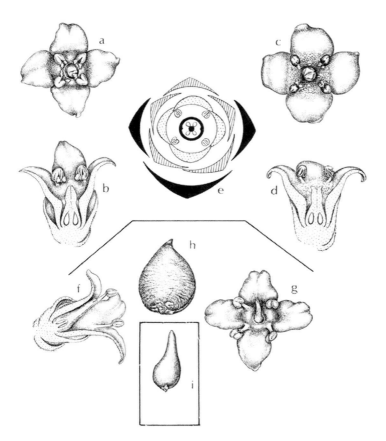

Figure B.13. *Avicennia* spp. (Avicenniaceae) flowers and fruit. (a)–(e) *A. marina* var. *australasica* flowers (x6): (a) Flower from above; male stage, pollen extruded, stigmas appressed. (b) Same flower in L.S. (c) Flower from above; female stage, pollen discharged, anthers turning black, stigmas diverged. (d) Same flower in L.S. (e) Floral diagram. (f)–(g) *A. officinalis* flowers (X3): (f) Flower in L.S. (g) Flower from above, just prior to pollen release. (h)–(i) Fruits (x3/4): (h) Fruit of *A. officinalis*. (i) Fruit of *A. alba*. Material (a)–(d) cultivated at Fairchild Tropical Garden from New Zealand collections; (f)–(i) from Sungei Santubong, Kuching, Sarawak (AMJ).

Fruit a one-seeded, leathery capsule, ellipsoidal to flattened ovoid, with a longer or shorter persistent stylar beak (cf. Plate 5C, E); pericarp thin, brown (Plate 5C), green (Plate 5E), or grayish green. Seed solitary, testa thin or fragmentary, ephemeral, endosperm absent; seed consisting mainly of ripe embryo with two fleshy cotyledons folded, one abaxially, the other adaxially, around the short plumular axis, the radicle blunt and exposed, with a distal tuft of fine hairs. Cotyledons sometimes purple as in *A. officinalis* and populations of *A. marina*. Germination epigeal.

This general account is amplified later with a more detailed description of functional features in relation to pollination; specific differences are described later.

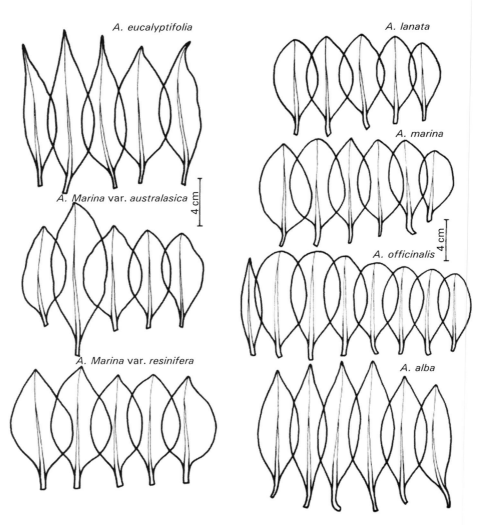

Figure B.14. *Avicennia.* Left: species of eastern Australia and New Zealand (*A. marina* var. resinifera) showing relative constancy of leaf shape. Right: Malaysian species showing range of leaf shape and relative constancy of leaf shape for individual species.

Leaf Morphology and Systematics

The keys in Moldenke (1960) emphasized leaf morphology as a diagnostic feature. Examples from IWP species (Fig. B.14) and AEP species (Fig. B.15) give some indication of the range. Malaysian species are best distinguished by a combination of leaf shape (e.g., Plate 8C) as well as color and density of the indumentum. Such characters are less useful for identification in the herbarium.

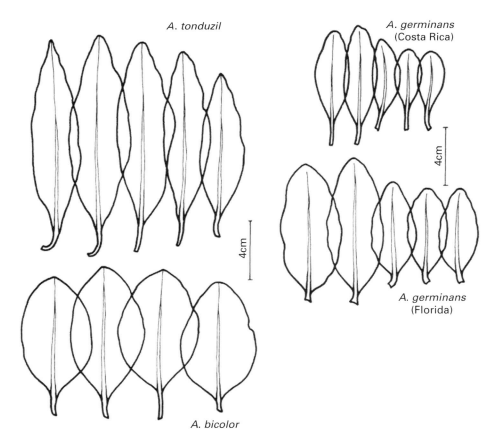

Figure B.15. *Avicennia* species from Caribbean and Central America. *A. tonduzii* is a form of *A. bicolor* with linear leaves. Size range for *A. germinans* represents variation due to soil conditions (scrub form in Costa Rica, tree form in Florida).

Architecture and Growth

The tree conforms essentially to Attims' model but with irregular substitution of flowering axes; branching is usually discontinuous, but may be continuous in vigorous saplings. No resting terminal buds are developed, but the terminal meristem seems capable of extended inactivity (as in populations at high latitudes in both hemispheres or during drier seasons nearer the equator). In the inactive condition, the bud is protected by the closely appressed margins of the petiolar grooves of the ultimate leaf pairs; axillary meristems are protected individually in the same way by older leaves. Axes vary much in the expression of their vigor; young saplings may have internodes 20–30 cm long, in depauperate shoots the internodes are scarcely developed. Shoot extension is often nonsynchronous in different parts of the plant.

Branching is, at first, always by syllepsis. On vigorous leaders or sapling shoots, all nodes may develop branches in a regular acropetal sequence; but on distal, less vigorous shoots, branching is diffuse with intervening nodes supporting unexpanded (reserve) meristems. Prolepsis characterizes all other branching, such as that in the reiterative development of reserve buds or stump sprouts on badly damaged trees. A distinctive form of clonal development is by the rooting of reclining trunks in old open-grown trees, as in *A. marina* var. *australis*.

The architectural capabilities of trees are thus somewhat versatile; tall, single-trunked, broad-crowned trees occur in pure stands on good substrates; on adjacent open sites with poor soils, such as salt flats, the tree is represented by low, dense-crowned shrubs. This shrubby habit occurs exclusively in stands at the limit of the trees' range (e.g., in central Florida, southern Victoria, Australia, and the North Island of New Zealand).

Root System

Two kinds of aerial roots develop. The least common are pendulous aerial roots, which in the saplings can lead to the development of a system of stilt-like roots. These are particularly characteristic of *A. officinalis* (Fig. 8.1; Plate 13E) and sometimes of *A. marina*. Similar aerial roots can develop on low branches that curve near the substrate, or at other levels, in response to injury. Chai (1982) demonstrates a support function for them in leaning trees. Aerial roots have been observed in American species (e.g., *A. germinans*) but are sufficiently infrequent to be treated as an anomaly. In one instance, it is implied that they develop as a response to oil pollution (Snedaker et al., 1981).

More familiar are the extensive erect pencil-like *pneumatophores*, up to 30 cm tall, which typically taper to a bluntly pointed tip, rarely branch, and have a somewhat corky texture. These pneumatophores arise from horizontal cable roots, which in turn originate at the base of the trunk and extend over distances of many meters (Plate 11E). Pneumatophores are of limited length, remain slender, and undergo little secondary growth, in contrast to those of *Sonneratia* (cf. Plate 13E with Plates 11B and 12F). Descending lateral roots, so-called anchoring roots, arise more or less alternately with the pneumatophores from the cable roots. These are not long but usually become somewhat longer than pneumatophores. Both kinds of these second-order roots produce, within the substrate, a fine system of third- and higher-order roots, which increase the absorptive area considerably.

Reproductive Biology

Inflorescence. The flower-bearing axes have a characteristic physiognomy that is not easily described using standard terminology, but it may be regarded as a panicle with leaves progressively reduced distally in size and number. In the flowering seasons, branches toward the outside of the crown (but not the more vigorous leaders) produce determinate axes, which all end in the basic units of the inflorescence, each of which may be described as a flower cluster. Usually, the lower inflorescence branches are subtended by foliage leaves, but distally the subtending structure may be a narrow leaf with characteristically inrolled margins. On vigorous shoots, the lower axils may bear two flowering axes, one inserted above the other and apparently arising from separate axillary meristems. Because a whole branch complex is determinate, a renewal system must develop from a meristem at a much lower level, but this substitution growth is not precise in its expression. Inflorescence morphology can be diagnostic; for example, the spicate, extended inflorescence of *A. alba* contrasted with the globose inflorescence units of *A. officinalis*.

Floral biology. The subdivision of the genus is based on the three contrasted types of floral organization described previously, which in turn seem correlated with contrasted types of floral mechanism. Protandry is universal, but the floral mechanism seems directed toward different-sized visitors in different species. The arrangement of bracteoles and calyx lobes is similar in all species.

In the *Germinans* type, the flower is appreciably zygomorphic (Fig. B.12e). The stamens lie under the upper, somewhat enlarged corolla lobe; the lower lobe provides a platform for insect visitors so that pollen is most likely to be deposited on a visitor's back in the initial phase when the anthers dehisce. Subsequently, the anthers shrivel somewhat and the two lobes of the stigma, which are at first appressed, diverge to expose their inner face, which appears to be the site of pollen reception. In any flower cluster, there tends to be only one open flower at a time. Flower visitors are short-tongued insects, and visits by honeybees are sufficiently extensive for *Avicennia* to be a good source of honey ("mangrove honey"), as on the Florida Keys.

In the *Officinalis* type, the stamens are exserted and the style is toward the adaxial side of the corolla (Fig. B.13f, g). The floral mechanism is comparable to that in the *Germinans* type.

In the *Marina* type, the corolla is actinomorphic (Fig. B.13a–d) and the stamens are either sessile or have very short filaments. The pollen is displayed when the corolla lobes reflex; it is extruded introrsely through short longitudinal slits in each theca; these gape only slightly and the pollen remains adherent, presumably removed only by contact with an insect. In such shallow flowers with short stamens, the site of pollen deposition is likely to be the insect's abdomen. When pollen is finally dispersed, the stamens turn black and the stigma lobes diverge and become receptive. In this type, protandry is more obvious, as the flower phases are more distinctly separate. Flower visitors are again recorded as bees, and the plant is considered to be a source of honey, as in *A. marina* var. *australis* in New Zealand and *A. germinans* on the Florida Keys.

In all three types, the floral reward is nectar, which accumulates in some quantity in the corolla tube. Its origin is almost certainly the minute glands on the corolla surface, because they visibly exude fluid; however, there are similar glands on the outside of the tube.

Floral evolution. *Avicenna* seems to provide a good example of an infrequent evolutionary sequence in the flowers of angiosperms in which an ancestral zygomorphic (monosymmetric) condition is ancestral to an actinomorphic (polysymmetric) condition, the reversal of the usual trend (Endress, 1999; Hileman, 2014). During floral development (Borg and Schönenberger, 2011), there is a change from five to four petals by fusion of two adjacent members of the corolla, as in the contrast of *A. germinans* with *A. marina* (cf. Figs. B.12 and B. 13). A consequence is seen in floral anatomy, in which the larger adaxial petal is supplied by two and not one vascular bundle, a condition maintained, seemingly, in all species. (P.B.T., unpublished observations.) Floral features are said by Borg and Schönenberger (2011) to support an affinity between *Avicennia* and Thunbergioideae in the extended family Acanthaceae.

Seedling Morphology and Establishment

The fruit is incipiently viviparous. The embryo is fully developed at seed maturity as an unextended seedling that will germinate promptly on release (Clarke and Myerscough, 1991). The fruit is still shed as a unit, but the pericarp splits immediately or even at the time of detachment; because the testa is absent or lost with the fruit wall, the unit of dispersal is the embryo, which roots immediately on becoming stationary. The size and shape of the embryo are somewhat diagnostic for different species. The cotyledons are thick (up to 3 mm), fleshy, and folded in opposite directions, the outer cotyledon wholly enclosing the inner (Fig. B.12j). The cotyledons may also differ appreciably in shape; the outer is broadly emarginate or even bilobed, and the inner is rounded. The radicle and hypocotyl are well developed, with the radicle densely hairy and protruding slightly even in the fruit. The distinction between embryos with and without a prominent plumule, a diagnostic character used by some authors, seems obscure and has not been used in this account. The plumule is always short when the embryo is shed. In dispersal, seedlings may float or sink. The differences seemingly are related to density, as established by Steinke (1986), where "sinkers" have a lower proportion of intercellular spaces in the cotyledons than "floaters."

Germination is epigeal, the hypocotyl extends; the radicle may also contribute to extension, because the hairy region becomes longer, notably in *A. officinalis*. The shape of these hypocotylar hairs can be diagnostic; they are hooked in *A. alba*, so that seedlings may become entangled (Fig. B.16). Lateral root primordia become evident on the radicle apex; these develop and anchor the seedling. The cotyledons do not yet unfold so that there is a pronounced twist to the hypocotyl just below the cotyledonary node. The hypocotyl untwists and straightens so that the cotyledons are elevated and the plumule begins to elongate. The cotyledons expand during seedling extension but do not become completely flattened. The first pair of plumular leaves has rather short blades and is borne at the end of a long internode.

Wood Structure

A conspicuous feature of the wood of *Avicennia* is the uniform but anomalous growth rings, which consist of bands of xylem, including radial multiples of wide vessels,

Figure B.16. *Avicennia alba* (Avicenniaceae). Seedlings of this species have characteristic hooked hypocotyl hairs so that groups of entangled individuals are common; Ponggol, Singapore.

alternating with bands of conjunctive tissue, which include strands of phloem (Fig. B.11). Separating each band is a narrow band of sclerenchyma, two or three cells wide. The bands are quite regular, but there is radial connection between them at intervals where the conjunctive tissue of adjacent rings becomes continuous. These connections are most frequent in the region of branch insertions. Zamski (1979) has shown that the phloem strands anastomose extensively in a tangential plane but that there is also continuity in a radial plane through these anastomosing connections. He suggests that phloem is active (conducting) in at least the four outermost rings.

It is now generally agreed (Studholme and Philipson, 1966; Zamski, 1979) that the rings are the products of division of successive cambia. The first cambium is formed in the usual way for dicotyledons with secondary thickening, and it cuts off xylem internally and phloem externally. It ceases to function after a limited period of activity, although it may remain locally as a nondividing layer on the inner side of phloem strands. The second cambium is formed from the innermost cortical layers; the third cambium partly from cortical cells and partly from outer parenchyma derivatives of the previous cambium; and all later cambia are derived entirely from the outermost parenchyma layer produced by each previous cambium.

In the development of each ring, the cambium produces a broad band of xylem to the inside and six to 10 layers of parenchyma cells to the outside. The xylem differentiates immediately, but the parenchyma differentiates centripetally, with differentiation being completed only after the cambium has ceased to function. The product of the first differentiation to the outside of the cambium is the layer of sclereids; the phloem strands appear later by cell divisions within the conjunctive parenchyma. Cambial initials remain inactive on the inside of each phloem strand, but the rest of the cambium differentiates as conjunctive parenchyma. The new cambium is then initiated by dedifferentiation of parenchyma cells immediately outside the sclerenchyma layer. Sclereids and the new xylem are always separated by two or three parenchyma layers.

Early studies suggested that the rings had a regular chronology in their appearance, either annually or semi-annually, but the detailed investigation by Gill (1971) shows that they are correlated directly with the stem diameter in a linear way and their formation is therefore under direct endogenous control. This can be demonstrated because the base of branches and the internode immediately above their insertion develop different numbers of growth rings, even though they are contemporaneous in development, because branches are sylleptic. The fact that the rings are non-annual can therefore be observed in a single specimen, and provides an interesting exercise in deductive reasoning for students. In consequence, the number of rings cannot be used to determine precise age in *Avicennia*. Zamski (1981) observed that there is no correlation between the ring number at any level and the number of branches above that level, so that ring formation is not directly dependent on branch development. Most of this information relates to *A. germinans* and *A. marina* but is probably generally applicable, such is the evident uniformity of the wood structure in the whole genus. A dendrometer study of the variation in radial increment through successive cambia by Robert et al. (2014) describes the "patchyness" of external influences on these changes, but the authors leave the causal nature of their observations unresolved.

The functional significance of this type of anomalous wood structure is not clear, although the vascular tissues of branches are highly integrated with that of stems through the higher frequency of anastomoses between rings in the region of their insertion (Zamski, 1979). The anomalous structure of the wood of *Avicennia* renders it unsuitable for constructional timber, as the phloem decays relatively rapidly to leave extensive pores. The xylem, however, is persistent and quite hard.

Leaf and Hair Anatomy

The microscopic anatomy of the leaf of *Avicennia* seems very consistent and has been described by numerous authors (Areschoug, 1902; Mullan, 1931a,b; Baylis, 1940; Trochain and Dulau, 1942; Fahn and Shimony, 1977). It is characteristic of the leaf of many mangroves in its markedly dorsiventral symmetry (Plate 9E), with a well-developed colorless hypodermis below the adaxial epidermis, abundant terminal tracheids (Plate 9G), and few bundle sheath fibers (Fig. 6.2D). A feature is the glandular hairs on the upper and lower surfaces that secrete salt (Fig. B.12d). Those on the lower surface are obscured by a dense layer of three- or four-celled uniseriate capitate nonsecreting hairs (Plate 9E). These two basic hair types occur in various guises on most parts of the plant. For example, glandular hairs in the flower apparently secrete nectar; capitate hairs on the calyx become enlarged and somewhat T-shaped; those on the seedling of *A. alba* are hooked. There is some suggestion in the literature that the Old World species (e.g., *A. marina*) differ from the New World species (e.g., *A. germinans*) in having three-celled and not four-celled stalks to the capitate hairs. I have not been able to confirm this suggestion. There is, however, appreciable variation in leaf anatomy among individuals of the same species in such features as epidermal wall thickness, density of glands, number of cells per gland, and density of uniseriate hairs, as reported for *A. germinans* by Schnetter (1978). Factors apparently responsible for this variation include soil salinity and degree of shading. Extremely shaded plants that grow in sites with low salinity had large, nearly hairless leaves with very thin external walls and a poorly developed mesophyll palisade. This kind of variation must be considered when comparative studies are done for systematic purposes.

General Key to Species of *Avicennia*

(From the subsequent presentation, it should be apparent that this treatment is a work in progress, but it does provide a guide to the characters used by a succession of earlier authors. Clearly the problems that will be recognized can only be resolved by extensive fieldwork, together with more detailed morphological analysis and the application of modern genetic and molecular methods.)

1A.   Corolla orange-yellow, glabrous within. Plants of the ***eastern (IWP) mangroves*** – in the area of East Africa, Indo-Malaya, and the Western Pacific. 2

1B.   Corolla white, creamy white or with a yellow throat, rarely yellow, glabrous or hairy within. Plants of the ***western (AEP) mangroves*** – West Africa and tropical America. 6

Eastern (IWP) Mangroves

2A.   Flowers 6–10 mm in diameter when expanded, appreciably zygomorphic, stamens exserted, 3–4 mm long, the filament exceeding the anther. Ovary about 5 mm long, oblique, hairy throughout. Flower heads capitate, globose, the units 5–7 mm in diameter; bracts with the margin fimbriate and becoming black with age. Leaves usually rounded apically, but not glaucous white below. Fruit 2–2.5 cm long (Fig. B.13h), abruptly narrowed to a short beak, russet brown (Plate 5C). . . . . *A. officinalis*

2B.   Flowers 3–4 mm in diameter when expanded, actinomorphic, stamens included, 1–2 mm long, the filament about equalling the anther. Ovary about 3 mm long, erect, usually hairy only on the upper half. Flower heads not as wide as 5 mm, bracts without fimbriate or black margins. Leaves pointed or rounded apically, sometimes glaucous white beneath. Fruit various, not russet, sometimes beaked. 3

3A.   Leaves usually rounded apically. 4

3B.   Leaves usually pointed apically. 5

4A.   Young twigs, petioles, midrib, leaf undersurface and upper surface of expanding leaves conspicuously but finely woolly hairy, giving a characteristic yellowish white color. Ovary hairy throughout, the apical hairs obscuring the stigma. Fruit rounded apically or with a short beak, covered with a dense woolly indumentum. . . . . *A. lanata*

4B.   Young twigs, etc., with a yellow but not woolly hairy indumentum. Ovary with a median ring of hairs not obscuring the stigma. Fruit rounded or at most shortly beaked, grayish green, not yellowish. Fruit abruptly narrowed to a short beak, silvery gray-green. . . . . *A. marina*

5A.   Leaves silvery gray or white beneath; inflorescence spicate. Fruit up to 4 cm long, grayish green, gradually narrowed to an extended beak (Fig. B.13i). Bark grayish; lenticellate smooth but often blackened by a sooty mould. . . . .*A. alba*

5B.   Leaves greenish yellow beneath, inflorescence capitate, never spicate. Fruit shorter than 3 cm, greenish yellow, without a conspicuous beak. Bark little flaky but peeling in patches, mottled yellow-brown or green. . . . . *A. eucalyptifolia*

Western (AEP) Mangroves

6A.   Corolla conspicuously hairy within, appreciably zygomorphic, stamens slightly to appreciably unequal. 7

6B.   Corolla glabrous within, only slightly zygomorphic; stamens equal and included. Restricted to Windward Islands, Trinidad, and northern South America to Uruguay. . . . . *A. schaueriana*

7A.   Flowers 10–15 mm long and almost as wide at anthesis, stamens exserted and only slightly unequal; style long, exserted from calyx when corolla is shed. Fruit distinctly beaked, glaucous, surface rough but not pitted. Leaf ovate to elliptic at most. Widely distributed from tropical America to West Africa. . . . . *A. germinans*

7B.   Flowers 5–6 mm long and about as wide at anthesis; stamens included and markedly unequal, with a short filament (0.5 mm) in inner pair and a longer filament (1 mm) in outer pair; style short, either deciduous or not exserted from calyx when corolla is shed. Fruit blunt, globose, greenish yellow (Plate 5E), surface irregularly pitted. Leaf blade oblong, elongate, often less than three times as long as wide. Restricted distribution in Costa Rica, Panama, and adjacent Colombia. . . . . *A. bicolor*

Hybrids in *Avicennia*

The existence of hybrids in *Avicennia* has been examined in AEP populations by Mori et al. (2015), revealing new phylogeographic patterns. Where *A. germinans* and *A. schaueriana* grow together, there is genetic evidence for hybridization, but this is absent where *A. schaueriana* grows in isolation, as on the Brazilian coast. As might be expected, West African and western Atlantic populations are the most distinctive.

Avicennia in Australia

Nomenclatural and identification problems in *Avicennia* are illustrated well by Duke (1991a) and summarized in Duke (2006). His conclusions are based on extensive field study and can therefore be considered authoritative. He distinguishes two species,

A. integra Duke (endemic to Australia) and *A. marina* wherein three varieties are named. Characters and geographic distribution are as follows:

1A. Leaf apex rounded, leaf width relatively constant; corolla 7–11 mm wide; calyx margin entire; style long slender; fruit somewhat elongate; radicle of dispersed seedling continuously hairy. Endemic to 15 riverine estuaries of the Northern Territory. . . . . *A. integra*

1B. Leaf apex pointed, leaf width variable; corolla 3–7 mm wide; calyx margin ciliate; style short broad; fruit rounded; radicle of dispersed fruit only hairy on the narrow collar. Widely distributed from East Africa north to the Middle East, throughout southeast Asia to the islands of the western Pacific and north to Japan (Okinawa). . . . . *A. marina*

In Australia, the following varieties have been recognised:

2A. Bark rough, fissured; calyx mostly pubescent; corolla width 3–7 mm; style "intermediate" (to the same height as the stamens); leaf ovate, elliptic. Restricted to southeastern Australia and northern New Zealand. . . . .*A. marina* var *australasica* (Walp.) Moldenke ex Duke 1991

2B. Bark smooth or flaky; calyx with a wide glabrous margin; style above or below the level of the stamens. 3

3A. Bark smooth, little flaky, corolla 3–5 mm wide; style long, exserted beyond the stamens; leaf narrowly lanceolate. Restricted to tropical northern Australia, from West Australia to Queensland, but elsewhere in Indonesia and the Philippines. . . . . *A. marina* var *eucalyptifolia*

(This taxon clearly corresponds to *A. eucalyptifolia* in the general key given here.)

3B. Bark conspicuously flaky; corolla 5–6 mm wide; style short, included within the stamens. Widely distributed within and outside Australia, from the Red Sea, Persian Gulf, East Africa. . . . . *A. marina* var *marina*

Ecology

Variation in the characters used earlier may correlate as much with the unusual wide ecological tolerance of *Avicennia* species as with genetics. Variables encountered include temperature, degree of tidal inundation, water availability, salinity, and substrate type. These are factors that are only analyzable from field and extended study combined with genetic analysis.

Species of East Africa and the Indo-Pacific (IWP)

Avicennia integra Duke 1988 [*Austral. Syst. Bot.* 4: 315]

This is distinguished from *A. marina* in the characters cited in the earlier key. It has the most limited distribution of any species in the genus. It is said to be closest to *A. officinalis.*

> **Avicennia officinalis** L. 1753 [*Sp. Pl.* 1: 110]
> *A. tomentosa* Willd. 1800 [*Sp. Pl.* 3(1):395]

This species has a wide range from India through Indo-Malaya to New Guinea and the Philippines. It is one of the more distinctive *Avicennia* species of the eastern tropics and is recognized by its large orange-yellow flowers that are 10–15 mm in diameter and therefore larger than any other of the eastern group. The flowers have a rancid or fetid smell and are slightly zygomorphic, as the adaxial (posterior) lobe of the corolla is the broadest and usually shallowly bilobed, a feature that recalls the corolla of the white-flowered New World *A. germinans.* In addition, the ovary is slightly oblique, with the style below the adaxial corolla lobe and not in the center of the flower. Unlike *A. germinans*, however, the inner surface of the corolla is glabrous. The tips of the petals unfold in an irregular way, blackening somewhat with age. The stamens are 3–4 mm long, exserted, and prominently displayed with well-developed filaments that exceed the anthers. The ovary is 0.5 mm long and densely hairy throughout, except the tip of the well-developed style, which is 3–4 mm long and tapered to the unequal stigma lobes. The flower heads are rather short (1–1.5 cm long and 0.5 cm wide) and are always more or less globular (cf. *A. alba*), but they have up to 12 flowers per unit. The bracts and bracteoles are black tipped with a fringed margin. The fruit is densely hairy, about 3 cm long, somewhat longer than wide, and with a short apical beak. The hypocotyl of the seedling often becomes hairy throughout. The cotyledons can be either purple or green. The bark is smooth, lenticellate, light colored, and not fissured.

Leaf shape is rather constant, being ovate or broadly elliptic to obovate, and the apex is rounded, but this shape is not always sufficient to distinguish it from *A. marina.* However, *A. marina* has a whitish, not greenish yellow, underleaf. Frequently in *A. officinalis*, the indumentum fails to develop in patches or even over the whole lower leaf surface. The seedling leaves are narrow with an acute apex, and this seedling character is illustrated in an exaggerated form by Biswas (1934, under "*A. marina*"); it also occurs in *A. alba.*

Aerial stilt roots seem more commonly developed in this species than in other *Avicennia* species.

Nomenclature

This species is the type of the genus (one of the two described by Linnaeus), but as Moldenke (1960) points out, the name "*A. officinalis*" has been widely misapplied by herbarium and other workers, and is a major contribution to the nomenclatural confusion with which species of this genus are invested. In the field, however, species can be distinguished with little difficulty. Examples of earlier confusion are the misapplication of the name "*A. officinalis*" to *A. marina* and the application of the name "*A. officinalis*" to one or more species of *Avicennia* in the New World, even though the true *A. officinalis* does not occur there. The ultimate is the reduction of all species and varieties to synonymy under this one name. This confusion seems to have originated with Linnaeus himself; having described two species, the Asian *A. officinalis* and the New World *A. germinans*, in the first edition of the *Species Plantarum*; he later extended his concept of the former to include both species, implying that *A. germinans* could be reduced to synonymy. His first assessment, however, was correct, because we now know that the two species are morphologically dissimilar as well as geographically separate.

Where good illustrations are available, it is sometimes possible to correct an author's nomenclature; thus Biswas (1934) describes *A. marina* as "*A. officinalis*" and the real *A. officinalis* as "*A. tomentosa*" (a common early mistake of Indian authors); his *A. alba* was correctly named. On the other hand, identification using illustrations of leaf shape alone may not be reliable.

As a consequence of this kind of confusion, often based on the exclusive use of herbarium specimens, descriptions by some authors are virtually worthless in the absence of voucher specimens, and few modern workers are likely to be in a position to verify identifications, even if specimens exist. Taxonomic work on tropical plants with a pantropical distribution requires resources not readily available to most botanists. The situation is particularly difficult for ecologists who may have no reliable basis for comparative study; geographical distribution still remains a valuable diagnostic character.

Avicennia marina (Forsk.) Vierh. 1907
[*Denkschr. Akad. Wiss. Wien Math.-Nat.* 71: 435]
A. intermedia Griff. 1846 [*Trans. Linn. Soc. Lond. Bot.* 20:pl. 1]
A. mindanaense Elmer 1915 [*Leafl. Philipp. Bot.* 8: 2868]
A full bibliography is given by Moldenke (1960).

When the name is used in the most inclusive sense, this species has the broadest distribution, both latitudinally and longitudinally, of the genus (indeed of any mangrove), with a range from East Africa and the Red Sea (the type locality), along tropical and subtropical coasts of the Indian Ocean to the South China Sea, throughout much of Australia into Polynesia as far as Fiji, and south to the North Island of New Zealand. The maps in Duke (1991a) and Spalding et al. (2010) are very generalized. The genetic diversity of this species has been extensively investigated (Maguire et al., 2000a, b, 2002).

In this broad sense, this species may be recognized by its regular orange-yellow flowers, not more than 6 mm in diameter with four equal lobes of the corolla that are glabrous within (except for minute glandular hairs). The stamens are short, scarcely 2 mm long, and not exserted, the filament about equalling the length of the anther. The ovary is rather short (2–3 mm), usually with a short indistinct glabrous style and two equal stigmas that diverge only in late anthesis. The ovary itself is glabrous below but hairy above as far as the style. The flowers are sweetly scented. The fruit is nearly spherical to slightly ovoid in outline but is always about as long as it is wide (2 cm); the short beak evident early in development is almost completely obscure in the mature fruit. The seedling hypocotyl is hairy only at its base, in the region of the short radicle. The leaves are rounded apically, more or less ovate, and whitish beneath. The bark is rather distinctive, being flaky, mottled greenish yellow, and peeling in patches. The flower heads are capitate and remain so, but with the units rather distant. The name *A. intermedia* Griff., sometimes used (incorrectly), is descriptively appropriate where *A. marina* co-occurs with *A. officinalis* and *A. alba*, as it has the leaf shape of the former and the indumentum of the latter.

Nomenclatural History

Retained here from the first edition of this book is the list of names used by Bakhuizen van den Brink (1921) and Moldenke (1960, 1967) that illustrates the range of geographic variation within *A. marina*. It is a burden that future workers may wish to abandon.

1. *Avicennia marina* (Forsk.) Vierh. var. *typica* Bakhuizen = *A. marina* (Forsk.) Vierh. var. *marina* of Moldenke.

    In the opinion of Bakhuizen van den Brink, this has a limited range in East Africa and Arabia (Red Sea), which is the type locality (Yemen). However, Moldenke's view is much broader, and the variety in his interpretation ranges from East Africa through Indo-Malaya to Australia and New Caledonia.

2. *Avicennia marina* var. *resinifera* (Forst.) Bakhuizen used in much the same sense by both authors. This variety is distinguished by its compact inflorescence units with two to 12 flowers per unit, the elliptic oblong leaf blades with an acute or acuminate apex, and the ovary hairy only in the upper half. It is distributed in the Western Pacific; for instance, Australia, the North Island of New Zealand, New Caledonia, New Guinea, and the Philippines. Again, Moldenke ascribes to it a wider distribution than does Bakhuizen and includes localities as far west as Sumatra in its range.

    The uncertainty of these designations is suggested by the inclusion of New Caledonia in the range of both the typical form and var. *resinifera*. There is only one form in New Caledonia.

3. *Avicennia marina* var. *intermedia* (Griff.) Bakhuizen is included within *A. marina* var. *marina* by Moldenke. Bakhuizen records it throughout the Malay Archipelago, from the Malay Peninsula to New Guinea and the Philippines. It is said to be distinguished by its leaf shape (ovate-rotund, with an obtuse base and short petiole, less than 1/5 cm long) and the characteristically small fruits. Bakhuizen based his description on living material.

4. *Avicennia marina* var. *rumphiana* (Hall. f.) Bakhuizen is accepted by Moldenke. It is said to be distinguished by its long petioles (1.5–3 cm) and the rounded leaf apex. This has much the same range as the preceding variety. In addition, Moldenke recognizes the following new varieties:

5.   *Avicennia marina* var. *anomala* Moldenke 1940 (*Phytologia* 1:141); recognized by its "attenuated" inflorescences with scattered or opposite pairs of flowers and recorded for only Low Island, Port Douglas, and Queensland.

6.   *Avicennia marina* var. *acutissima* Stapf & Moldenke, ex Moldenke 1940 (*Phytologia* 1:411); recognized by "its decidedly sharp-acute or acuminate leaf apex." It is restricted, according to its author, to the vicinity of Bombay.

7.   *Avicennia marina* var. *australasica* (Walp.) Moldenke has been accepted by Duke (1991a) and presumably should now be used for the New Zealand populations.

In view of the extreme uncertainty of the status of other varieties of *A. marina*, it seems inappropriate to recognize presumed forms of such limited occurrence as categories 5 and 6.

> **Avicennia alba** Blume 1826 [*Bijdr. Fl. Ned. Ind.* 14: 82I]
> *A. marina* (Forsk.) Vierh. var. *alba* (Blume) Bakh. 1921 [*Bull. Jard. Bot. Buitenz.* Series 3(3): 205, 207–2019]

A. alba has a wide distribution from India to Indochina, through the Malay Archipelago to the Philippines, New Guinea, and the Solomon Islands. It is distinguished within the *A. marina* complex by the long spicate distal flower units that are 1.5–3 cm long, with 10 to 30 flowers per unit and the lower flower pairs usually quite distant. The flowers are of the *Marina* type, the ovary scarcely 2 mm long, and the style absent; the middle of the ovary is conspicuously hairy, but the lower part includes only glandular hairs. The leaves of *A. alba* are characteristically lanceolate, pointed at the apex (in contrast to the rounded apex of *A. officinalis* and *A. marina*), and with a white under surface.

A good field character is the conical shape of the fruit (and consequently seed), which is extended into a pronounced beak, especially in the early stages of development (Fig. B.13i). The bark is slightly roughened, but not fissured, and is somewhat brown. A frequent character, which may be diagnostic, is the development of a sooty mould on older stem parts.

Several observers (Watson, 1928; Chai, 1982) comment on the tendency of drifted seedlings in this species to become clumped, apparently because the hooked hairs of the hypocotylar axis of different seedlings become entangled (Fig. B.16). The hairs are also useful diagnostically.

> **Avicennia lanata** Ridley 1920 [*J. Fed. Malay State Mus.* 10: 151–152]
> *A. officinalis* L. var. *spathulata* Kuntze 1891 [*Rev. Gen. Pl.* 2: 502]

This species is readily distinguished in the field by the white woolly indumentum that covers leaves and younger stem parts so that, even from a distance, the crown is characteristically whitish. Flowers are of the *Marina* type but with very hairy outer petals and calyx. The ovary is about 2.5 mm long and hairy; the hairs project distally and envelope the base of the short style, which has a bifid stigma. This species is not listed in Spalding et al. (2010).

According to Ridley, *A. lanata* is distinguished from *A. officinalis* by its recurved (not erect) calyx lobes (corolla lobes according to Moldenke, because the calyx lobes are always erect in all species), the filaments the same length as the anthers (not longer), the short (not long) style, and the agreeable (not disagreeable) floral fragrance. The flowers of the two species are possibly the same size, but this is nowhere reported unequivocably, and Moldenke speculates that this species is, in fact, close to *A. marina*.

The species has been recorded only for the states of Malacca and Pahang in the Malay Peninsula, and more recently for extreme western Sarawak (Chai, 1982). There are several records for Singapore, but most sites seem to have been destroyed by real estate development.

> *Avicennia eucalyptifolia* (Zipp. ex Miq.) Moldenke 1960 [*Phytologia* 7(3): 162–165; see Zippel ex Miquel, *Fl. Ned. Ind.* 2: 912]
> *A. officinalis* var. *eucalyptifolia* Valet. 1907 [*Bull. Dep. Agric. Ind. Neerl.* 10: 53]

This taxon has been recognized by a number of authors entirely on the basis of its ovate-lanceolate to lanceolate leaves, which may be three to five times as long as they are wide, or if less, with the leaf apex gradually narrowed to a point. Moldenke regarded this feature as sufficient for recognizing a new species. *A. marina* is largely separated by its rounded or at most bluntly or abruptly pointed leaves. However, the two seem to have identical flowers and fruits.

A distinguishing character may be the smooth bark, which is grayish, light brown to greenish yellow, and flaking in patches. *A. marina* seems to have similar bark characters in part of its range, however. Thus, Semeniuk et al. (1978) illustrate, for the bark of *A. marina*, what might in Queensland be thought of as that of *A. eucalyptifolia*. They suggest that *A. eucalyptifolia* is merely a northern variant of *A. marina* but do not describe its bark. It has this status in Duke (2006).

> *Avicennia balanophora* Stapf and Moldenke ex Moldenke 1940 [*Phytologia* 1: 409–410]

The status of this species is dubious and is not accepted by Duke (2006). Moldenke based it on material from Queensland (type F. Mueller s.n.), and its locality is said to be the mouth of the Brisbane River. It is also recorded from Keppel Isles (MacGillivray s. n.). It is said to be recognizable by its oblong, acorn-like fruit and the corolla limb growing to 6 mm wide during anthesis. However, search of the mouth of the Brisbane River has not revealed any *Avicennia* except *A. marina*, with which this species may be equated in the absence of further evidence to the contrary. This type locality is now, in part, the site of an international airport.

Species of the New World and West Africa (AEP)

> *Avicennia germinans* (L.) Stearn 1958
> [*Kew Bull.* 1958: 34–36]
> *A. nitida* Jacq. 1760 [*Enum. Syst. Pl. Carib.* 25]
> *A. tomentosa* Jacq. 1760 [*Enum. Syst. Pl. Carib.* 25]

The full synonymy of this species is long and complex and has been provided in full by Moldenke (1960), but the resolution by Stearn is now generally accepted (see also Compère 1963).

A. germinans is widespread from West Africa to southern Florida, and the Bahamas throughout the Caribbean to Mexico, Central America, and the Pacific coast of South America to Peru and the Galapagos; its distribution on the Pacific coast of the Americas seems discontinuous. It is recorded on the Atlantic coast of South America to northern Brazil, but some of this distribution may be in error because it has not always been clearly distinguished from *A. schaueriana*.

Genetic diversity between Caribbean and Pacific populations has been investigated by Dodd et al. (2002) in relation to their separation by the Central American Isthmus. A further discussion is found in Cerón-Souza et al., 2005, 2012).

In its vegetative morphology, *A. germinans* is characterized by its ovate leaves, usually with a blunt apex; the lower leaf surface and the twigs are conspicuously glaucous white in contrast to the dark green of the upper leaf surface. The leaves are 5–8 cm by 2–4 cm with a short petiole. The flowers are white and conspicuously zygomorphic, with the stamens long, equal, and exserted; the petals are conspicuously hairy inside; the fruits are somewhat variable in size (2–3 cm long) and always with a short lateral beak.

Nomenclature. This species was made known to European science by material collected by Patrick Browne in Jamaica that became incorporated into the Linnaean herbarium and used by Linnaeus, together with material from Venezuela, for the description of *Bontia germinans* in the *Systema Naturae*. In the second edition of the *Species Plantarum* (1763, 2:891), Linnaeus expanded his use of the name to include material from India that he had described as *Avicennia officinalis* in the first edition (1753, 1: 110). This was later recognized to be a distinct species; because the confusion was soon evident, even to Linnaeus' contemporaries, the name *A. nitida* of Jacquin was adopted for the New World species and became well established until Stearn pointed out the need for clarification.

Avicennia africana Palisot de Beauvois 1805
[*Flore Oware Benin* 1: 79-80, pl. 47]

This name is applied widely to material from West Africa, but in the absence of any clear-cut and consistent morphological distinction, it may be better to broaden the concept of *A. germinans* and include plants from both sides of the Atlantic. Schauer (in Moldenke, 1960) suggests that there may be a difference in the color of the dried leaves and that the African species has narrower and more elongated leaves. It is not included in Spalding et al., 2010, hence it has been seen simply as a geographic entity.

Avicennia bicolor Standley 1923
[*J. Wash. Acad. Sci.* 13: 354]
A. tonduzii Moldenke 1938 [*Phytologia* 1: 273–274]

A. bicolor has a limited distribution and is recorded for the Pacific coast of Costa Rica (Jiménez and Soto, 1985) to Colombia, notably at the mouth of the Buenaventura River. When growing with *A. germinans*, this species can be recognized by the dense, dark green crown, which forms a discontinuous understorey to the more diffuse, grayish crowns of *A. germinans*, which overtops it. Otherwise, it can form pure stands behind the *Rhizophora* zone, with trees of appreciable stature. Its flowers are distinctly zygomorphic and like those of *A. germinans* in having the petals hairy inside, but they are much smaller, scarcely 5–6 mm in diameter, and expanded. The white corolla sometimes has a yellow throat, hence the name, but it can be almost pure yellow in some populations. A distinctive feature is the inequality of the stamens; they are inserted at the same level in the flower, but the outer pair has filaments at least 1 mm long, compared with the inner pair where filaments are scarcely 0.5 mm long. The fruit is more rounded than that of *A. germinans* and without a prominent beak.

Although described by Moldenke as "very distinct," *A. tonduzii* seems to be only a variant of *A. bicolor*, distinguished by the narrow leaves (Fig. B.15) and the paniculate assemblages of floral axes with the individual flower pairs distant from each other. Material from Panama to southern Mexico (Chiapas) has been referred to it.

Avicennia schaueriana Stapf and Leechman ex Moldenke 1939
[*Lilloa* 4: 336]

This species has a more or less continuous distribution from as far north as the Turks and Caicos Islands to the lower Lesser Antilles and the Atlantic coast of northern South America from Guyana south to Uruguay. It is based originally on material sent by Leechman from Guyana (British Guiana) to Kew, where it was studied by Stapf, who wrote a lengthy unpublished account (quoted by Moldenke in full) of the contrasted sets of specimens at his disposal under the names "*A. nitida*" (= *A. germinans*) and "*A. tomentosa*" (whose origins he discusses), and for which he suggests the name that was later formally applied by Moldenke.

The flowers are larger than those of *A. bicolor*, the corolla being 10–12 mm long and almost as wide, and so they approach those of *A. germinans* in size, but the inner face of the corolla is glabrous or at most slightly hairy. The lobes are narrow and are not reflexed and thus enclose the equal stamens with filaments 1.5–2 mm long. The ovary is uniformly hairy and not beaked so that it is not exserted or only scarcely exserted when the corolla falls, unlike *A. germinans*. The fruit is a pale sap green, seldom with a purple tinge, and is flatter and more pointed than that of *A. germinans*.

Family: Bataceae (Batidaceae)

A family of uncertain affinity containing the single genus *Batis*, a characteristic coastal plant throughout the tropics in salt flats, often extending into the mangal fringes.

Batis P. Browne. 1756 [*Civ. Nat. Hist. Jamaica*: 351]

A subshrubby genus of two species with linear succulent leaves and spreading, prostrate, or scrambling woody stems. Plants are dioecious with reduced fleshy spicate inflorescences.

Batis maritima L. 1759 [*Syst. Nat. Ed.* 10: 1289]
See Cafferty & Monroe 1999 [*Novon* 9(4): 490] (Fig. B.17)

Leaves opposite, sessile, and subcylindrical to about 3 cm long. Spikes in leaf axils are up to 1 cm long with fleshy bracts (male); or in the female flower, the carpels are fleshy and fused together. The flowers have no perianth; the male includes four stamens, the female a fleshy pistil with four apically inserted ovules. The fruit consists of the fleshy aggregate of flowers with the numerous included seeds, and is dispersed by water or possibly animals.The plant is strictly a halophyte but is common in hypersaline areas adjacent to mangal, and is commonly associated with more salt-tolerant species, for example, *Avicennia* in South Florida.

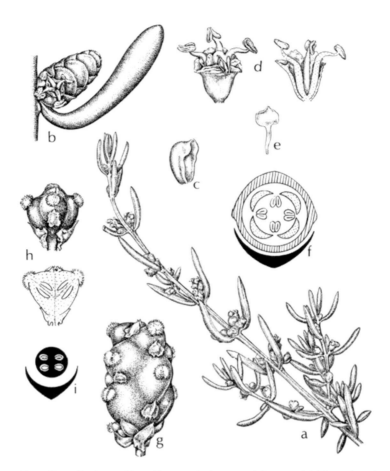

Figure B.17. *Batis maritima* (Bataceae) shoots and flowers. (a) Flowering shoot (x1/2). (b) Male inflorescence (x3). (c) Single male flower in bud (x6). (d) Male flower at anthesis (x6); left: side view; right: L.S. (e) Petal (x6). (f) Floral diagram of male flower. (g) Female inflorescence (x3). (h) Aggregate of female flowers (x3); upper: side view; lower: L.S. (i) Floral diagram of single female flower. (From Correll and Correll, 1982.)

The second species (*B. argillaceus*) has a limited distribution in New Guinea and northern Australia.

Family: Bignoniaceae

A tropical family of trees and lianes usually with opposite pinnately compound leaves; the flowers large and conspicuous with a tubular, trumpet-shaped corolla. A number of genera are recorded as having species that are mangrove associates; *Dolichandrone* is an Old World genus of trees; four genera are New World: *Anemopaegma* and *Phryganocydia* are viny, and *Amphitecna* and *Tabebuia* are trees. These genera show no close relationships within the family.

Dolichandrone Fenzl. ex Seeman 1862
[*Ann. Mag. Nat. Hist.* Series 3, 10: 31]

A genus of about 10 species with a range from East Africa to New Caledonia. *Dolichandrone spathacea* has the widest range of any species in the genus and is a frequent constituent of back mangal, but only in areas flooded by the highest tides. Most typically, it grows in swampy or beach communities such as dunes and dune slacks or on riverbanks.

Dolichandrone spathacea (L.f.) K. Schumann 1889
[*Fl. Kais.-Wilhelm Land*: 123]
(Figs. B.18,19; Plate 7K)
See Chatterjee, D. (1948, *Bull. Bot. Soc. Bengal* 2: 67) for a full synonymy. An account is provided by van Steenis (1977); see also van Steenis (1928).

This species is found from southern India throughout Malesia to New Caledonia. It is distinguished from *D. serrulata* of India and Ceylon, which has similar long-tubed flowers, by its entire (not serrate) leaflets. In Australia, it is restricted to far north Queensland; the leaflets there are said on occasions to be reduced and needle-like (Duke, 2006).

Evergreen or briefly deciduous trees up to 20 m high with conspicuous perfect flowers; bark gray to dark brown, somewhat fissured in older trees. Twigs thick, conspicuously lenticellate with prominent and rather irregular pairs of leaf scars; bud scales not developed. Leaves opposite, decussate, imparipinnate, 20–30 cm long, petiole to 6 cm, with two to four (usually three) pairs of leaflets and a somewhat large terminal leaflet, petioles and young leaves often reddish. Leaflets 5–10 cm by 3–5 cm, ovate to ovate-lanceolate, narrowed gradually to the pointed apex, base abruptly narrowed to a short (1 cm) petiolule; margin entire, often somewhat undulate, glabrous, or at most, minutely hairy when young; lateral leaflets often asymmetric, with the

Figure B.18. *Dolichandrone spathacea* (Bignoniaceae) leaves and flowers. (a) Flowering shoot (x1/2). (b) Domatia (x6) in angle between midrib and main vein on lower leaf surface. (c) Leaf outline (x1/6). (d) Abaxial glands in tip of calyx (x3/2); inset: detail of glands (x12). (e) Flower bud just before anthesis (x3/2). (f) Surface glands on calyx (x12). (Material cultivated at Fairchild Tropical Garden, Miami, FL.)

basiscopic half narrowest. Domatia with a hairy fringe in abaxial angles between larger veins and midrib (Fig. B.18b).

Flowers large, perfect, zygomorphic, solitary in the axil of minute scale-like bracts and forming terminal racemes of two to six flowers at the end of leafy twigs. Flowers develop in acropetal order; pedicel thick, about 2 cm long with bracteole inserted at about the midpoint. Calyx green, inflated tubular, 6–8 cm long, split adaxially, spathe-like and recurved at anthesis, sometimes caducous; apex with a blunt mucro and a

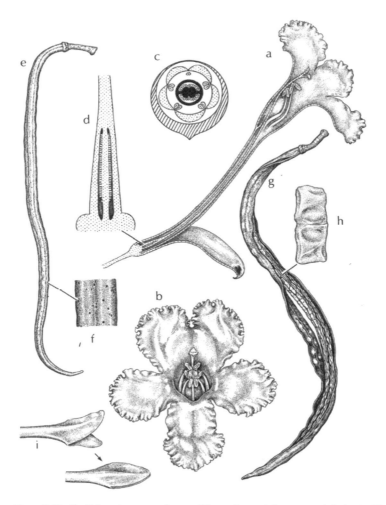

Figure B.19. *Dolichandrone spathacea* (Bignoniaceae) flowers and fruit. (a) Flower in L.S. (x1/2). (b) Flower from front (x1/2). (c) Floral diagram. (d) Ovary in L.S. (x3). (e) Unripe fruit (x1/4). (f) Surface of unripe fruit (x1). (g) Dehisced fruit (x1/4). (h) Single seed (x3/2). (i) Stigma of receptive flower (x3) in diverged (unstimulated) and closed (stimulated) position. (Material cultivated at Fairchild Tropical Garden, Miami, FL.)

purple glandular patch on the abaxial side. Corolla at first green, maturing to white, tubular, trumpet-shaped, 15–20 cm long, the tube only 7–8 mm wide but about 10 cm long, abruptly enlarged to the five-fringed lobes, spreading and about 12 cm in diameter at maturity, the three abaxial lobes 4–5 cm by 3–4 cm, the two adaxial lobes somewhat shorter, with a shorter median sinus between them. Stamens four, with the fifth adaxial stamen represented by a filiform vestige; fertile stamens inserted in the throat of the tube and enclosed by the corolla lobes, filaments about 3 cm long, anthers medifixed and horizontal. Ovary slender, cylindrical, on a short, narrow basal disk, tapered upward into the filiform style, stigma bilobed, peltate, extending beyond the stamens. Ovary bilocular, scarcely 1 cm long, with numerous ovules on the transversely axile placenta. Fruit a flattened, pendulous linear capsule, 50 cm or longer, drying white, with two valves splitting to reveal the central replum. Seeds numerous, packed in regular series, white, about 1.5 cm long, oblong-rectangular with a soft, corky winged testa; endosperm absent. Germination epigeal; cotyledons bilobed.

Growth and Reproduction

Shoot construction. The species is deciduous in some climates; for example, cultivated specimens in South Florida lose their leaves for several weeks in winter. The tree is also recorded as deciduous within its natural range (Chai, 1982), but the periodicity is not noted; van Steenis (1977) comments that it may "stand leafless in the dry season for many months." Shoot growth is somewhat articulate, with reduced leaves at intervals, but without the formation of organized, scaly, resting buds, despite the deciduous state. Distal shoots are sympodial through flowering, with the terminal raceme substituted by one or two lateral buds that develop by prolepsis. The architecture of the tree is not regular, but in open situations it becomes broad-crowned with a short trunk. Buttresses and aerial roots are not developed.

Floral biology. The flowers develop rapidly in acropetal order and expand in a few days from a stage where they are buds 2–3 cm long enclosed by the inflated calyx to the pre-flowering stage, after which the corolla reaches its maximum length of about 20 cm. Van Steenis (1977) emphasizes that the calyx is at first filled with water, a condition found in other Bignoniaceae, and which he describes as one of the uniquely tropical features of plants. Flowers usually develop one at a time within a single inflorescence, but because there are many inflorescences at different stages on a single tree, the flowering period is extended. The corolla expands in the early evening and each flower usually lasts only one day, with the corolla falling on the morning of the following day. When the flower opens, a pervasive scent is produced, and nectar, presumably produced by the floral disk, accumulates copiously in the base of the corolla tube; the anthers dehisce. The stigmas protruding beyond the anthers diverge and become receptive. They are sensitive and close within three or four seconds when the inner receptive face is touched, as by the head of a suitable flower visitor (Fig. B.19i). Because a visitor carrying pollen touches the stigma first, it is likely to receive foreign pollen, but the flower's own pollen, picked up as the visitor withdraws, is not likely to be deposited because the closed stigma lobes now conceal the receptive surface. Plants appear to be self-compatible so that self-pollination of different flowers on one tree is possible; isolated trees may set fruit. Pollination is presumably by long-tongued nocturnal animals, probably hawk moths, but no observations have been made.

Fruits develop rapidly, as is characteristic of many Bignoniaceae, maturing within one or two months. The fruits elongate and are at first green and beanlike, but they eventually dry and turn white, and the valves separate by twisting to reveal the numerous oblong seeds. When released, these float readily and are dispersed by water, but they germinate immediately on being stranded. The corky wing replaces the usually thin wing of the wind-dispersed species found otherwise in the family and provides a good example of specialization in relation to habitat.

Extrafloral nectaries. In addition to the nectar produced within the corolla tube, there are small discoid nectaries on exposed surfaces. The most obvious are those that occupy the tip of the calyx tube (Fig. B.18d), but larger ones occur on the outside of the corolla lobe before it expands (Fig. B.18e, f). The function of these nectaries is not known, but van Steenis suggests that they are the source of a varnish-like material that covers the young flower bud.

Amphitecna Miers 1870 [*Trans. Linn. Soc.* 26: 163]

Like the related calabash (*Crescentia cujete* L.), this genus is unusual in the family in its simple, spirally arranged leaves and non-winged seeds.

Amphitecna latifolia (Mill.) A. Gentry 1976 [*Taxon* 25: 108]
Crescentia cucurbitina L. 1771 [*Mant. Pl.* 2: 250]
Enallagma latifolia (Mill.) Small 1913 [*Fl. Miami*: 171]
A complete synonymy is given by A. H. Gentry 1980 [*Fl. Neotrop. Monogr.* 25: 66]

A small to medium-sized tree with simple, obovate, spirally arranged leaves, 5–10 cm by 15–20 cm, inserted on a short woody cushion. The flowers are solitary or in pairs in leaf axils or on the older wood, with a tubular, somewhat zygomorphic, yellowish green corolla. The fruit is gourd-like, 6–8 cm in diameter, and has a hard shell enclosing many seeds in a slimy pulp. Both the fruit and seeds are dispersed by water.

Gentry (1982) includes this species as a mangrove associate and suggests that it is restricted to the Pacific but not the Atlantic coast of tropical America. Records of inland distributions may be due, in part, to its occasional cultivation, and it may not always have been distinguished from closely related species. Johnston (1949, p. 270) records it at the margins of mangrove swamps.

Anemopaegma Mart. ex DC. 1845 [*Prodromus* 9: 182]
A genus of vines with about 30 species in the New World.
Anemopaegma chrysoleucum (HBK) Sandw. 1938 [*Lilloa* 3: 459]

This and possibly other species of the genus have been recorded in association with mangroves in northern South America. It is recognized by the showy, trumpet-shaped flowers and the opposite compound leaves with the rachis extended into a tendril. The capsules are rounded, 2–3 cm in diameter, and have numerous flattened papery seeds.

Other Bignoniaceae

Gentry (1982) records *Phryganocydia phellosperma* (Hemsl.) A. Gentry as restricted to the Pacific side of Colombia in the tropical American mangroves. This species is distinguished from the common and widely distributed tendrillous vine with large purple flowers [*Phryganocydia corymbosa* (Vent.) Bureau] by its simple tendril, short ovoid fruit with corky, almost wingless seeds, and floral details (Sandwith, 1940).

It is seemingly restricted to coastal, swampy habitats, and might be termed a mangrove associate. Gentry also records *Tabebuia palustris* Hemsl. as a mangrove associate with a strictly Pacific distribution. It has a limited range from Costa Rica to Colombia. Johnston (1949, p. 274) records it as a shrub or small tree up to 4 m tall, in the drier marginal parts of mangrove swamps, an "unobtrusive plant and not very common." It may be recognized by the trifoliate leaves and conspicuous white flowers with a yellow throat. It is deciduous. However, Spalding et al. (2010) regard it as a "core mangrove" species.

Family: Bombacaceae (Malvaceae: Bombacoideae)

The genera in the traditional family Bombacaceae have become dispersed in the much-expanded Malvaceae (Alverson et al., 1999) The following genus now appears as a sister lineage to the traditional Malvaceae, itself now represented as a subfamily Malvoideae, but a more precise designation suggests Helicteroideae within a modified Durioneae, but even then somewhat isolated (Nyfeler and Baum, 2000).

> ***Camptostemon*** Masters 1872 [*Hook Icon. Pl.* 12: 18, Pl. 1119]
> *Cumingia* Vidal 1885 [*Cuming. Philip.*: 211]
> See also Masters 1874 [*J. Linn. Soc. Bot.* 14: 505]

A genus with two species of evergreen trees in back mangal from Borneo, the Moluccas, and northern Australia, to New Guinea and the Philippines. It is distinctive in its scaly pubescence, cupular epicalyx, and closed calyx, the five or more stamens united into an androecial tube and particularly the two- (rarely more) locular ovary with one or two ovules in each locule. The fruit is a two-valved capsule, with one or two hairy seeds, but without endosperm. The surface roots may bear knobby pneumatophores. Masters suggested that the genus formed a link between Bombacaceae and Malvaceae because of its stamen tube, which is not divided into separate phalanges. The geographical distribution of the two taxa has been clarified in recent years, especially in Australia (Duke, 2006), while Spalding et al. (2010) indicate they co-occur in Indonesia.

The two species, which may overlap in the middle part of their range, are distinguished as follows:

1A. Leaves glabrous above; stamens numerous (~20), dithecate ....
.... *C. schultzii* (northern Australia, Central Indonesia, and New Guinea)
1B. Leaves sparsely but uniformly lepidote above; stamens five, polythecate... ....
.... *C. philippinensis* (Borneo, Central Indonesia, and Philippines)

Camptostemon schultzii Masters 1872 [*Hook. Icon. Pl.* 12: 18, Pl. 1119] (Fig. B.20)

Evergreen hermaphroditic trees to 30 m tall, trunk flanged at the base, bark dark, scaly, unfissured; exposed roots with blunt rounded pneumatophores to 30 cm tall. Younger exposed surfaces (except upper surface of leaf) with a continuous indumentum of minute peltate scales. Twigs with prominent round leaf scars. Leaves spirally arranged, petiolate, the petiole 2–3 cm long, terete but finely grooved above, blade 8–10 (–15) cm

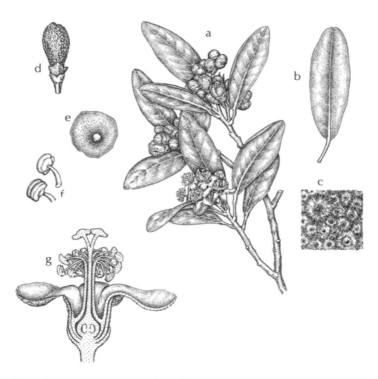

Figure B.20. *Camptostemon schultzii* (Bombacaceae) leaves and flowers. (a) Flowering shoot (x1/2). (b) Leaf (x1/2). (c) Scales on lower surface of leaf (x35). (d) Flower bud (x4) with bracteoles, calyx, and scaly petals. (e) Single peltate petal scale from below (x70). (f) Stamens (x8). (g) Flower in L.S. (x3). (Material from Missionary Bay, Hinchinbrook Island, Queensland; part (a) after Masters, 1872.)

by 2.5–3 (–4.5) cm oblong-ovate to broadly lanceolate, margin entire, apex acute to rounded or even slightly emarginate, base cuneate but always with an abrupt notch at the petiole insertion. Flowers in axillary three- to six-flowered cymes, as either regular dichasia or umbellate; peduncle 4–8 mm long, pedicels 3–4 mm long. Epicalyx cupulate, 2–3 mm long, with an irregular margin; calyx campanulate, about twice as long as epicalyx, more or less distinctly three-lobed, but margin becoming irregular with age. Petals five, imbricate, bluntly pointed to 6 mm long, white but densely scaly without. Stamens numerous, united into a narrow tube about the same length as the petals, free portion of filament 1–2 mm long, anthers bithecate. Ovary 2 mm long, superior, globular, scaly, included within inflated base of stamen tube, with two locules and one campylotropous ovule in each locule on an axile placenta. Style slender, 8 mm long, with two exserted flattened peltate stigmas, each obscurely three-lobed. Fruit scaly, a somewhat laterally flattened ellipsoidal capsule 15 mm by 8 mm, with persistent basal epicalyx, dehiscing loculicidally into two valves, each with a narrow persistent placental replum. Seeds two, conspicuously hairy and sometimes described as "cotton-like" or "kapok"; embryo green.

Ecological and Geographic Distribution

The species is more characteristic of open rocky shores than estuarine mangroves, and is commonly described as occurring on sandy beaches, but within the tidal range. On the coast between the Oriomo and Fly rivers in Papua New Guinea, it is described (*Brass* 6465) as a "dominant species of extensive swampy rain forests; 30 m or more." It is recorded for southern New Guinea and northern Australia, but also into Indonesia as far as Borneo.

Growth

Shoot construction. The shoots are not morphologically articulate, but extension is periodic, with the exposed but unexpanded leaf primordia of resting terminal buds protected by the scaly indumentum.

   Indumentum. The peltate scales that form the indumentum are about 200 μm in diameter, but there may be some dimorphism, as there are a few uniformly scattered, larger scales. These may represent a preliminary population that covers the leaf in early stages of development; after a period of rest, the majority of narrower scales may be produced. Scales on calyx and petals may correspond to the larger size class.

   Root system. Troll (1933a) mentions its similarity to *C. philippinense*. The above-ground roots are knobby, woody structures up to 30 cm tall, originating as slight irregularities of the extending horizontal roots, much as in *Bruguiera* but less prominent. They are augmented by secondary growth, the wood being very thin-walled and with few tracheary elements. They are not the site of feeding roots, as in *Bruguiera*. Troll says that a rôle as an aerating organ has still to be demonstrated because the cortex of the pneumatophore is not aerenchymatous.

   Camptostemon philippinense (Vidal) Becc. 1898 [*Malesia* 3: 273]
      See also K. Schumann (1890) in Engler and Prantl [*Die natürlichen Pflanzenfamilien*, ed.1, t. 3, ab. 6: 67]

This species differs from *C. schultzii* in the presence of lepidote scales on the upper as well as the lower surface of the leaves, and the five polythecate stamens. The ovary may more often have three or four instead of two locules; the stigma may then have more than two lobes, the number corresponding to the number of locules, with the capsule separating into a corresponding number of valves. The two species are said by Bakhuizen van den Brink (1924) also to differ in leaf shape (elliptic in C. *schultzii* as distinct from obovate-oblong to lanceolate in *C. philippinense*), but they both have a similar range of leaf size.

Nomenclature and Other Notes

The genus is based on *Camptostemon schultzii* (the specific name referring to the collection of *Schultz 511* Port Darwin, North Australia; the specimen had no fruit).

The genus *Cumingia* was erected by Vidal (1885, *Phan. Cuming. Philip.:*212, Fig. 1) in his description of plants collected by Cuming in the vicinity of Luzon in the Philippines. He did comment on its proximity to the *Camptostemon* of Masters, but he considered it sufficiently distinct in its "unilocular" anthers. He also noted the prior use of *Cumingia* Don (in Benth. and Hook. *Gen. Pl.* 3:69) in the synonymy of *Conanthera* Ruiz et Pavon, which he dismissed in view of the remoteness of this taxon (contrary to today's rules of botanical nomenclature), in his desire to commemorate the collector's name. Schumann retained the two genera in his treatment of the Bombacaceae but still without a description of the fruit of *C. schultzii*. Beccari recognized the two species as congeneric, in part from having better material to study, and made the necessary transfer of Vidal's species.

There is some confusion about the structure of the androecium, even though this is a primary factor in the distinction of species or even genera in the Bombacaceae. Masters described the anthers of *Camptostemon* as bilocular, and his illustration shows a typical tetrasporangiate anther that is medifixed. Vidal described the anther as "unilocular." In Schumann's account, the distinction is between polythecate (*Cumingia*) and dithecate (*Camptostemon*) anthers, although it is not clear from the illustration what is the precise difference. The illustration in Bakhuizen van den Brink (1924) suggests that *C. philippinensis* has a tetrasporangiate anther. The differences may relate to the number of microsporangia, but needs confirmation.

Family: Celastraceae

A large, mainly woody but commercially unimportant family with over 50 genera and 1000 species, having a wide distribution, particularly in the tropics. The following species has been recorded as a mangrove associate.

> ***Cassine*** L. 1737 [*Gen. Pl.*: 338; *Sp. Pl.*: 1753: 268]
> *Elaeodendron* Jacq. F. 1782 ex Jacq. [*Icon. Pl. Rar.* 1: 48]

A genus of about 80 species, mainly African. The status of the genus is discussed by Ding Hou (1963).

> ***Cassine viburnifolia*** (Juss.) Hou 1963 [*Fl. Males.* I, 6: 286]

A small tree to 8 m tall with smooth gray bark. Leaves opposite, simple, shortly petiolate; blade obovate, about 7 cm by 4 cm, undersurface light green to glaucous; margin obscurely crenate, initially with glands in each sinus. Petiole short (1–1.5 cm). Flowers small (2–3 mm diameter), in axillary long-stalked cymes toward the ends of the shoots, four or more, white; stamens alternate with the petals. Disk inside the stamens green; ovary conical, scarcely 1 mm long, with a sessile bilobed stigma. Fruit a one-seeded drupe ripening yellow, about 1 cm long, narrowed at the base, with corky rather than fleshy mesocarp; endocarp somewhat bony.

This species occurs in wet coastal communities, and has been recorded in mangrove swamps and along tidal rivers. Ridley (1930) notes that the fruit structure is unique in the genus and clearly adapted to water dispersal. It has a range from the Andaman Islands through the Malay Peninsula to northern Sumatra, Borneo, and into Sulawesi.

Family: Clusiaceae (Calophyllaceae)

A tropical group characterized by simple opposite leaves with finely parallel lateral veins and colored (usually yellow) latex. Some genera, including the following, have been separated from the traditional Clusiaceae as Calophyllaceae (Soltis et al., 2011). This represents the resurrection of an old name (*J. Agardh. Theoria Syst. Pl.*: 121 1858). The flowers have numerous stamens. One species is included here as a commonly encountered mangrove associate.

> *Calophyllum* L. 1753 [*Sp. Pl.* 1: 513]

A pantropical genus of about 200 species. The following, which is the type of the genus, has wide coastal distribution, typically in beach communities in the eastern tropics, with a range from East Africa to Polynesia, but extended artificially into the Pacific. It is a commonly planted coastal ornamental in the tropics, including the New World. The seed oil has been suggested as a biofuel.

> *Calophyllum inophyllum* L. 1753 [*Sp. Pl.*1: 513]
> A full synonymy and extensive citation are provided by Stevens (1980).
> (Fig. B.21; Plate 5I, 7F)

Trees grow to 30 m without buttresses or root modifications; bark dark, fissured. Latex yellow to whitish, sticky, from all cut surfaces. Shoots with naked terminal buds except for obscure indumentum; stems angular. Leaves glabrous opposite, simple, with a short petiole (1–2 cm); blade elliptic to oblong, about 15 cm by 8 cm, margin entire; apex always rounded, base cuneate or rounded. Lateral veins uniformly parallel. Inflorescence up to 15 cm long, axillary, usually a simple raceme with five to 15 flowers. Flowers about 1 cm in diameter, apparently bisexual, tetramerous with eight (to 13) tepals; two outer pairs decussately arranged, four inner in an imbricate whorl, little differentiated from outer. Stamens numerous, the filaments shortly connate below. Ovary globose with a simple style 4–8 mm long; stigma peltate; ovule solitary, basal, anatropous. Fruit a one-seeded globose drupe, 2–4 cm in diameter; the single embryo enclosed by a spongy layer and a thin stony layer derived from the testa (Corner, 1976). Germination hypogeal, the radicle both breaking the stony layer and pushing aside a basal plug.

    The tree is most characteristic on sandy soils and is a good example of a beach tree, but has been recorded to an altitude of 200 m. It therefore occurs only occasionally in association with true mangroves and always in transitional habitats. Stevens (1980)

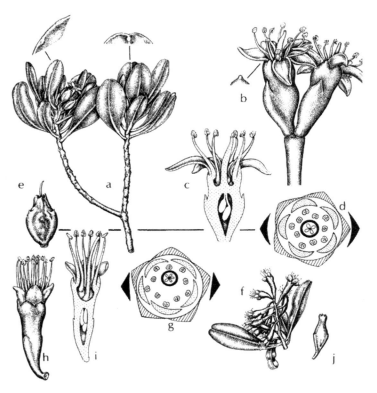

Figure B.23. *Lumnitzera* (Combretaceae) leaves, flowers, and fruit. (a)–(e) *Lumnitzera racemosa*:
(a) Habit with axillary flowers (x1/2); left inset: detail of leaf margin, right: detail of leaf tip.
(b) Two-flowered inflorescence (x3); inset: detail of tip of sepal (x4). (c) Flower in L.S. (x3).
(d) Floral diagram. (e) Immature fruit (x3/2). (f)–(i) *Lumnitzera littorea*: (f) Terminal axis (x1/4).
(g) Floral diagram. (h) Flower in side view (x3/2). (i) Flower in L.S. (x3). (j) Fruit (x1/2).
(Material (a)–(e) cultivated at Fairchild Tropical Garden, Miami, FL; (f)–(j) from Labu Lagoon,
Lae, Papua New Guinea.)

glabrous at maturity. Flowers in short terminal or axillary spikes, perfect, actinomorphic
or slightly zygomorphic, sessile or shortly pedicellate, a pair of short bracteoles inserted
on the calyx tube. Calyx lobes five, short, rounded, sometimes minutely hairy on the
margins or gland-tipped; petals five, free, red or white (rarely yellow or pink). Stamens
usually 10, but often fewer, inserted on the inner rim of the calyx cup; anthers versatile,
sometimes with a minute apical appendage. Ovary inferior, the calyx tube forming a
deep nectar cup. Style simple, filiform, persistent, and arising from the center of the
calyx cup (*L. racemosa*) or partially adnate to it on one side (*L. littorea*). Ovules two to
five (perhaps seven), pendulous from the upper part of the ovary cavity. Fruit a flattened
one-seeded pseudocarp (i.e., a drupe-like structure) with the sclerotic portion developed
from the inner tissues of the calyx tube. Petals caducous, style and calyx lobes
persistent, outer parts of dispersed (floating) fruits decaying to leave fibrous tissues of
fruit wall. Germination hypogeal.

Lumnitzera racemosa Willd. 1803 [*Neue Schr. Ges. Naturf. Fr. Bert.* 4: 187]
(Fig. B.23a–e; Plate 6C)

Flowers in short axillary spikes 1–2 cm long, petals white, spreading (or yellow in var. *lutea* Presl.); actinomorphic, the stamens somewhat exserted. Pneumatophores as looping lateral roots sometimes developed, but much less commonly than in *L. littorea*.

Architecture. *Lumnitzera racemosa* is a good example of Attims' model (see Fig. 5.1A). The sapling shows continuous or diffuse branching, with no evidence of articulations. The twigs are glabrous, reddish brown, with close-set regular leaf scars. Each leaf subtends a single but obscure axillary bud; branching is always by syllepsis and is continuous on vigorous shoots; flowers are axillary. Branches are orthotropic and repeat the construction of the parent axis, but with fewer laterals and these mostly toward the outside of the tree. Consequently, at an early stage of development the plant is much branched. Further development as a single-trunked tree involves the loss of branches. This is somewhat selective, and at a later stage, the tree shows a confused intermixture of dead and persistent branches. In open situations, the tree may retain the rounded, low-crowned shape, especially as lower branches can root basally when in contact with the substrate. Otherwise, where the tree suffers crown competition, the lower branches die and a single-trunked form with a narrow, conical crown appears.

The two species differ in the position of the inflorescences, and by definition *L. littorea* conforms to Scarrone's model. This is seen in the forking of distal axes. In younger stages, the terminal inflorescences are closely substituted by lateral branches and the resultant sympodial growth does not influence the crown form strongly.

Lumnitzera littorea (Jack) Voigt. 1845 [*Hort. Suburb. Calc.* 39]
(Fig. B.23f–i; Plate 6A)
L. coccinea Wight and Arnold 1834 [*Prodromus* 316]

Flowers in terminal spikes 2–3 cm long; petals red, erect; flowers slightly zygomorphic, the calyx tube curved and the style slightly adnate to the inner adaxial side of the tube. Stamens up to 10 mm long and prominently exserted at anthesis.

Pneumatophores usually well developed and with looped laterals that protrude up to 10 cm above the substrate, the loops remaining spongy and with little secondary thickening (see *Bruguiera* and *Ceriops*). This species may be glabrous at all times. Given that it has terminal inflorescences, their scars between branch forks are useful diagnostic features. It may form taller trees than the following.

Lumnitzera x rosea (Gaud.) Presl. 1834 [*Rep. Bot. Syst.* 1: 156]
Laguncularia rosea Gaud. 1827 [*Voy. Uranie* 1: 155]
Freycinet [*Voy. Bot.* 481, t. 105, f. 2]

A form with pink flowers (Plate 6B), interpreted by Tomlinson et al. (1978) as a hybrid *L. racemosa* x *L. littorea,* from specimens on Hinchinbrook Island, Queensland, and described in greater detail in Duke (2006), where it is also recorded for the Philippines

Figure B.24. *Conocarpus erectus* (Combretaceae) female flowers and fruit. (a) Flowering shoot (x1/2). (b)–(e) Female flowers (x10): (b) Side view with bract (x10). (c) Face view (x10). (d) L.S. (x10). (e) Floral diagram. (f)–(i) Fruits: (f) Fruiting heads (x1/2). (g) Single fruiting head (x2). (h) Detached fruit from front (x5). (i) Detached fruit from back (x5). (From Tomlinson, 2001.)

and Papua New Guinea. This form is intermediate in morphology between its putative parents, and is sterile. Its hybrid status has been verified from molecular evidence by Guo et al. (2011), who also found *L. racemosa* to be the maternal parent.

The hybrid is equated with *Laguncularia rosea* Gaud. on the basis of the description and illustrations by Gaudichaud-Beaupré. Unfortunately, the type is not identifiable because all flowers have been lost, although the specimen is clearly the one from which the illustration (t. 105) in Freycinet's *Voyage* is drawn. His specimen is said to come from the Philippines in a locality where presently only one species (*L. racemosa)* occurs. Furthermore, Gaudichaud-Beaupré described the flowers as axillary, but they may have been terminal, as in the Queensland form.

Reproduction

Floral biology. The differences in floral structure and inflorescence position reflect contrasted methods of pollination. *Lumnitzera littorea* (Plate 6A) seems to be pollinated predominantly by birds, specifically sunbirds and honey eaters, although bees and wasps are additional flower visitors. Bird pollination is suggested by the exposed position of the flowers, which are slightly zygomorphic to accommodate a probing beak; the petals are red and remain erect, effectively lengthening the tube somewhat and providing some protection from short-tongued insects. The more prominent anthers thus brush the bird's head beyond the beak. The calyx tube is relatively deep and nectar is abundant. In Australia, lorikeet parrots are said to be noisy but destructive nectar gatherers (Duke, 2006). *Lumnitzera racemosa* (Plate 6C), on the other hand, is visited by a variety of day-active wasps, bees, butterflies, and moths, for which the white spreading petals, actinomorphic flower, and somewhat shallower calyx cup are suited. Pollen is presented on the first day, with the stigma not becoming receptive until the second and subsequent days, so protandry exists. Plants, however, may be self-compatible, because single trees in cultivation set viable seed, and seed set in wild populations is often high, with all the flowers in a head setting fruit. Cross-pollination can be promoted by visits of more discriminate pollinators.

Fruits and seedlings. Although the fruit set seems high, a high percentage of fruits with aborted embryos occur so that seemingly mature fruits are commonly empty, sometimes also because the embryo has been eaten by a small grub that originates from eggs laid by the parent insect early in fruit development.

In normally formed fruits, the embryo is well protected by the layer of sclerenchyma within the outer corky or fleshy layers of the fruit wall. There is no evidence of vivipary. Fruits are dispersed by water and lose the softer outer layers, which exposes the sclerenchymatous fibers. In nature, most floating fruits lose their viability, but those taken directly from trees germinate fairly readily. The seedling is hypogeal, but without other obvious adaptive devices, which aids its establishment.

Ecological and Specific Isolation

Van Slooten (1924) and Exell (1954) have both produced maps showing a similar distribution of the two species in Malesia. Van Slooten suggested that the two species occupy different ecological sites, and this is confirmed by Tomlinson et al. (1978), who comment that stands in north Queensland, almost without exception, contain one species to the exclusion of the other and that the two species never intermix. According to Duke (2006), this difference can be accounted for in terms of tidal range, salinity differences, and moisture availability. This suggests that *Lumnitzera littorea* is better suited to less saline, well-drained sites, and is most vigorous on highly organic substrates, whereas *L. racemosa* is more resistant to saline conditions, as at the margin of bare salt pans, in association with *Ceriops* and *Avicennia.* In Queensland, *L. littorea* has a more extensive southerly range than *L. racemosa* (Byrnes, 1977).

It is interesting to consider why the two species remain distinct, with the acceptance of their hybridization, as this suggests that the reproductive barriers between them can be broken. The ecological isolation assists in the continued separation, which could be further strengthened if the pollinators were quite specific or had limited ranges. The evidence for this interpretation is circumstantial and presently slight.

Systematics and Nomenclature

The genus is based on Willdenow's recognition of it as distinct from *Laguncularia* in his description of *Lumnitzera racemosa*. Presl made the necessary combinations for a number of names included in *Laguncularia* by Gaudichaud-Beaupré, but of these, *Laguncularia lutea* (as *Lumnitzera lutea*) from Timor is now considered to be a variety of *L. racemosa* (van Steenis-Kruseman, 1950, p. 187). *Laguncularia coccinea* is a synonym of *L. littorea*. *Laguncularia purpurea* has been interpreted as a name used in error by Gaudichaud-Beaupré in the plate and legend of the entity he called *L. coccinea*. *Laguncularia rosea*, as we have seen, was resurrected by Tomlinson et al. (1978) simply to draw attention to the fact that Gaudichaud-Beaupré, in creating this name, could have been referring to a hybrid between the two now generally accepted species.

Gaudichaud-Beaupré used the presence or absence of calyx glands in his diagnoses, but in agreement with Exell (1954), this seems an inconstant character.

Conocarpus L. 1753 [*Gen. Pl. Ed.* 5: 81]

A genus of two species: one with a very restricted distribution in East Africa and the other described here and widely distributed in coastal communities in tropical America and West Africa, that is, throughout the range of the CEP community.

Conocarpus erectus L. 1753 [*Sp. Pl.* 1: 176]
(Figs. B.24, B.25; Plate 8J)

Dioecious trees, usually with several spreading trunks, or a low shrub. Bark rough and fissured. Branching frequent, but diffuse and irregular (Petitt's model). Shoots ortho-tropic, without articulations; leaves spirally arranged, scattered, or somewhat clustered distally. Twigs angled with a ridge beneath each leaf. Leaves (Fig. B.25c) either glabrous at maturity or persistently slightly to densely silver hairy (var. *sericeus*). Petiole short (to 1 cm) with a pair of circular glands, one on each side on the decurrent extension of the blade. Blade folded in bud; ovate-lanceolate, 4–9 cm by 2–3 cm, apex acute to acuminate, margin entire. Lower leaf surface with several domatia (minute shallow pits) in the angles between the midrib and the major veins (Fig. B.25d). Inflorescences terminal, compact, or rather diffuse little-branched panicles; the flowers clustered in terminal globose heads 5 mm in diameter, the branches of the inflorescence subtended by either foliage leaves or obscure bracts; female inflorescences usually somewhat more compact than male. Male heads with prominent stamens at anthesis. Each head with about 25 flowers, each flower subtended by a short (1–2 mm) hairy bracteole. Flowers somewhat dimorphic, each with an inferior ovary, hairy above, and five acute calyx lobes with sparse hairs; petals absent; disk well developed and hairy. Male flowers distinguished by 5–10 functional stamens, at first recurved but erect and about 3 mm long at anthesis; style simple, ovary narrow and commonly including one or two nonfunctional ovules. Female flowers with a well-developed laterally com-pressed ovary; staminodes usually 5 mm, 1–2 mm long and inconspicuous; style with a tuft of short sparse hairs below the stigma. Fruiting head of short laterally compressed

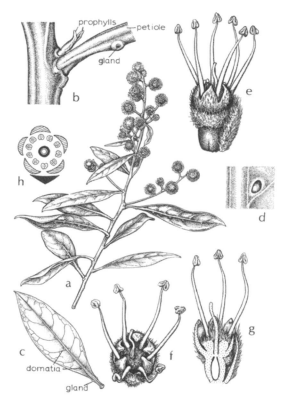

Figure B.25. *Conocarpus erectus* (Combretaceae) male leaves, flower, and fruit. (a) Flowering shoot (x1/2). (b) Detail of node (x3), dormant axillary bud enclosed by prophylls, one of a pair of basal petiolar glands (see part c). (c) Leaf from below (x1/2) showing glands and domatia. (d) Single domatium (x4). (e–h) Male flowers: (e) Side view (x10). (f) From above (x10). (g) L.S. (x10). (h) Floral diagram. (From Tomlinson, 2001.)

nutlets about 5 mm long, released by shattering of the head. Seeds small, angular; germination hypogeal.

Ecological Distribution

The species is frequently considered a "true" mangrove, but it is better regarded as a mangrove associate because it lacks any of the morphological and biological features (such as pneumatophores and vivipary) that characterize true mangroves; furthermore, it occurs in inland communities. The point is somewhat pedantic, however; the somewhat weedy tendencies of the tree partly account for its mangrove association. In the mangrove community, *Conocarpus* is a back-mangal constituent, but only within the limits of the highest tides. It is tolerant of high salinities and rather dry soils, but it also grows in or near fresh water. In South Florida, for instance, it can form almost pure stands on saline marly soils, but it also occupies low hammocks inland, commonly as a fringe to open ponds or artificial canals in the company of *Salix*, *Chrysobalanus*, and *Myrica*.

A form of this species that has silvery leaves because of a dense indumentum of long hairs is sometimes given a varietal name (var. *sericeus* Grisebach = *C. pubescens* Schumach.), and it is commonly cultivated in South Florida as "silver-leaved button-wood" in beach plantings. However, the silver-leaved condition is not genetically fixed (Semple, 1970).

Growth and Reproduction

Plants are ever-growing in the sense that they produce no protected terminal buds, but, at least in South Florida, they undergo an extended dormancy in the winter. Branching is by both prolepsis and syllepsis. Little is known about the reproductive biology. Seeds are dispersed by water because they float. The seed set by female trees is abundant, but either many seeds are aborted or germination is difficult, as seedlings are not common compared with the dense swards that are produced in most years by *Laguncularia*. A suggested contrast difference between these two genera is discussed earlier.

Terminalia L. 1767 [*Syst. Nat. Ed.* 12(2): 674]

A pantropical genus of about 150 species, including many commercially important timber species. Many species have two-winged fruits, the wings formed from a lateral extension of the fruit wall. In other species (including the following), the wing is represented by a narrow ridge, and the fruits are dispersed by water, creating a degree of reduction that parallels that in *Heritiera*. No species are mangrove constituents, but the following is a characteristic strand plant of tropical Asia that is widely naturalized in other parts of the tropics.

Terminalia catappa L. 1767 [*Syst. Nat. Ed.* 12, 2: 674] (Fig. B.26)
For a complete citation, see Exell (1954, p. 566).

An andromonoecious deciduous tree with broadly obovate leaves 10–20 (perhaps 30) cm long with a rounded or bluntly pointed apex. Flowers of two kinds, male and perfect, are borne in axillary spikes, the male distally and the perfect proximally. Each flower has a calyx of five lobes and 10 stamens with an inconspicuous hairy disc. Male flowers (Fig. B.26c-f) lack any conspicuous pistillode; perfect flowers (Fig. B.26g-i) include a well-developed ovary with two functional ovules and a slender style. The flowers are rendered relatively conspicuous by the yellowish calyx and massed stamens. Fruits develop in clusters at the base of the spikes. Each is almond shaped, 5–7 cm long, with a shallow lateral ridge. The mesocarp is fleshy and fibrous, the fruit floats, and the fibrous part rots. Germination is hypogeal.

Of interest is the architecture of this species (Aubréville's model), which is a "type" for the characteristic shoot physiognomy known as "*Terminalia*" branching in tropical

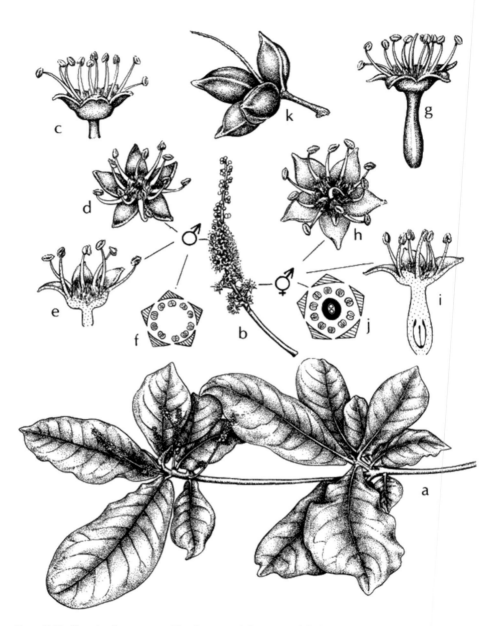

Figure B.26. *Terminalia catappa* (Combretaceae) flowers and fruit. (a) Flowering shoot (x1/4) representing two sympodial units produced by apposition growth. (b) Flower spike (x1/2), perfect flowers at base, male flowers distally. (c–e) Male flower (x3): (c) From side. (d) From above. (e) L.S. (f) Floral diagram of male flower. (g–i) Perfect flower (x3): (g) From side. (h) From above. (i) L.S. (j) Floral diagram of perfect flower. (k) Fruit (x3/8). (From Tomlinson, 2001.)

trees, represented diagrammatically in Fig. 5.2. Appreciation of the form is ancient; its architecture was illustrated by Rheede (1678). The trunk axis is orthotropic and bears spirally arranged leaves. After one or more flushes of growth, a tier of sylleptically produced branches establishes a series of lateral axes with obligate plagiotropy and apposition growth. This results in a sympodial system in which each unit extends horizontally for a limited distance and then turns erect, but remains as a functional short shoot with a rosette of closely set, spirally arranged leaves (Fig. B.26a). The system is extended further by one or two renewal shoots originating in a terminal rosette so as to repeat the process. The total result is a trunk with widely spaced rosettes of leaves in a regular hexagonal pattern of horizontal axes. Trees are deciduous, and in some parts of the range, lose their leaves regularly twice a year, with the leaves turning yellow and then red as they senesce. In this condition, the symmetry of the branching pattern is striking.

The shoot system has been investigated extensively both directly and by computer simulation (Fisher and Honda, 1977, 1979a, b; Fisher, 1978,). These authors have shown that this branch system minimizes mutual shading between rosettes arranged in an axis system with minimal path length. Fisher's (1978) precise description cele-brates the 300th anniversary of Rheede's first illustration. The same physiognomy is achieved by *Bruguiera* and *Rhizophora*, with the former also representing Aubréville's model and the latter Attims' model, as described elsewhere. In all these examples, the regular symmetry of the basic architecture is confounded by reiterative responses in damaged trees. It remains reasonable to accept the original architecture as a highly adaptive mode of branch expression. This is probably equally effective in a tree in open sites, like *Terminalia catappa*, as in a tree in the forest understorey, like *Bruguiera* saplings that grow in the shade of parents.

Family: Ebenaceae

A sizeable family of three genera and about 500 species, mostly in *Diospyros*. A few species occur in warm temperate regions (e.g., *Diospyros virginiana* L., persimmon), but the family is primarily tropical. Ebony is primarily the wood of *D. ebenum* Koen.

Diospyros L. 1753 [*Sp. Pl.*: 1057]
Maba J. R. & G. Forst. 1775 [*Char. Gen. Pl. Ed.* 2: 61]

A pantropical and subtropical genus of mainly dioecious trees with over 400 species, often difficult to distinguish; apomixis is suspected in some forms. The genus *Maba* now is usually included in *Diospyros*. The following elements may include populations in back-mangal or beach communities, and have been described as mangroves by some recent authors.

Diospyros ferrea (Willd.) Bakhuizen 1932 [*Gard. Bull. Straits Settl.* 7: 162]
See also 1937, *Bull. Jard. Bot.Buitenz.* 3, 15(3): 1–515 and 1941, ibid. 3, 15(4): 428–444

A moderate-sized evergreen dioecious tree, up to 25 m tall, with dull gray, finely fissured bark; buttresses scarcely developed. Trunk axis with spirally arranged leaves; branches distributed continuously to discontinuously on the trunk (either Massart's or Roux's model), strongly plagiotropic with distichously arranged leaves. Leaves on branches ovate-oblong, 6–12 cm by 2.5–4 cm, with a short (to 5 mm) petiole; apex acute to rounded or sometimes emarginate, base rounded cuneate. Flowers hypogynous, unisexual in few-flowered condensed axillary cymes, the female inflorescences with fewest flowers. Male flowers subsessile, calyx campanulate, three-lobed, corolla white or pale yellow, three-lobed, stamens six to 12 (rarely more), pistillode slender, densely hairy. Female flowers similar, the calyx and corolla three- (to five-) lobed; staminodes absent; ovary ovoid, three-locular, each locule with one or two ovules; style simple, short. Fruit an ellipsoid or globose one- to six-seeded berry, about 1 cm wide, with the calyx persistent and cupuliform. Seeds oblong, with a thin testa, endosperm present, the embryo straight.

The species is exceedingly polymorphic and occupies diverse habitats, from back mangroves to lowland montane rain forests, with a geographic range from East Africa, throughout the Asian tropics to Polynesia and northern Australia. The modern tendency is to segregate local species on the basis of careful study (Smith, 1981, p. 730), but identification of specimens in areas where there has been no field study is still exceedingly difficult. For example, Fosberg (1939) recognizes 12 nameable taxa in his treatment of the Hawaiian forms. Elsewhere, the mangrove forms have been referred to var. *littorea* (R. Br.) Bakh. and var. *geminata* (R. Br.) Bakh.

In northernmost Australia and adjacent Papua New Guinea, the tree has been described under the name *Diospyros littorea* (R. Br.) Kosterman 1977 and occurs in the mid to high intertidal with *Brugeria gymnorrhiza* and *Xylocarpus granatum*. It flowers over a very limited period in November, with fruits maturing in the following September (Duke, 2006).

Growth rings. The species in North Queensland has been the subject of a detailed study correlating growth rings with climate (Duke et al., 1981). A specimen about 12 m high and 10 cm in diameter at a height of 1.3 m was shown to have distinct growth rings differentiated by differences in fiber wall thickness. The width of these was shown to have unusual associations with rainfall. First, there is a general correlation with extended changes in average annual rainfall, so that a period of decreasing rainfall is matched by a period of decreasing ring width. Second, the rings are not annual but the numbers come close to an average production of seven rings every four years (i.e., 1.77 rings per year). The authors account for this in terms of the erratic rainfall in this part of Australia, and emphasize that although this complicates the problem of determining the age of mangrove trees, it indicates that it is not always insurmountable. This study remains one of the few in which (1) growth rings of a standard anatomical kind have been shown clearly in mangroves, and (2) the growth rings have some correlation with climate (in this instance, rainfall).

Family: Euphorbiaceae

One of the largest plant families with over 7000 species, most commonly tropical. The family is very diverse; the tribes delimit the most natural units (Pax and Hoffmann, 1931). Three relatively unrelated genera are considered here: two Old World and the other New World (Webster, 2014).

1A. Plants without milky latex, fruit a capsule 1.5–2 cm in diameter with 12–15 locules. . . . . *Glochidion*

1B. Plants with a watery or milky latex; fruit fleshy or three-lobed. 2

2A. Old World, dioecious; fruit a three-lobed schizocarp. . . . . *Excoecaria*

2B. New World, monoecious; fruit indehiscent, resembling a small apple. . . . . *Hippomane*

Excoecaria L. 1759 [*Syst. Nat. Ed.* 10: 1288 and *Sp. Pl.* 2: 1451, 1763]

A genus of 35 to 40 species in tropical Africa and Asia to the Western Pacific. The following species, which is the type of the genus, occurs in mangroves but also occasionally in inland stations; two other species may be considered as mangrove associates. Webster (2014) indicates that *E. agallocha* is distinctive among species in the genus.

Excoecaria agallocha L. 1759 [*Syst. Nat. Ed.* 10: 1288] (Fig. B.27; Plate 5G)
[*Stillingia* sect. *Excocaeria agallocha* (L.) Baillon 1858 [*Et. Gen. Euphorb.* 518.]

From India, and Sri Lanka to Hainan and the Ryu-Kyu Islands, through Malesia and Papuasia, including tropical Australia, and into the Pacific as far as Niue and Samoa.

A dioecious tree to 15 m high, with abundant white latex, sometimes branched from the base and somewhat shrubby. Bark gray, becoming fissured; lenticels prominent on younger twigs. Leaves spirally arranged, sometimes somewhat clustered toward the ends of erect shoots. Shoots orthotropic, with infrequent diffuse branching (Attims' model); articulations indistinct, terminal buds inconspicuous with, at most, an envelope of persistent stipules. Stipules minute, ephemeral as lateral triangular scales on each side of petiole. Leaves simple, coriaceous or somewhat fleshy, with a terete petiole 1–2 cm long, blade ovate-elliptic to obovate, up to 9 cm long and 6 cm wide. Apex rounded, slightly emarginate, or at most, bluntly acuminate. Margin inconspicuously notched, with a minute gland in the

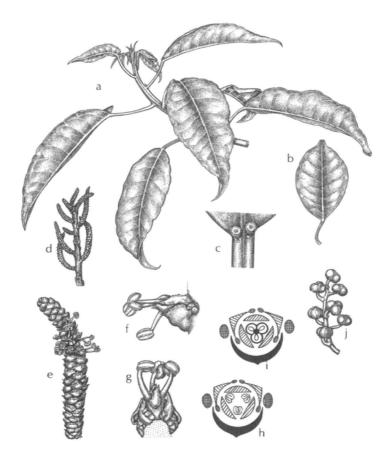

Figure B.27. *Excoecaria agallocha* (Euphorbiaceae) leaves, flowers, and fruit. (a) Leafy shoot (x1/2). (b) Single leaf (x1/2). (c) Detail of pair of glands at leaf base (x4). (d) Deciduous axis with immature axillary male spikes (x3/4). (e) Male spike at early anthesis (x4). (f) Single male flower from the side (x8). (g) Male flower from above, including axis in T.S. (x 8). (h) Floral diagram of male flower. (i) Floral diagram of female flower. (j) Immature fruiting head (x3/4). (Vegetative shoot from Bako, Sarawak; female flowers from Bootless Bay, Papua New Guinea; male flowers from Kinoya, Fiji.)

notch in young leaves. Basal glands two (to four), usually one on each side of blade at its insertion on the petiole; the glands usually circular and often associated with additional outgrowths like the marginal glands. Each leaf subtending one (to two) axillary bud with a series of minute, decussately arranged, woody bud scales.

Inflorescences axillary, pale green, 3–7 cm long, initiated as catkin-like structures within the leaf-bearing portion of the shoot, but sometimes persistent and not expanding until the subtending leaf falls. Male inflorescence up to 7 cm long at maturity, with a series of spirally arranged, often glandular bracts, each bract subtending a male flower. Male flower almost sessile with three narrow laciniate tepals (calyx) below the three yellow stamens, each anther bilocular, basifixed to almost versatile and longitudinally dehiscent. Pistillode absent. Filament initially obscure, but rapidly extending to 5 mm at maturity. Female inflorescence usually shorter than male, bracts glandular; flowers initially sessile and with a pair of basal bracteoles, pedicel extending to 5 mm in fruit. Tepals three, initially somewhat cupulate, wider than those in male flowers; staminodes absent. Ovary trilocular with three short, spreading or recurved simple styles, each locule with a single basal ovule. Fruit three-lobed about 7 mm in diameter, dehiscing into three cocci (schizocarp) to release the solitary seeds; the pericarp somewhat leathery but not fleshy. Seeds black, about 3 mm in diameter. Endosperm absent. Germination epigeal; cotyledons somewhat cuneiform.

Varietal distinction has been made based on the leaf tip, degree of marginal serration, and petiole length. In Australia, two taxa are more or less separated into Western versus Eastern regions (Duke, 2006).

1. *E. agallocha* L. var *agallocha.*
    Leaf tip more or less acuminate, margin often serrate, petiole mostly >18 mm. (Northern Territory, Queensland, New South Wales.)
2. *E. agallocha* L. var *ovalis* (Endl.) Mull. Arg. 1860
    Leaf tip more or less blunt, margin mostly entire, petiole <17 mm. (Western Australia, Northern Territory.)

Field Identification

The tree is easily recognized by the catkin-like inflorescences (Fig. B.27d) and the copious white latex that exudes from wounds. The leaves have distinctly crenate margins (Fig. B.27b). The laticifers are abundant in the bark (Santos, 1932). This latex is irritating, and contact with it, especially around the mouth and eyes, should be avoided. However, the latex is water soluble (as in *Hippomane*). The simple petiolate leaves, obscurely notched margins of the ovate blade, and especially the basal pair of glands (Fig. B.27c), are also distinctive.

Geographic and Ecological Distribution

This species is recorded in the Asian tropics throughout much of the eastern range of the genus *Excoecaria,* but is not known for certain to occur in Africa. It is a characteristic associate of the mangrove community but only in open sites in the back mangal. It occurs more commonly on exposed beaches and in sandy estuaries and is also recorded in disturbed or open sites to an elevation of 400 m, thus it cannot be regarded as an exclusively mangrove component. In keeping with this, the plant shows no obvious morphological adaptation to mangal; the root system is not obviously specialized.

Growth and Reproductive Biology

The tree grows intermittently and irregularly. There are obvious articulations associated with the inconspicuous resting terminal buds. The plant may be briefly deciduous in dry seasons, with the senescent leaves becoming orange or even scarlet. Branching is diffuse, irregular, and by prolepsis. Axillary buds are present, either solitary or sometimes in vertical pairs, but most remain inconspicuous and undeveloped. Reiteration can occur, leading to somewhat shrubby forms. Inflorescences are developed on younger shoots, usually from the upper of the axillary bud pair (but sometimes both), and seemingly independent of the time of shoot extension. The inflorescences may remain unextended for a long time, but the final extension at anthesis is rapid.

Floral biology. Plants are typically unisexual, but there are records of male catkins with a few basal female flowers, a condition that recalls the bisexual catkins of related monoecious genera (e.g., *Ateramnus* and *Hippomane*). Flowers are pollinated by insects because the pollen is sticky; bees are particularly common flower visitors and may be the chief pollinators. They are attracted by the yellow nectar-secreting glands, which characteristically occur at the margin of the catkin bracts. These glands may be homologous with the glands on the foliage leaf blades, the function of which is not known. Leaf glands are larger and more or less circular; inflorescence glands are irregular and almost peltate. The distribution of glands in association with the inflorescence scales is somewhat irregular, and additional glands within the bract axil may represent those belonging to bracteoles, if the suggested homology is correct.

Taxonomy and Nomenclature

Excoecaria is a typical member of the subfamily Euphorbioideae in the simple leaves, absence of petals, trilocular ovary with one ovule per loculus, inarticulate laticifers with white latex, and biglandular bracts on the inflorescence. It is included in Webster (1975, 2014) within the tribe Hippomaneae, which contains a number of genera with similar inflorescences, including *Hippomane* itself, which has a coastal distribution. *Excoecaria* is distinguished from members of its tribe by a combination of characters, including the dioecious condition, axillary inflorescences, male flowers with only (two to) three stamens, and the absence from the seed of a caruncle. The persistent columella of the schizocarp fruit lacks any horizontal projection at its base, as is found in the related genus *Stillingia*.

As with many common plants of the Asian tropics, the species *Excoecaria agallocha* was first described by Rumphius in *Herbarium Amboinensis* in 1741 (as *Arbor excoecans*); the spelling *Excaecaria* is used by a number of authors. The remaining species of the genus occupy inland sites and are not necessarily restricted to the coast. There is no recent treatment of the genus that discusses the evolutionary relationships of *E. agallocha* with other members of the genus. The most similar inland species seems to be *E. philippinensis* with much larger leaves, up to 20 cm long, and longer (up to 10 cm) lax inflorescences. However, Airy-Shaw (1975) comments that generic limits in the tribe Hippomaneae are not clear, and the distinction between *Excoecaria*, *Stillingia*, and *Sapium* has not been maintained by all authors (Baillon, 1875). *Excoecaria* itself is the oldest generic name of the three.

> **Excoecaria dallachyana** (Baill.) Benth. 1873 [*Benth. Fl. Austr.* 6: 153]
> *E. agallocha* var. *dallachyana* Baill, 1866 [*Adansonia* 6: 324]

This taxon is described from south Queensland and northern New South Wales, but seems chiefly distinguished by its upland habitat in scrubby vegetation, for example, Sandiland Ranges (Pax and Hoffman, 1912, p. 168). The relation between *E. agallocha* and *E. dallachyana* would merit comparative study if, as seems likely, the latter is simply an inland form of the former. Within the species, a number of varieties have also been named, based on leaf dimensions and the size of the reproductive parts (Pax and Hoffmann, 1912).

> **Excoecaria indica** (Willd.) Muell.-Arg. 1863 [*Linnaea* 32: 123]

Airy-Shaw (1975, p. 114) indicates that this species occupies "primary *Nypa* forest in salt water, in seasonal swamps, on tidal riverbanks and seashores, on black soil or yellow clay alluvium from sea-level up to 10 m alt." It ranges from south and east India, Southeast Asia, Malesia, and the Solomon Islands, but not the Philippines. It is distinguished from *E. agallocha* by its thorny trunk, regularly crenulate, almost lanceolate leaves, and globose smooth (not lobed) capsular fruit, up to 3 cm in diameter, which dries black.

> **Hippomane** L. 1753 [*Sp. Pl.* 2:1191 and 1754 *Gen. Pl. Ed.* 5: 499]

A genus of perhaps three species of the New World tropics. The genus is in the same tribe as the eastern *Excoecaria*, but is distinguished by the terminal inflorescence, monoecious condition, multilocular ovary, and several-seeded indehiscent fruit. The following species is widely distributed; two other species are endemic to Hispaniola.

> **Hippomane mancinella** L. 1753 [*Sp. Pl.* 2: 1191] (Fig. B.28)

This species ("manchineel") is a characteristic constituent of seashore communities throughout the Caribbean (including the Florida Keys), Mexico, and northern South America, extending to the Galapagos and Revillagigedo islands (Webster, 1967). It is not a mangrove, but is mentioned here because it may be encountered, and its white but water-soluble latex is poisonous (Howard, 1981). Its fruits (Fig. B.28j) resemble small apples and are very dangerous. The Latin name means, literally, "the little apple that makes horses mad." This species is a common but scattered constituent of drier, sandy, coastal communities within its range, and it may be encountered in back-mangal or adjacent beach communities in the New World.

Hippomane mancinella forms a low, much-branched, usually wide-spreading tree or sprawling shrub. It is recognized by its terminal inflorescences and the small, several-seeded, apple-like fruits that are borne later in the crotch of the branch forks. These fruits develop from a multilocular ovary, each locule with a single ovule; there are three

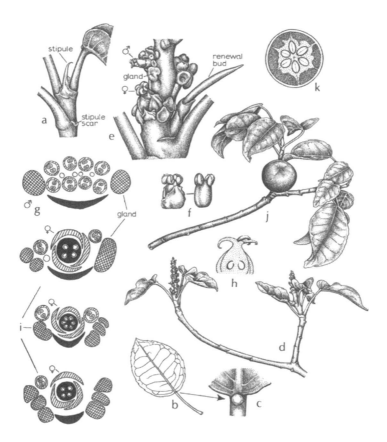

Figure B.28. *Hippomane mancinella* (Euphorbiaceae) shoots, flowers, and fruits. (a) Terminal bud (x3) with stipule and stipule scar. (b) Outline of foliage leaf (x1/2). (c) Detail of gland at base of leaf (x4). (d) Shoot with terminal inflorescences (x1/2). (e) Base of flowering shoot with male and female flowers (x3), renewal bud in axil of leaf to right. (f) Male flower (x6) with (left) and without (right) perianth. (g) Diagram of male flowering unit (glands cross-hatched). (h) Female flower in L.S. (x 5). (i) Diagrams of lower flowering units with three examples of flower arrangement (glands cross-hatched). (j) Fruiting branch (x1/2). (k) Fruit in T.S. (x1/2). (From Tomlinson, 2001.)

or more recurved stigmas. The leaves are stipulate, long petiolate, and have a single conspicuous gland at the base of the ovate blade that has a shallowly toothed margin (Fig. B.28b, c). The inflorescence bears unisexual flowers, with male flowers in axillary aggregates above, and one or more female flowers below, often also associated with further male flowers. The structure and arrangement of the flowers are somewhat variable (Fig. B.28g, i), and the conspicuous yellow glands associated with each flower cluster are presumably attractive to pollinators. Perhaps because of its sinister reputation, the plant has not been studied carefully; details of the pollination mechanism are not known and the method of seed dispersal is also unknown. Germination has apparently not been studied.

Glochidion J. R. & G. Forst 1776 [*Char. Gen.* 113, t. 57]

A large genus of about 300 species, mainly in tropical Asia, distinguished by its broadly ovate, glabrous leaves with rounded apex, and flowers with a multilocular ovary, producing a multilocellate fruit. The following coastal species is recorded in back-mangal communities, but it also has a wide distribution inland, including "elfin woodland" (var. *culminicola*, Airy-Shaw, 1975).

Glochidion littorale Blume 1825 [*Bidjr.* 585]

A shrub or small tree growing to 6 m with brown or gray, somewhat flaky or fissured bark. Leaves spirally arranged, glabrous, obovate with a rounded apex, about 5–7 cm by 2.5–4 cm, narrowed basally to the short (2–3 mm), somewhat fleshy petiole, margin entire or slightly crenulate. Fruit a globose capsule about 1.5–2 cm in diameter with numerous (12 to 15) locules, becoming lobed before dehiscence; green, flushed crimson and splitting between the lobes to reveal the two red seeds per locule.

A species with a wide distribution in India, Sri Lanka, Indochina, and West Malesia. Airy-Shaw (1975) distinguishes a number of varieties, of which the following are coastal in their distribution: var. *littorale*, the usual form, with the leaf apex rounded and the pedicels of the female flowers short; and var. *caudatum*, with the leaf apex pointed and the female pedicels long.

Family: Fabaceae (Leguminosae)

One of the largest tropical families, it is divided into three main subfamilies, Mimosoideae, Caesalpinioideae, and Papilionoideae (sometimes recognized as separate families), but a characteristic fruit, the legume (pea pod), occurs in all three. Although no member of the family grows in the front-mangal communities, the following genera are characteristic mangrove associates with a wide distribution. Four climbers are mentioned here: they are the more frequent viny species with this habit in mangrove associate communities.

Artificial key to legumes treated in this account

1A. Flowers more or less regular, unlike those of sweet pea. Leaves usually compound. . . . . .2 (subfamily Caesalpinioideae)

1B. Flowers zygomorphic, like those of either sweet pea, or if regular, plants with simple leaves. . . . . .4 (subfamily Papilionoideae)

      [*N.B.*: The most consistent feature distinguishing these two subfamilies is somewhat technical and contrasts the petals in bud that overlap either from below upward (imbricate-ascending: Caesalpinioideae) or from above downward (imbricate-descending: subfamily Papilionoideae).]

2A. Trees, unarmed. 3

2B. Climbers with numerous recurved prickles. . . . .*Caesalpinia*

3A. Leaves uni- or bijugate with two or four leaflets; fruit about 3–4 cm long; Old World. . . . .*Cynometra*

3B. Leaves pinnate, usually with two pairs of leaflets; fruit large, up to 25 cm long; New World . . . .*Mora*

4A. Vines with twining stems. 5

4B. Trees or shrubs; if climbing, without twining stems. 6

5A. Inflorescences obviously paniculate, flowers borne on extended second-order axes, corolla green and white. . . . .*Aganope*

5B. Inflorescences in contracted spike-like panicles, the flowers on short second-order (often dichasial) axes, corolla white or pink. . . . .*Derris*

6A. Shrubs or semi-scandent plants with grapnel-like short shoots. . . . . . .*Dalbergia*

6B. Trees, plants always self-supporting. 7

7A. Leaves simple. . . . .*lnocarpus*

7B. Leaves pinnately compound. 8

8A. Leaf equal pinnate with two to four leaflets. . . . .*Intsia*

8B. Leaf unequal pinnate with five to seven leaflets. . . . .*Pongamia*

Subfamily Caesalpinioideae

Cynometra L. 1753 [*Sp. Pl.*: 382; *Gen. Pl. Ed.* 5, 1754: 179]

A pantropical genus of about 70 species of trees or shrubs, mainly restricted to moist lowland forest. The genus is included within the subfamily in the very large tribe Detarieae DC 1825 (Cowan and Polhill, 1981), and is considered to be a somewhat "basal" (i.e., unspecialized) type from which more specialized genera may be derived. It is distinguished by the uni- or bijugate leaves, small bud scales, short inflorescence axis, four calyx segments, usually 10 stamens, usually five short petals, and thick, one-seeded fruit. The following account is based on the monograph by Knapp-van Meeuwen (1970), where a complete synonymy is given.

The two species distinguished in the following key occur somewhat infrequently in back mangal but are also recorded inland to an altitude of 400 m. A related species, *C. cauliflora* L., is the type of the genus, but known only in cultivation although possibly originating in East Malesia, and is grown for its edible fruit ("nam-nam").

1A. Style straight, in line with the dorsal part of the ovary, fruit without a prominent lateral beak. Ovary wall glabrous inside. Sepals not curved distally when reflexed at anthesis. . . . .*C. ramiflora*
     (From India throughout Southeast Asia and Malesia to the Pacific but not Australia)

1B. Style bent, basally oblique to the dorsal line of the ovary, becoming a prominent lateral beak in fruit (Fig. B.29e). Ovary wall hairy inside. Sepals curved distally when reflexed at anthesis. . . . .*C. iripa*
     (With much the same distribution but much more localized, occurring in North Queensland)

Cynometra ramiflora L. 1753 [*Sp. Pl.*: 382]

Trees to 25 m high with smooth, thin, gray bark; the trunk fluted somewhat at the base but not to more than 1 m. Leaves distichous; shoots somewhat zigzag, developing rapidly from lateral buds with numerous overlapping distichous bud scales, the shoot at first limp and pendulous with white or reddish leaves. Stipules intrapetiolar, narrow, caducous, and leaving no scar. Leaves glabrous, essentially paripinnate but with a short (1–2 cm or longer) axis and one (to two) pair of opposite leaflets, thus usually unijugate, less commonly bijugate, the lower pair of leaflets often very small. Leaflets 10–13 cm by 4–6 cm, but commonly smaller, asymmetrical, ovate, lanceolate, the apex usually acute or bluntly rounded, margin entire, the acroscopic side almost straight, the basiscopic side curved. Petiole and leaflets each with a short (1–2 mm) corky or fleshy pulvinus.

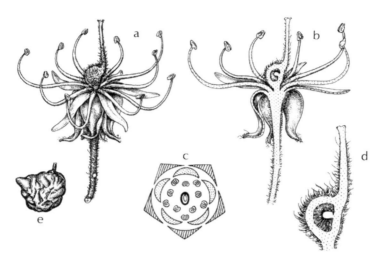

Figure B.29. *Cynometra iripa* (Fabaceae-Caesalpinioideae) flower and fruit. (a) Single flower (x4). (b) Flower in L.S. (x4). (c) Floral diagram. (d) Ovary in L.S. (x9). (e) Mature fruit (x1). (From Sunday Creek River, Deluge Inlet, and Hinchinbrook Island, Queensland; Duke and Bunt, 1979.)

Flowers (Fig. B29) perfect, slightly zygomorphic, in short, axillary, hairy racemes, sometimes with two inflorescences per node, racemes up to 1 cm long but usually almost capitate. Flowers about 15 per axis, each subtended by a short (to 10 mm), narrow, scarious bract, the lower bracts often empty. Pedicels 6–15 mm long with one or more basal bracteoles. Floral envelope white, but turning brown, the calyx forming a shallow, slightly asymmetric hypanthium about 1 mm deep; calyx deeply four-lobed above, the lobes narrow, 3–4 mm long, at first imbricate but becoming reflexed. Petals five, free, to 5–8 mm long, lanceolate, narrower than the calyx lobes. Stamens 10, rarely fewer (down to eight) or more (up to 13), free with small anthers on slender curved filaments up to 7 mm long. Ovary inserted slightly eccentrically with a short (1 mm) stalk, asymmetric-elliptical, hairy, with a slender style 3–4 mm long and a somewhat capitate stigma. Ovule one (to two), inserted dorsally. Fruits one or more on each raceme, each with a prominent beak (Fig. B.29e), one-seeded, elliptic to semi-orbicular, about 3–4 cm by 2–3 cm, brown with a roughened, wrinkled, often hairy surface; indehiscent, distributed by water currents. Seed without endosperm. Germination epigeal.

The species is somewhat variable and much of the synonymy refers to local forms. A number of varietal or subspecific names has also been proposed, but only one, var. *bifoliolata*, as in *C. ramiflora* var. *bifoliolata* (Merr.) van Meeuwen (= *C. bifoliolata* Merr. 1917, *Philipp. J.* Sci. *Bot.*12: 272), is accepted by Knapp-van Meeuwen. It is distinguished by its petiolules, 5–8 mm long. The most clearly related species seems to be *C. glomerulata* Gagnep., which is distinguished by the consistently bijugate leaves, with a short (less than 2 cm) axis, the acuminate-cuspidate leaflets, and the rhombic outline of the pod. It is not likely to be encountered, given that it has a limited distribution in Laos and is known only from three collections.

Cynometra iripa Kostel 1853 [*Allg. Med. Pharm.* Fl. IV 1341] (Fig. B.29)

As discussed in Duke (2006), *Cynometra ramiflora* has been most frequently confused with *C. iripa*, which apart from being similar morphologically, can also occur in back mangal. According to Knapp-van Meeuwen (1970), all material from Australia referred to as *C. ramiflora* should be identified as *C. iripa*, so that Australia can no longer be included in the range of *C. ramiflora*. The two are distinguished primarily by the characters used in the key.

Caesalpinia L. 1753 [*Sp. Pl.*: 380]

A number of species in this genus of about 150 species (which can include segregate genera like *Guilandina*, *Mezoneuron*, and *Poinciana*) are climbers and often spiny. Two of these with twice-compound leaves extend into beach vegetation and are frequent mangrove associates; one is pantropical, the other Asian. Although they have been confused and so given rise to a complex synonymy (Dandy and Exell, 1938; Hattink, 1974), they are readily distinguished, as follows:

1A.   Stipules conspicuous, pinnate; leaflets opposite or subopposite, 16 to 24 per pinna, base unequal. Inflorescence always axillary; flowers unisexual. Fruit armed with rigid spines, seeds gray. . . . .*Caesalpinia bonduc* (pantropical)
1B.   Stipules obscure; leaflets always opposite, two to four pairs per pinna, base equal. Inflorescences often terminal; flowers bisexual. Fruit unarmed; seeds black. . . . .*Caesalpinia crista* (Indo-Malaya)

Caesalpinia bonduc (L.) Roxb. 1832 [*Fl. Ind.* 2: 362, emend Dandy & Exell 1938 in *J. Bot. Lond.* 179] (Figs. B.30, B.31; Plate 7H)
A full synonymy is given in Hattink (1974).

A coarse scrambling vine. Leaves twice pinnate, up to 1 m long, with six to 11 pairs of pinnae, leaflets 16 to 24 per pinna, 2–4 cm long, asymmetric. Rachis and stem with hooked prickles; stems with additional numerous stout to soft prickles. Stipules conspicuous, pinnate. Inflorescences lateral, supra-axillary, often several per axil and serial, branched, axes up to 50 cm long. Flowers unisexual by abortion, petals yellow with reddish streaks; stamens 10, anthers hairy. Fruit one- to two-seeded, ellipsoid, 6–9 cm by 3–4 cm, with numerous stout spines (Fig. B.31g), ultimately dehiscent. Seeds ovoid, smooth, gray.

This species is essentially pantropical; it is widely distributed partly because seeds can float and retain their viability in water for extended times, over two years. It is familiar in a diversity of coastal communities, including back mangal, especially in disturbed sites, but also occurs inland, chiefly in secondary forests to an altitude of about 850 m.

A similar and also widely distributed species, which may be confused with it and can also occur in beach vegetation, is *C. major* (Medik) Dandy & Exell, distinguished by its simple stipules, if they are present at all, fewer (six to 14) symmetrical leaflets per pinna, longer pedicels, and yellow seeds; the ovary has four ovules (not two, like *C. bonduc*).

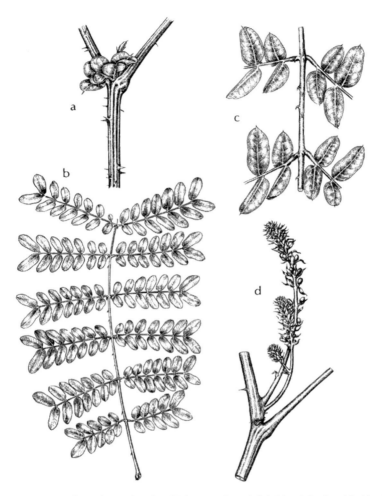

Figure B.30. *Caesalpinia bonduc* (Fabaceae-Caesalpinioideae) leaf and habit. (a) Node with pinnate stipules (x1). (b) Leaf (x1/4). (c) Part of leaf with recurved spines (x1/2). (d) Node with axillary complex of young inflorescences (x1/2). (Material from mangal at Fairchild Tropical Garden, Miami, FL.)

Caesalpinia crista L. 1753 [*Sp. Pl.*: 380]
Caesalpinia nuga (L.) Ait. 1811 [*Hort. Kew,* ed. 2, 3: 32]

Hattink (1974) provides a full synonymy; see also Dandy and Exell (1938). A climber to 15 m. Leaves bipinnate, to 30 cm long, with two to four pairs of pinna, leaflets per pinna about four to six, opposite. Stipules obscure or absent. Rachis and stem armed with recurved prickles. Racemes either axillary or aggregated into terminal inflorescences. Flowers bisexual, petals yellow; stamens 10, with woolly filaments. Fruit ellipsoidal, 4–7 cm by 2–3 cm, flat, beaked with one (to two) seed. Seeds ovoid and black.

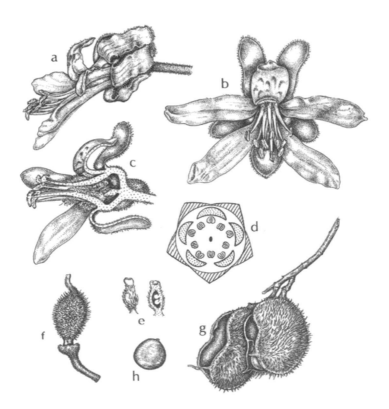

Figure B.31. *Caesalpinia bonduc* (Fabaceae-Caesalpinioideae) flower and fruit. (a) Male flower from side (x3). (b) Male flower from front (x3). (c) Male flower in L.S. (x3), ovary aborted. (d) Floral diagram of male flower. (e) Functional ovary of female flower (x3). (f) Young fruit (x3/2). (g) Mature fruit (x1/3). (h) Seed (x1/3). (Material from mangal at Fairchild Tropical Garden, Miami, FL.)

Commonly recorded as a back-mangal constituent from India and Sri Lanka through most of Southeast Asia to the Ryukyu Islands, Queensland, and New Caledonia.

Mora Schomb. ex Benth. 1839 [*Trans. Linn. Soc.* 18:210, t. 16]
Dimorphandra Schaft [in Sprengel 1827 *Syst.* 4; App. 404]

A genus of trees with six species in tropical South America and the West Indies. *Mora excelsa* Benth. is a common and often dominant swamp species in Trinidad and northern South America, where it may extend into the mangrove fringes. The following species is recorded as growing with *Pelliciera*, *Rhizophora*, and *Acrostichum* on the Pacific coast of Colombia, Ecuador, and Central America.

Mora oleifera (Triana) Ducke 1925 [*Arch. Jard. Bot. Rio de Janeiro* 4: 45]
Dimorphandra oleifera Triana ex Hemsl. 1885 [*Bot. Voy. Challenger* 3: 301]
M. megistosperma Britt. & Rose 1930 [*North Am. Flora* 23(4): 218]

A tall buttressed tree with large, alternately pinnately compound leaves usually with two pairs of leaflets, and long dense pendulous spikes of caesalpinioid flowers. There are five stamens opposite the petals alternating with five staminodes. Leaflets are 10–25 cm by 5–10 cm, ovate, acuminate. The fruits are massive one-seeded, woody structures up to 25 cm long, the seed flattened and about 12 cm in diameter, with the largest embryo of any dicotyledon.

Subfamily Papilionoideae

> ***Inocarpus*** J. R. & G. Forst. 1776 [*Char. Gen.* 66, t. 33]
> *Inocarpus* is a name conserved against *Aniotum* Parkinson 1773 [*J. Voy. Endeavour*, 39]

A genus of perhaps three species in Malesia and the Pacific Islands, distinguished from related genera (e.g., the American *Etaballia)* by its fruit, the almost equal petals, and the stamens free or only basally joined. Verdcourt (1979) comments that the systematic position of these two genera within the Papilionoideae is uncertain. One species, *Inocarpus fagifer*, readily recognized in the field by its red sap, is recorded on river-banks subjected to tidal influence, brackish swamps, and sandy foreshores, but is essentially a lowland swamp forest species. The seeds are roasted or boiled and eaten, and the young leaves are said to be edible. The tree is sometimes planted, presumably for its edible seeds. It is one of the very few examples of a plant in the mangrove associations that provides human food, as its older name *I. edulis* indicates. In English-speaking parts of the Pacific it is known as "Tahitian chestnut."

> ***Inocarpus fagifer*** (Parkinson) Fosberg 1941 [*J. Wash. Acad. Sci.* 31: 95]
> *I. edulis* J.R. & G. Forst. 1776 [*Char. Gen.* 66, t. 33]

In the original account, the spelling *fagiferus* was used, but *fagifer* is orthographically correct (Verdcourt, personal communication).

The tree grows to 30 m tall with a fluted trunk, exuding a red sap from cut stem surfaces. Leaves alternate, simple, oblong to oblong-lanceolate, about 20 cm by 7 cm, but up to 30 cm or longer, apex usually rounded, base cordate; petiole short (1 cm). Flowers scented, in axillary spikes to 12 cm long, the spikes either single or in clusters on a common stalk, often on older wood. Calyx pinkish white; petals white or yellow, 1–1.5 cm long, recurved at the tips. Petals almost equal and without characteristic differentiation usual in papilionoid flower. Stamens 10, in two series, alternately long and short. Fruit one-seeded, indehiscent, irregularly globose, flattened, 5-10 cm in diameter, and variously keeled, ribbed, or smooth.

The precise distribution of the plant is uncertain, as its range may have been artificially extended through cultivation. Corner (1939) concluded from its abundance in seemingly natural situations that it was native, at least to Johore in the Malay Peninsula, and that part of its eastward distribution was the result of human migration and exploration. Presently it occupies the full range of the genus. A closely related species (or possibly simply a variety: *Inocarpus papuanus* Kosterm.) has a limited distribution in New Guinea. It is distinguished by its smaller, inedible fruit and the usually pubescent undersurface of the leaf, and occurs exclusively in rain forest.

Intsia Du Petit-Thouars 1806 [*Gen. Nova Madag.:* 22]

A genus of perhaps three species along coasts and on islands of the Indian and Pacific Oceans. It is distinguished by its three fertile stamens and exarillate seeds (three to four) in the flat, oblong, woody fruit. The following widely distributed species is recorded as a mangrove associate.

Intsia bijuga (Colebr.) O. Kuntze 1891 [*Rev. Gen. Pl.* 1: 192] (Plate 7J)

A tree up to 40 m tall with a long bole, but little buttressed, crown spreading, foliage deciduous. Leaves paripinnate with (two to) four leaflets; leaflets ovate, about 10–12 cm by 4–5 cm, apex obtuse, glabrous except for hairs on midrib beneath. Flowers numerous in dense terminal, finely hairy spikes or panicles. Sepals four, unequal, up to 10 cm long. Petal solitary, clawed, 2–3 cm long, at first white but turning red or orange. Stamens red (Plate 7J), three plus seven staminodes. Fruits flat, oblong, to 20 cm by 7 cm, seeds one or more, black, flattened with a white fleshy funicle that turns brown.

The timber has been extensively used in canoe building.

Pongamia Vent. 1803 [*Jard Malm.* 1, t. 28]

An older genus of probably two species sometimes included as a section *Pongamia* (Adans.) J. Bennet in the larger genus *Derris*. More recent evidence suggests it should be included in the even larger genus *Milletia* Wight and Arn. It is recognized by its panicles of white or pink flowers and round one-seeded fruits.

The following species is widespread in tropical Asia in coastal environments, including back mangal, but it is also recorded inland to altitudes of over 500 m. It is commonly planted elsewhere in the tropics in coastal areas because it is resistant to salt and exposure and may be found as a street tree.

Pongamia pinnata (L.) Pierre 1899 [*Fl. For. Cochinch.* sub. t. 385]
P. glabra Vent. 1803 [*Jard. Mlim.* I, t. 28]
(Fig. B32; Plate 7G)

A tree growing to 25 m without buttresses, the crown in open-grown trees typically spreading with dense foliage and producing deep shade; briefly deciduous in seasonal climates. Leaves with five to seven ovate-elliptic glabrous leaflets in the order of 10–15 cm by 5 cm but variable in size. Inflorescences 20–25 cm long as lax panicles from the axils of distal leaves on each shoot. The ovary is unusual in having only one or two ovules; if two, the distal ovule is the one that always aborts (Arathi et al., 1999). Fruit oblong-ellipsoid, beaked, 4–6 cm by 2–3 cm, seed always single, compressed, with a smooth coat.

Pongamia velutina (C. T. White) Verdcourt 1977 [*Kew Bull.* 32: 250; cf. Verdcourt 1979, pp. 311–314)]

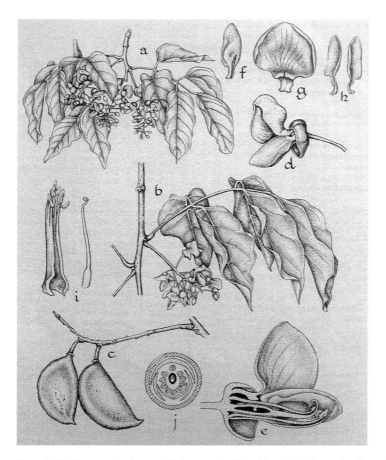

Figure B.32. *Pongamia pinnata* (Fabaceae-Paplionoideae) (a) Flowering branch (x1/2). (b) Leaf with axillary flowering branch (x2/3). (c) Fruit (x2/3). (d) Flower (x2). (e) Flower in LS (x4). (f) Keel (x2). (g) standard (x2). (h) Wings (x2). (i) Stamens (x4), nine fused (left) and one free (right). (j) Floral diagram. (From Tomlinson, 2001.)

This species, known from a very limited distribution in Papua New Guinea (Madang District and Central District), is distinguished by the velvety lower surface of the leaflets and the dense gray spreading hairs on the calyx. The fruit is said to be dehiscent, whereas in *P. pinnata* the fruit is always indehiscent, somewhat larger, and perhaps never so elongated.

Dalbergia L.f. 1782 [nom. cons. *Suppl.* 52: 316]

A pantropical genus of about 100 species; the tree forms provide many valuable timbers. At least two viny species are commonly recorded in mangrove communities.

Dalbergia ecastophyllum (L.) Taub. 1894 [Engler and Prantl, *Nat. Pflanzenfam.* 3(3): 335]

Figure B.33. *Dalbergia amerimnion* (Fabaceae-Paplionoideae) shoot and flowers. (a) Apical region of major axis (x1/2) to show sylleptic branching and characteristic zigzag configuration of extending internodes. (b) Determinate lateral branch (x1/2) modified as a woody tendril. (c) Floral diagram. (d) Flower from side (x4). (e) Flower in L.S. (x4). (f) Standard petal (x4). (g) Wing petal (x4). (h) Keel petals (x4). (Material cultivated at Fairchild Tropical Garden, Miami, FL.)

A shrub or small tree with reclining or scandent branches. Leaves with a short (to 1 cm) petiole, the flowers in axillary clusters. Legume 2-3 cm long and usually one-seeded. The species has a wide distribution in the Caribbean and South America, with an extension to West Africa. In the Bahamas, it is recorded as "frequent in mangrove swamps" (Correll and Correll, 1982).

> **Dalbergia amerimnion** Benth. 1860 [*J. Linn. Soc.* 4 Suppl: 36]
> (Fig. B.33)

Resembles the previous species closely but is distinguished by its longer petioles (to 2 cm), more numerous ovules, and the frequently two-seeded fruits. It is more characteristically scandent and supported by curled lateral determinate shoots that function as grapnels (Fig. B.33b). The extended ends of the main axes are characteristically zigzag. This species is restricted to the New World.

> **Aganope** Miq. 1885 [*Fl. Ind. Bat.* 1(1): 151]

A genus of six species ranging from West and Central Africa to India, China, the Philippines, and New Guinea. For a discussion of the distinction between *Aganope* and *Derris,* see Polhill (1971). Recent treatments have renamed it *Ostryocarpus* Hook. F.

> **Aganope heptaphylla** (L.) Polhill 1971 [*Kew Bull.* 25: 268]

Woody climbers or scrambling shrub growing to 15 m; shoots hairy when young. Leaflets five to seven, elliptic or oblong elliptic, 5–10 cm by 3–6 cm. Flowers about 1.5 cm long in long narrow inflorescences up to 30 cm long. Standard green, wings white. Fruit elongate, up to 20 cm by 3 cm, flat, often constricted between the two to six flat seeds.

This species is recorded in mangrove swamps and associated communities, with a range from Sri Lanka through Bengal, South China, Malaysia, Indonesia, and the Philippine Islands to New Guinea. It may grow inland but has a decidedly coastal distribution. Verdcourt (1979) records that it sometimes grows intertwined with *Derris trifoliata*, with which it has been confused; mixed collections have been made under a single number.

> **Derris** Lour. 1790 [*Fl. Cochinch.*: 433]

A genus of about 40 species in the tropics and subtropics, mainly Asian; many provide fish poisons and insecticides.

> **Derris trifoliata** Lour. 1790 [*Fl. Cochinch.*: 433]
> *Dalbergia heterophylla* Willd. 1800 [Sp. *Pl.* 3: 901]
> *Derris uliginosa* Benth. in Miq. [*Pl. Jungh:* 252]

Erect shrub or a rambling climber to 15 m or more, spreading by root suckers. Leaflets three (up to five or seven), ovate to elliptic-lanceolate, 10–12 cm by 4–5 cm but variable. Inflorescence slender, about 20 cm long; flowers about 1 cm long, corolla white or pale pink. Fruit elliptic, 3–4 cm by 2 cm, seeds two or three, wrinkled.

This species has a coastal distribution from East Africa, Madagascar, and throughout tropical and subtropical Asia to tropical Australia. It is recorded in coastal communities such as beaches, strand vegetation, and coastal swamps, and is a frequent constituent of back mangal throughout its range. It is distinguished from the frequently associated *Aganope heptaphylla*, with which it is sometimes confused, by its fewer glabrous leaflets and smaller flowers and fruits. A good field character is the dark red, strongly ridged younger stems with prominent lenticels.

Other Legumes

Gentry (1982) includes *Muellera moniliformis* L. f. as a mangrove associate common to the Atlantic and Pacific mangal of the New World. It is a small tree, most characteristic of riverbanks, with imparipinnate leaves 15–25 cm long and five to 11 rather delicate pulvinate leaflets. The fruit is a lomentum; that is, the pod is constricted between the seeds. It is most commonly recorded well away from the sea.

Family: Flacourtiaceae

A diverse tropical family of woody plants with more than 1000 species in 90 genera. The family has many disparate elements and is seemingly unnatural.

Scolopia Schreber 1789 [*Gen.*:335 nom. cons.]

A genus of about 40 species with a distribution from West Africa eastward to the Solomon Islands and eastern Australia.

Scolopia macrophylla (Wight and Arnold) Clos 1857 [*Ann. Sci. Nat.* Series 4, 8: 253]

For a detailed account, including full synonymy, see Sleumer (1972).

A small tree growing to 10 m with smooth grayish-pink bark, the trunk described as having spirally arranged simple spines. Leaves spirally arranged but appearing two-ranked on the branches. Petiole slender, 5–10 mm long, scarcely 1 mm wide. Blade simple, ovate, and about 8–11 cm by 4–6 cm, either tapered gradually to a point or rounded apically; margin minutely and regularly toothed; base acuminate with two orange glands at the insertion of the blade. Flowers in axillary spikes up to 6 cm long, sometimes aggregated distally to make a loose terminal panicle; flowers white, about 4 mm in diameter, at first green and then becoming red or orange, style persistent and up to 4 mm long; base enclosed by persistent perianth.

This tree is typically a species of riversides or marshes to an altitude of almost 1000 m but is recorded in "inland mangroves with consolidated soil and numerous crab mounds" (Anderson and Chai, Sarawak Forestry Department #S29344). It has a distribution from Indochina and the Malay Peninsula through Sumatra and Borneo to Java.

Family: Goodeniaceae

A family of 11 genera and about 400 species, mostly in Australia. *Scaevola*, a genus of some 130 species, is also concentrated in Australasia but has a pantropical distribution by virtue of two species, one Old World and the other New World, which are coastal beach plants. Carolin (1960) describes its method of pollen presentation together with those of related families.

> **Scaevola** L. 1771 [nom. cons. *Mant. Pl.* 2: 145] (Fig. B.34; Plates 5B, 7I)

The two species considered here are erect, shrubby plants with succulent leaves, pithy stems, and axillary cymes of conspicuous flowers with undulate petal margins. They are both typical of beach communities, especially sand dunes, where they can form extensive colonies apparently by subterranean branching of the stems. They sometimes occur in mangrove communities, but only in sandy, well-drained areas. They are recognized by their white zygomorphic flowers, with the corolla split completely down one side to expose the curved style. Fruits are fleshy, with one or two seeds. Germination epigeal, the cotyledons fleshy.

1A. Leaves usually larger than 10 cm; calyx segments obvious, pointed; petals often with violet stripes on the inside; fruit white at maturity. Asian tropics. . . . . *S. taccada*

1B. Leaves usually shorter than 10 cm; calyx segments not obvious or at most short and rounded; petals without internal markings; fruit black at maturity. American tropics. . . . . *S. plumieri*

> **Scaevola plumieri** (L.) Vahl. 1796 [*Symb. Bot.* 2: 36] (Fig. B.34)

This species ranges from Florida throughout the West Indies to Central and South America. It is also known as "probably the most common dune plant on the Natal coast" of South Africa (Steinke and Lambert, 1986), where it has been much studied as a key stabilizer of the foredunes (Barker et al., 2003; Knevel and Lubka, 2004). Results show that although it spreads clonally within localized populations, there is genetic exchange by sexual means among these populations as a source of diversification. If this species is correctly identified in this locale, it is a curious disjunction from American populations, perhaps as an artificial introduction.

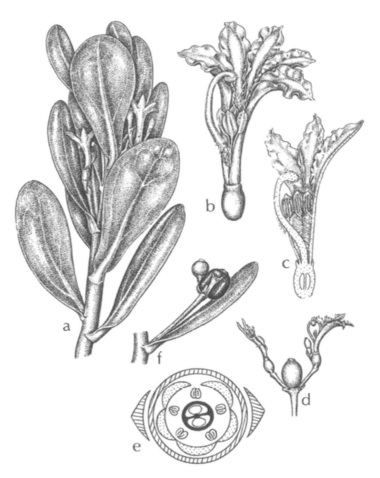

Figure B.34. *Scaevola plumieri* (Goodeniaceae) flowers, fruit and foliage. (a) Apex of flowering shoot (x1/2). (b) Flower from side (x3/2). (c) Flower in L.S. (x3/2). (d) Part of flowering shoot with young fruit (x1/2). (e) Floral diagram. (f) Mature fruit on axillary branch (x1/2). (From Correll and Correll, 1982.)

> ***Scaevola taccada*** (Gaertn.) Roxb. 1814 [*Hort. Bengal.* 15]
> See also Green 1991[*Taxon* 40: 118–122] (Plates 5B, 7I)

This species has a wide range from the Malay Peninsula to the South Pacific and Hawaii. Two types of seeds are known, resulting in a curious seed dimorphism (Emura et al., 2014). The C-morph has a corky testa so that seeds float. The NC-morph has a pulpy testa, attractive to birds but not floating. This may account for differences in the frequency of the two morphs across differing substrates;basically, beaches versus cliffs.

The species has been introduced to the New World and occasionally escapes from cultivation (as in South Florida). In the Bahamas, *Scaevola taccada* var. *sericea* is listed as an invasive species (Eshbaugh, 2014).

Family: Lecythidaceae

A small tropical family of 15 genera and 325 species with large, showy flowers. The most familiar product is Brazil "'nuts," the seed of *Bertholletia excelsa* H.B.K.

Barringtonia J. R. & G. Forst. 1776 [*Char. Gen.* 75, t. 28]

A genus of about 40 species distinguished by its large simple leaves clustered at the ends of the branches and the large angled fruit. It has a distribution from East Africa to Polynesia. Two species are characteristic of coastal communities in the Asian tropics, and others may be encountered in back-mangal communities. A full discussion is provided by Payens (1967). The following species are those most likely to be encountered in cultivation for the attractive flowers and fruits. The flowers open at night and are pollinated by night-flying animals. The bark and crushed fruit of both of them contain saponins and are used as a fish poison (Tattersfield et al., 1940). The fruit of *B. edulis* Seem. is edible.

1A.　Leaf margin always entire. Racemes usually terminal, erect, relatively short (to 8 cm) with few (to 10) flowers; pedicels 5–9 cm long. Flowers white, petals 6–8 cm long, filaments white, red-tipped. Fruit cubical, 10–15 cm in diameter. . . . . *B. asiatica*

1B.　Leaf margin usually toothed. Racemes terminal or from axils of fallen leaves on older wood, pendulous, long (more than 10 cm), many-flowered; pedicels at most 5 cm long. Flowers pink, petals at most 3 cm long, filaments red, white, or yellow-tipped. Fruit narrow, longer than wide (5–7 cm by 3–4 cm). . . . . *B. racemosa*

Barringtonia asiatica (L.) Kurz 1876 [*J. Asiat. Soc. Bengal* 45: 131, see also 46: 70]
　　(Fig. B.35; Plate 7A, B)

A tree growing to perhaps 20 m with a short, buttressed bole and dense crown. Bark rough and thick in older trees. Leaves obovate, somewhat fleshy, 30–40 cm by 12–30 cm, apex rounded, base slightly auriculate, petiole short, stout. Flowers in short, erect terminal racemes, showy (Plate 7A), up to 10 cm in diameter with four white petals. Fruit (Plate 7B) massive with two persistent calyx lobes, floating by virtue of the fibrous pericarp.

　　This is a characteristic strand plant of the Indo-Malayan and Polynesian region. Brass (field notes of Brass 2610) describes it as "the largest of the beach trees."

Figure B.35. *Barringtonia asiatica* (Lecythidaceae). Germinated seeds. Bako National Park, Sarawak.

Barringtonia racemosa (L.) Spreng. 1826 [*Syst. Veg.* 3: 127]

The synonymy of this species is long and complex, and Payens (1967), who has monographed the genus, provides an extensive list but says that it is impossible to interpret pre-Linnean names with certainty, so that the extensive early synonymy serves no useful purpose in citation.

A usually small tree, but up to 10 m; leaves obovate-lanceolate, 20–30 cm by 5–8 cm, tapered below to the short, somewhat fleshy petiole scarcely 1 cm long. Flowers on slender pendulous spikes 40–50 cm long; the flowers about 2 cm long with a slender filiform style. Fruit narrow, pointed at each end and four-angled or grooved.

This tree has a distribution in Indo-Malaya to Polynesia and Queensland and as far north as the Ryukyus; it is variously recorded as a mangrove and is abundant along tidal rivers and in areas subject to tide and salinity. It also occurs in and on the edge of peat swamp forests and on hillsides to altitudes of 200 m.

Barringtonia conoidea Griff. 1854 [*Notul.* 4: 656]

A species with a more limited distribution than the previous two, recorded as growing along riverbanks but only seaward at the limit of saline influence. It is distinguished from *B. racemosa* by its few-flowered but still pendulous raceme that is 5–10 cm long, but most characteristically by the top-shaped fruit about 7 cm long with eight projecting basal flanges. A suggested field character (Chai, 1982) is that the leaves of *B. conoidea* wither yellow, whereas those of *B. racemosa* wither red.

Seedling and Embryo in *Barringtonia*

The embryo is unusual, has a distinctive method of germination, and bears some comparison with viviparous seedlings (Ng, 2014). The embryo is solid and undifferentiated and the mature seed is usually interpreted as lacking endosperm. The shoot apex is enclosed by a series of spirally arranged minute scales, but there is no recognizable pair of cotyledons. The radicular end is blunt and undifferentiated. The bulk of the axis can be interpreted as a hypocotyl. On germination (Fig. B36), the plumule elongates and the scales expand, with a transition to the normal foliage leaves. Scales apparently subtend buds, which can grow out and form a replacement shoot if the apex is damaged. The "radicular" end produces roots that may be entirely adventitious, although one seems dominant.

   Although a comparison with Rhizophoraceae has been made, suggesting a close relationship (Miers, 1880), Payens does not agree. The embryology and development in Lecythidaceae seem quite diverse but provide no obvious clues to the morphology of this seedling.

Family: Lythraceae

A cosmopolitan, mainly herbaceous family of 25 genera and 500 species. The petals often crumpled in bud. Two genera unrelated in the family, one Old World, the other New World, and represented in mangrove communities. They are distinguished as follows:

1A. Shrubby or subarborescent, Old World; widely distributed . . . . *Pemphis*
1B. Herbaceous, New World; localized . . . . *Crenea*

> **Pemphis** J.R. & J.G. Forst. 1776 [*Char. Gen. Pl.*: 67, t. 34]
> See Koehne 1880–5 [*Engl. Bot. Jahrb.* 1–7: 132] and 1903 [*Engl. Das Pflanzenreich* 4, 216: 185]

In modern analyses, the genus may be found in Malvaceae. A tropical genus of coastal shrubs or at most small trees, ranging from East Africa (but curiously not Madagascar) through southeastern Asia to northern Australia, Polynesia, and northward to Hong Kong but with an apparent disjunction in Malaysia. The wide range of *P. acidula* seems related to its ecological status, somewhat intermediate between a strand plant and a mangrove. The genus is distinguished by its perfect hexamerous distylous flowers, a campanulate calyx persistent in fruit, the stamens united basally into a short tube, and the globose, many-seeded capsule with circumcissile dehiscence. Germination epigeal.

> *Two species are recognized*:

1A. Coastal, evergreen. Plant without black punctate dots. Flowers dimorphic (heterostylous). Petals white, stamens 12. Style long, slender, capitate; ovules inserted only on base of placenta . . . . *P. acidula*
1B. Inland, deciduous. Leaves and calyx with black punctate dots. Flowers monomorphic (homostylous). Petals rose colored, stamens ~18. Style short or very short, with a conspicuous capitate stigma; ovules inserted throughout placenta . . . .*P. madagascariensis*

> **Pemphis acidula** J. R. & J. G. Forst. 1776 [*Char. Gen. Pl.*: 68, t. 34]
> (Fig. B.36)

Small tree 7–10 m high, or more usually a low shrub 1–2 m high, of strands and rocky foreshores and more exposed mangrove associations. Bark light gray to brown, becoming deeply fissured with age and shredding into long curling strips. Young twigs angular, densely hairy as on the leaves, irregularly branched by syllepsis. Each node

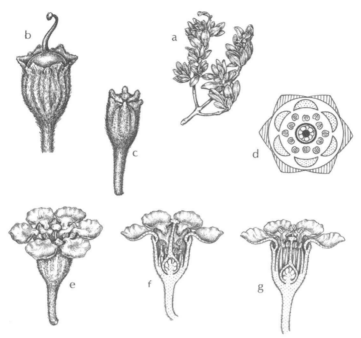

Figure B.36. *Pemphis acidula* (Lythraceae) shoot, flowers, and fruit. (a) Flowering shoot (x3/4). (b) Fruit (x3/2). (c) Flower bud (x3). (d) Floral diagram. (e) Flower from side (x3). (f) Flower, "pin" form in L.S. (x3). (g) Flower, "thrum" form in L.S. (x3). (Material from Idler's Bay, Papua New Guinea.)

supporting an axillary bud, the bud with a lateral series of minute, ephemeral linear colleters. Leaves decussate, sometimes fleshy (to 2 mm thick), very shortly petiolate (1–2 mm), 10–30 mm by 3–10 mm, narrowly ovate to obvate, apex bluntly pointed to rounded, margin entire, base shortly cuneate; both surfaces densely covered with stiff white hairs. Flowers axillary, solitary, less commonly in few-flowered bracteolate sessile cymes. Flowers hexamerous, perfect, perigynous, distylous; pedicel 5–10 mm long. Calyx 5–8 mm long, becoming somewhat larger and persisting in fruit, greenish red campanulate 12-angled or ribbed with six pointed triangular lobes, alternating with six thickened accessory lobes, one in each sinus, the calyx lobes valvate and at first form a flat lid to the calyx cup. Petals six, inserted on the mouth of the calyx cup below each sinus, shortly clawed, 5–8 mm long, crumpled in bud, ephemeral. Stamens 12 (–18) in two series inserted on lower third of calyx cup, the antesepalous stamens longer and inserted higher, filament 2–4 mm long, anthers dorsifixed, dehiscence introrse. Ovary superior, globular, 2 mm in diameter, unilocular but with three (to four) obscure basal dissepiments, placentation essentially free central. Ovules numerous, crowded, orthotropous. Style simple, stigma capitate. Short-styled (about 1 mm) flowers with stamens enclosing stigma, long-styled (about 4 mm) flowers with stigma exceeding stamens. Fruit enclosed in somewhat enlarged calyx, a spherical capsule (4–5 mm in diameter), with persistent style, dehiscence circumcissile; seeds numerous

(20–30), flattened, angular, with a corky margin or wing and narrowed below. Embryo minute, straight; endosperm present.

Growth and Reproduction

Architecture. This appears to correspond to Attims' model because the shoots are not articulate, although there may be some periodicity of extension correlated with the frequent but diffuse or irregular branching. Although often recorded as a small tree, the characteristic habit of this plant is a diffuse low-growing shrub. Every node seems capable of supporting a vegetative branch, as even flowering nodes have an additional (supernumerary) vegetative bud.

Floral biology. Floral dimorphism in *P. acidula* has been recognized at least from the time of Koehne's monograph [1880–1885; see also Koehne (1903, cited earlier)], but a more modern understanding of the significance of this in relation to the genetics of incompatability is only recent.

Lewis and Rao (1971) indicated distyly is not a local phenomenon because they recorded it in both Malaysia and East Africa. Gill and Kyauka (1977) made a quantitative study of populations in Tanzania. They describe the two forms in the traditional terminology of pin (long styled) and thrum (short styled), although the two sets of stamens (inner and outer whorl) and the stigma occupy three levels. The stigma then stands at a level either above the two sets of stamens (pin, Fig. B.36f) or between them (thrum, Fig. B.36g). The relative levels are such that the style of one flower form corresponds approximately to the level of at least one set of stamens of the other form. Styles are either 1 mm or 4–5 mm long, stamen length is either 3 mm or 4–5 mm. These authors conclude that the proportion of pin to thrum in natural populations does not differ significantly from a 1: 1 ratio.

Pollen grains from the two forms differ appreciably in size and rate of germination on artificial media, with thrum pollen growing faster. Stigmatic papillae are larger in thrum (16–18 μm) than in pin (12–14 μm) flowers. Artificial crosses between pin and thrum result in normal seed sets (20–30 seeds per capsule), whereas selfed flowers set no seed. This efficient outbreeding mechanism is maintained by a differential or disruptive selection mechanism, which has eliminated intermediate forms in favor of the two extremes.

> **Pemphis madagascariensis** (Baker) Koehne 1903 [*Engl. Bot. Jahrb.* 29: 164]
> *P. madagascariensis* (Baker) H. Perrier 1954 [*Fl. Madagas. Fam.* 147–222]

This endemic Madagascan species differs from *P. acidula* in the characters cited in the key. According to Perrier, this species occurs only inland in an inter-montane region of Madagascar. This contrast between species, one of which is a mangrove, is unusual. On the basis of pollen and floral features, it has been regarded as a distinct genus *Koehneria* S. A. Graham, Tobe & Baas.

> **Crenea** Aubl. 1775 [*Pl. Guia.* 1: 523, t. 209]

A genus of three species restricted to Trinidad and northern South America. All seem to have an association with salt water and the following species is more or less restricted to the muddy tidal flats of the Pacific coast of Colombia (Gentry, 1982). It could well be described as a mangrove, but is exclusively herbaceous and has no morphological adaptations to the mangal environment.

>**Crenea patentinervis** (Koehne) Standley 1947 [*Publ. Field Mus. Nat. Hist., Chicago Bot. Ser.* 23: 218]
>
>*C. surinamensis* (L.f.) Koehne ssp. *patentinervis* [Koehne, 1882, *Bot. Jahrb.* 3: 320]

A herb growing to about 50 cm with basal stolons. Leaves opposite, simple obovate-spathulate (3–4 cm by 0.5–1 cm), with a blunt, rounded apex. Flowers white, axillary. Capsules about 5 mm in diameter.

Family: Malvaceae: (Malvoideae)

Here three genera are treated in the traditional but narrow sense of Malvaceae, whereas molecular analysis reveals a broadened inclusion of a monophyletic group that also includes members of Bombacaceae, Sterculiaceae, and Tiliaceae, which family names are retained here for continuity with the older literature. These appear largely as subfamilies in the modern treatment, which recalls the older designation for this assemblage as an order Malvales. Strictly, the taxa in the following description are included in Malvaceae *sens lat* as Malvaceae: subfamily Malvoideae (Judd and Manchester, 1997; Baum et al., 1998; Alverson et al., 1999).

A large, mainly tropical family of trees, shrubs, or especially in temperate regions, woody herbs; including about 100 genera and over 1800 species. Characterized by simple alternate stipulate leaves and often stellate hairs or a scaly indumentum. The flowers are often large and showy and commonly have an epicalyx of free or fused scales below the calyx; the numerous stamens are fused to form a central staminal column and have spiny pollen; the ovary is superior with five or more locules; the fruit is usually dehiscent as a capsule or schizocarp. Members of the family often show a consistent overlapping of petals, described as contorted, either clockwise or anticlockwise. This may or may not be consistent within an individual or a species. Fig. B.37e shows what could be described as a left-handed and Fig. B.38d as a right-handed flower.

Species in two genera are characteristic trees of tropical seashores and are often associated with mangroves but never occur on strict mangal substrates. They both have very similar conspicuous flowers that are yellow on the evening they open but turn purple by the end of the following day. A third genus is added from the New World.

These taxa may be distinguished as follows:

1A. Shrubs, tropical America; flowers scarcely 2 cm in diameter.................
.... *Pavonia spicata* and *P. rhizophorae*
1B. Trees, pantropical (at least in cultivation); flowers 5–7 cm in diameter........2
2A. Calyx conspicuously five-toothed, persistent in fruit; flowers usually with a continuous maroon eye; receptacle without yellow latex. Capsule regularly loculicidally dehiscent and gaping widely. Seeds smooth. Leaves usually with nine to 11 main veins and always one to three glands on the abaxial surface near the petiole insertion; blade with an indumentum of stellate hairs..................
.... formerly *Hibiscus tiiaceus*, now *Talipariti tiliaceum*

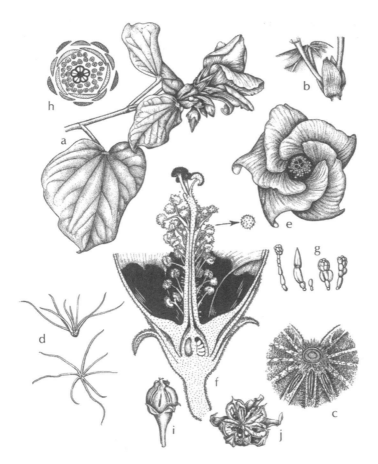

Figure B.37. *Talipariti tiliaceum* (Malvaceae) shoot, flowers, and fruit. (a) Flowering shoot (x3/8). (b) Leaf insertion with stipules (x1/2). (c) Leaf blade insertion from below (x3/2), glands on major veins. (d) Stellate hairs from leaves (x15). (e) Flower from above (x1/2);.(f) Base of flower in L.S. (x3/2); inset: pollen grain (x35). (g) Glandular hairs from the style (x30). (h) Floral diagram, epicalyx cross-hatched. (i) Unripe fruit (x1/2). (j) Dehisced fruit (x1/2). (From Tomlinson, 2001.)

2B.   Calyx entire or at most with minute teeth; flowers with an eye of five separate dark spots; receptacle with yellow latex. Capsule shattering irregularly, or if dehiscent, not gaping widely. Seeds finely hairy. Leaves usually with seven main veins, often lacking glands on the lower surface; blade with an indumentum of scales. . . . . *Thespesia populnea* [and *T. populneoides*]

Talipariti (L.) A. St.-Hil. 1828 [*Fl. Bras. Merid.* 1(7): 256]

The name change from the long-used *Hibiscus tiliaceus* is discussed extensively by Fryxell (2001). In recognizing *Talipariti* as a segregate genus from *Hibiscus*, he refers to an early name *Pariti* based on a Tamil name meaning "cotton," but which is invalid.

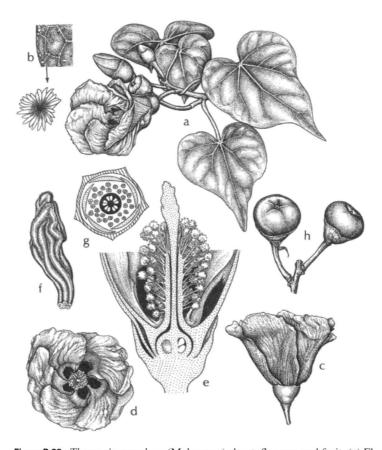

Figure B.38. *Thespesia populnea* (Malvaceae) shoot, flowers, and fruit. (a) Flowering shoot (x1/2). (b) Details of leaf surface (x6) and enlarged single scale (x90). (c) Flower from side (x1/2). (d) Flower from above (x1/2). (e) Base of flower in L.S. (x3/2). (f) Stigma (x3). (g) Floral diagram. (h) Fruits (x1/2), one epicalyx scale persistent to left. (From Tomlinson, 2001.)

The addition of the prefix *Pari* also changes the emphasis to mean "mucilage," which is perhaps more appropriate. Fryxell lists the characters in which the two genera differ, perhaps most obviously in the fruit.

> *Talipariti tiliaceum* (L.) A. St.-Hil. 1828 [*Fl. Bras. Merid.* 1(7): 256]
> *Hibiscus tiliaceus L.* [*Sp. Pl.*: 694] 1753 (Fig. B.37; Plate 7D)

This is a spreading tree growing to a height of 10–15 m in coastal communities, especially adjacent to beaches. It is recorded inland up to altitudes of 800 m but is widely planted, and its precise natural ecological and geographic distribution is uncertain. A closely related species, *T. elatum* Sw., is distinguished by the deciduous (not persistent) calyx and epicalyx.

The tree may be described as ever-growing, and flowers are present throughout much of the year, singly in the axils of foliage leaves, with the base of the long peduncle

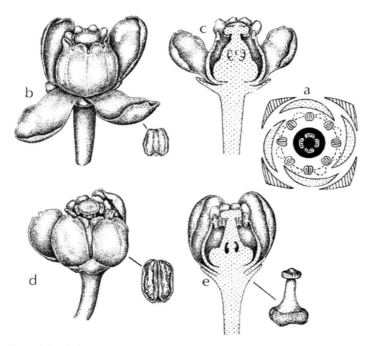

Figure B.41. *Xylocarpus granatum* (Meliaceae) flowers. (a) Floral diagram, scales on stamen tube dotted. (b) Female flower from side (x4); inset: nonfunctional stamen. (c) Female flower in L.S. (x4). (d) Male flower from side (x4); inset: functional stamen. (e) Male flower in L.S. (x4); inset: nonfunctional ovary (x3). (Material from Semetan, Sarawak.)

Further illustrations of the two types of flower in *Swietenia mahagoni* are in Tomlinson, 2001 (p. 189).

> *Xylocarpus granatum* König 1784 [*Naturforscher* 20: 2]
> *Carapa obovata* Blume 1825 [*Bijdr.* 179]
> *Xylocarpus obovatus* (Blume) Juss. 1830 [*Mem. Mus. Paris* 19: 244]
> *Xylocarpus benadirensis* Mattei 1908 [*Boll. Ort. Bot. Palermo* 7: 99]
> (Figs. B.40 & 41)

Vegetative Growth

Germination is initiated by extension of the hypocotyl so that the radicle is extruded from the seed (Fig. B.40d). Cotyledons stay within the seed, which may remain attached to the seedling for a long time (Fig. B.40g). The plumule grows upward through the cleft between the paired cotyledons, which are fused within the seed, and extends rapidly so as to reach a height of 50 cm in a few weeks (Fig. B.40g). It bears a series of spirally arranged scale leaves, which are progressively more distant. At the same time, the radicle extends forms, at least initially, a distinct tap root with subordinate laterals.

The first leaves are simple, but there is a fairly rapid transition to compound leaves with two, four, or six leaflets. Subsequently, the number of leaflets is reduced to four or even two on the older parts. Leaves are paripinnate, the rachis being golden brown and continuing beyond the insertion of the distal pair of leaflets as a minute appendage. Leaflets have a short corky petiolule, the same color as the rachis; this petiolule affects the primary positioning of the leaflet but there are no obvious later periodic leaflet movements. Leaflets vary somewhat in shape and size, up to a maximum of 12 cm long and 6 cm wide at the broadest part, but they may be less than half these dimensions in depauperate specimens. Typically, the apex is rounded or blunt and sometimes even slightly emarginate, but leaflets with a more or less acute apex are not uncommon. The leaflet thickness varies considerably according to the situation and is often quite succulent, up to 1.5 mm thick.

Growth of the shoot system is rhythmic and corresponds precisely to Rauh's model (Fig. 5.2A; Hallé et al., 1978). The resting bud is enclosed in short, thick, brown bud scales, which may retain a vestigial leaf blade. The bud scales leave a series of close-set scars so that the successive increments are clearly recognized. Trees may retain leaves throughout the year, even in seasonal climates, but Chai (1982) reports this species (and *X. mekongensis*) as deciduous in the nonseasonal climate of Sarawak.

Branching follows a period of seedling growth and is the result of the proleptic development of axillary buds at the end of the previous growth flush. Reserve buds are produced in abundance, as all leaves, including most of the bud scales, subtend a dormant bud. Few of these expand during the normal development of the tree. Saplings recover rapidly after damage, and depauperate plants may develop several trunks. The persistence of these reserve axillary buds probably accounts for the marked ability of older trees to sucker basally when they are damaged. Suckers essentially repeat the morphology of seedlings but have only a short series of basal scale leaves.

Normally trees remain single trunked. The bark is very characteristic, smooth pale greenish or yellowish brown, and peeling in irregular patches so that the trunk is blotchy (Plate 4E). Lenticels are not conspicuous. Buttress development appears fairly late and is usually preceded by the first evidence of the plate-like surface roots in trees about 5 m high. These have not been studied in detail but are commented upon by all authors who have seen the tree *in situ*. They can develop to the extent of becoming daunting obstacles (Plate 13C).

Reproductive Biology

In the absence of extended measurements on phenology, little precise information about the periodicity of shoot extension and flowering is available. Noamesi (1958) records this species as flowering throughout the year on the basis of herbarium specimens, but this is probably not generally true because of the pronounced rhythmic growth and synchrony in shoot development over the whole plant. Chai (1982) records quite precise flowering and fruiting periods for both species. Inflorescences appear mainly from the axils of leaves at the base of the current shoot increment, but they also develop directly from older previously dormant buds or on woody shoots. In this situation, they may appear to be terminal. This habit may be important in view of the large fruit, which requires a stout twig to support it. The tree cannot be described as ramiflorous, however.

Inflorescences consist of a central axis supporting spirally arranged inconspicuous and ephemeral bracts. Each bract usually subtends a three-flowered cyme with even smaller bracteoles. The irregular inflorescence structure in this species, emphasized by Noamesi, is the result of zigzag development of the axis and the frequent abortion of flowers in the cymes. In a regular cyme, the terminal flower seems most often to be female, the lateral (or distal flower of the whole inflorescence) to be male. Flowers expand rather irregularly (usually the terminal flower of a cyme opens first) and are quite strongly but pleasantly scented. Bees are recorded as flower visitors, and one could expect flowers to be pollinated by short-tongued insects that insert their mouth parts between the stigmatic disk and the stamen tube in search of nectar produced by the ovary disk. In withdrawing this organ, they would retain some pollen (male flowers) or deposit existing pollen on the under surface of the stigma (female flowers). Minute drops of nectar originate on each wart of the disk and accumulate in the floral cup.

Fruit develops rapidly and normally only one fruit appears from a single inflorescence. As the fruit enlarges, a short massive axis is produced, becoming pendulous with increasing weight. The mature fruit is the size of a good melon and weighs 2 or 3 kg. The four sutures of the future woody valves are evident as four shallow equidistant furrows. The fruit (Plate 5D) mainly shatters on impact after its fall from the tree, with the valves breaking away from the fruit stalk, releasing the several seeds. Fruits are always well filled, as if there is no abortion of ovules, and Noamesi records up to 20 seeds, but eight to 10 is a more usual number.

Seeds are angular and fitted completely into the fruit cavity; the outer convex surface is pressed against the inner fruit wall, and the angular, often pyramidal internal facets are fitted together in a completely close-packed arrangement. Seeds have a thick corky testa; the embryo has an undifferentiated cotyledonary mass with the plumule and radicle sometimes on one internally facing facet but more usually on the outer convex face. The site of the radicle is evident as a small boss. Seeds may start to germinate by extrusion of the radicle while still floating. No information is available about the longevity of seeds. Seedlings are quite common in *Xylocarpus* communities so that maintenance of populations as well as dispersal seems effective.

Xylocarpus mekongensis Pierre 1897 [*Fl. For. Cochinch.* plate 359B]
(Figs. 5.7, B.40, B.41)
Carapa moluccensis Watson 1928 [*Malay. For.* Rec. 6: 70, 75, tabs. 34 and 35. Non-Lam]
Xylocarpus gangeticus (Prain) Parkinson 1934 [*Indian For.* 60: 140]
X. australasicus Ridley 1938 [*Kew Bull.* 1938: 291; orig. as *X. australasica*]

This species is probably the "*Xylocarpus moluccensis*" of many non-taxonomic works, as for example in Duke (2006).

Apart from the diagnostic details outlined earlier, this species is morphologically and biologically very similar to *X. granatum*. The most striking differences are the bark texture and the presence of well-developed peglike pneumatophores. The name *X. australasicus* has been used in several recent accounts for plants from Western Australia and New Guinea, presumably with the acceptance of Ridley's Queensland species. However, *australiasica* is incorrect there and should be corrected to *australis* (P.G. Wilson, personal communication).

The species is more consistently deciduous than other species and in Queensland for example, it characteristically loses its leaves for one to two months in the cool, drier winter months (Duke, 2006).

The erect, conical pneumatophores up to 30 cm tall originate from the extended horizontal "cable roots," which grow away from the base of the trunk. They are entirely secondary and represent a localized outgrowth of the vascular cambium (Plate 12E). They can be forked or clustered. They may originate in slight loops where the apex of the parent root is displaced upward. This explains the concavity in their base. The cable roots in addition produce branch roots, which may descend or grow horizontally to proliferate the root system. New horizontal roots originate (presumably adventitiously) from the base of the older pneumatophores. The ultimate branches of the descending roots are very fine and lack root hairs. The surfaces of older pneumatophores become rough with numerous lenticels.

Xylocarpus moluccensis (Lamk.) Roem. 1846 [*Syn. Hesper. Fasc.* 1: 124]

This is regarded as a non-mangrove species, although its habitat may frequently impinge on mangrove swamps. I have had no opportunity to investigate this plant. It seems largely to be distinguished by its usually broadly ovate pointed leaflets. The bark is described by Parkinson (1934) as "grey, rough with longitudinal fissures and peeling in flakes."

Family: Myristicaceae

A tropical woody family of 18 genera and 300 species characterized by the dioecious condition, simple entire, alternate, exstipulate leaves, the 10 stamens united into a central column in male flowers, and the ovary of a single superior uniovulate carpel in female flowers. The one-seeded fruits are fleshy but dehisce to release the large arillate seed, which has ruminate endosperm. The family includes the commercial nutmeg, *Myristica fragrans* Houtt, native to the Moluccas.

Myristica Gronov. 1755 [*Fl. Or.*: 141]

A genus of understorey trees with about 70 species widely distributed in the Asian tropics. The following species occurs in back mangal with a distribution entirely restricted to the islands of New Guinea and New Britain. The account is taken largely from Sinclair (1968). In the field, the following species is recognized by its basal aerial stilt roots, alternate oblong leaves, copious red exudate, and nutmeg-like fruits.

Myristica hollrungii Warb. 1897 [*Monogr. Myrist.* 490, t: 19, f. 1–2]
See also Dewilde 1994 Blumea 39: 341–350

An evergreen dioecious tree growing to 36 m high with horizontal whorled branches suggesting Massart's model. Aerial roots forming basal stilts. Bark dark grayish brown, finely fissured vertically and flaking. Sap red, exuding copiously from cut surfaces. Leaves spirally arranged on the orthotropic (trunk) axis, two-ranked on the plagiotropic (branch) axes. Stems with distant leaves and a continuous line on each side of the stem along the internode connecting successive leaves. Leaves glabrous, thinly coriaceous, 20–35 cm by 5–13 cm, with a short slender petiole 1.5–2 cm long, the blade oblong or oblong-lanceolate, the apex acute or acuminate, the base rounded, to subcordate. Veins 16 to 22 on each side, parallel and equidistant, slightly prominent beneath. Inflorescences axillary; axis short, thick, producing flowers distally and acropetally over an extended period. Male flowers 5 mm long, shortly stalked, subglobose with a pointed apex, with a single lateral bracteole 1–4 mm long. Perianth cupular, with minute hairs, yellow outside, white within, shortly three-lobed apically at anthesis. Stamens 10 united in a central column with a short apiculus and a short basal, sterile, hairy stalk; pistillode absent. Female flower similar but on a shorter pedicel, the perianth lobes very short, ovary ovoid with a single basal anatropous ovule and a short bilobed stigma enclosing the receptive surface; staminodes absent. Fruit 3–4 cm by 2–3 cm, on a short thick stalk and with a pointed apex. Pericarp firm, orange glabrous at maturity, seed brown, surrounded by the laciniate aril that is orange-red at maturity.

Distribution and Ecology

Sinclair (1968) describes the tree as one of the most common wild nutmegs in New Guinea. Within its range it is regarded by several authors as a characteristic back-mangal constituent (Percival and Womersley, 1975), but it is recorded from wet places inland and has been collected at altitudes up to 900 m, where the stilt roots are little developed or even absent.

Growth and Reproductive Biology

There is no information in the literature on the biological aspects of this tree.

Systematics and Nomenclature

A full description and discussion of the nomenclature of this species are provided by Sinclair (1968). There has been some confusion with related species, which was compounded by the destruction of many types in the Berlin Herbarium in World War II, but problems have been resolved by Sinclair. He distinguished *M. hollrungii* from the related *M. subalulata* by its total lack of ant-inhabited swellings on the shoot and the absence of a distinct intermodal wing. Where these characters are not clear in herbarium specimens, the two species have sometimes been confused.

Family: Myrtaceae

This large, essentially tropical family of 130 genera and over 4000 species includes one monotypic genus associated with mangrove communities. The family has been traditionally divided into two major subfamilies; Leptospermoideae with capsular fruits and Myrtoideae with fleshy fruits, but modern treatments have abandoned this approach (Wilson, 2011).

Osbornia F. Mueller 1862 [*Fragmenta* 3: 30]

A genus ranging from tropical Australia to the Philippines with outliers to the east and west. In the field, its single species is recognized by its opposite, rather small, obovate, and aromatic leaves with pellucid dots, and the spongy gray striated bark.

Osbornia octodonta F. Muell. loc. cit. (Fig. B.42, Plate 4D)

Small aromatic evergreen, hermaphroditic trees, or more commonly low shrubs, with a gray, soft, spongy flaking and quite thick bark (Plate 4D), usually on more exposed sites in association with mangroves, never within the dense mangrove community itself. Plants usually with several irregular slender trunks, not forming a regular architecture. Leaves opposite, decussate on four-angled twigs; leaf obovate, usually 2–4 cm by 1.5 cm, but larger on more vigorous shoots; margin entire, apex rounded or sometimes slightly emarginated. Petiole 1–3 mm, reddish, grooved adaxially, with an adaxial ridge closing the groove basally. Colleters present as an inconspicuous palisade within this basal groove (Fig. B.42c). Stipules absent. Terminal bud scales absent; axillary and terminal buds protected within the pocket of the grooved petioles of a subtending leaf or terminal leaf pair, the pocket closed by the adaxial ridges. Leaves aromatic when crushed, with numerous scattered pellucid or translucent dots. Lamina anatomy described as isobilateral, with a palisade below both surfaces (Wilson, 2011). Flowers axillary, either solitary or in three-flowered cymes (dichasia: Fig. B.42e); dichasia otherwise appearing to be terminal, sometimes with a pair of reduced leaves immediately below. Inflorescence or flower with a basal pair of somewhat woody prophylls 1–2 mm long, subsequent bracteoles 3 mm long, scarious, ephemeral. Flowers (Fig. B.42h) about 5 mm long with a well-developed calyx cup 2 mm deep, ending in eight equivalent rounded lobes, the calyx densely silver hairy within and without, tapered below, appearing (and often described as) white. Petals absent. Stamens yellow, numerous (up to 48), protruding as a cluster at maturity, each slender and 2–3 mm long, inserted on the mouth of the calyx cup; anther minute (0.5 mm), the connective extended into a blunt appendage. Ovary inferior, apparently unilocular but

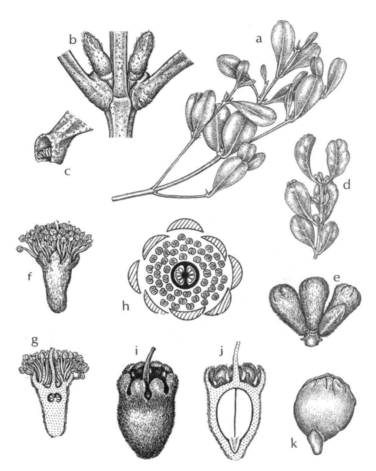

Figure B.42. *Osbornia octodonta* (Myrtaceae). (a) Leafy shoot (x1/2). (b) Node with axillary flower buds (x4). (c) Detail of notched leaf base with colleters (x4). (d) Shoot with flower buds (x1/2). (e) Dichasium of flower buds (x4). (f) Open flower (x3). (g) Flower in L.S. (x3). (h) Floral diagram. (i) Fruit (x3). (j) Fruit in L.S. (x3). (k) Embryo (x4). (Material from Kuching River, Sarawak.)

with several (up to 15) ovules on two basal, obscurely axile placentae (Fig. B.42g, h). Style simple, tapered, 3–4 mm long, and protruding to the same height as the extended stamens. Fruit a globose, leathery berry about 5 by 7 (up to 10) mm, somewhat larger than and included within the calyx tube. Seeds one (to two). Endosperm absent. Embryo straight, top-shaped, cotyledons hemispherical, much exceeding the short hypocotyl (Fig. B.42j). Germination epigeal.

Growth

The habit of the plant is irregular, leading to the development of a low shrub or bushy tree with a dense, rounded crown. Individual shoots are infrequently but irregularly branched with frequent sympodial development as a result of abortion of the terminal bud, or in older plants, the development of terminal flowers. The shoots grow rhythmically, but because bud scales are not formed, there are no obvious articulations. Branching may be correlated with resting periods. Frequently there are two buds at a node, with the lower becoming reproductive and the upper remaining vegetative. The plant is well illustrated in Duke, 2006.

Reproductive Biology

There is no information about pollination. The flowers are aromatic and made conspicuous by their silvery indumentum and expanded stamens; insect pollination is likely. The glandular appendage on the stamen may function as an osmophore. The fruits apparently do not dehisce, but are detached and float as a unit.

Root System

This remains unspecialized; the main roots may become exposed by erosion of the substrate and develop obscure buttresses. Larger-diameter and exposed roots develop a spongy bark, like that of the stem. Percival and Womersley (1975) record the presence of pneumatophores; the aerial roots tend to loop much as in *Lumnitzera*, to which they probably refer.

Systematic and Other Notes

The position of *Osbornia* within the Myrtaceae is isolated. There was uncertainty into which of the two traditional subfamilies it should be placed. Briggs and Johnson (1979) included it as a monogeneric group "*Osbornia alliance*" within the subfamily Myrtoideae, and indicate this in their phyletic diagram by showing it attached basally, originating more or less directly within the ancestral myrtoid stock. Schmid (1980) includes *Osbornia* within the subfamily Leptospermoideae because of its non-fleshy fruit. Wood anatomy, pollen morphology, and floral anatomy are said to support this assignation.

This historical problem resides in the texture of the fruit, which is leathery but not fleshy as in the typical myrtoid fashion. At the same time, the fruit is not dehiscent as is usual (with some exceptions) in the leptospermoid group. Nevertheless, *Osbornia* was included in the Leptospermoideae (tribe Backhousinae) by the early monographer Niedenzu, a tradition followed subsequently by other authors until this opinion was challenged by Briggs and Johnson. They indicate that features of the leaves, flowers, and embryo suggest Myrtoideae, without elaborating details. They argue for an ancestral succulent fruit, as in the Myrtoideae, with the fleshy condition being lost in the change from presumed animal dispersal to dispersal by floating in seawater. In addition, the fruit is most commonly one-seeded, with the embryo filling the cavity. The pericarp expands to accommodate this, but increases little in thickness.

In modern literature (Wilson, 2011), the original typological division into the two contrasted subfamilies has been restructured, based on molecular evidence, with the creation of a series of tribes. *Osbornia* appears as the monotypic tribe Osbornieae, indicative of its isolated status and suggestive of an early divergent group.

Floral Morphology

Although the flower of *Osbornia* is usually described as apetalous, with eight calyx lobes, Briggs and Johnson (1979) refer to this octomery as "false"; they consider that the calyx and corolla are scarcely distinguishable and simply simulate a single whorl. This opinion is reiterated by Schmid (1980); neither statement is substantiated in detail. In material that I have examined, the lobes overlap in a single spiral, all are about the same size, and there is no textural difference between them. Schmid also comments that *Osbornia* has an incompletely divided ovary with the ovules morphologically axile but basally located on the septum; this configuration can be seen in a transverse section just above the base of the loculus.

Geographic Distribution

The distribution map by van Steenis (1936) shows an apparent discontinuity in the distribution of this plant between a western area (Philippines, North Borneo, Sulawesi, and eastern Java) and an eastern area (north Queensland and adjacent Papua New Guinea).

Family: Pellicieraceae

Also called Pelliceriaceae and *Pelliceria* (see under "History and Systematics").

A monotypic family; the genus *Pelliciera* has traditionally been placed in the Theaceae or Ternstroemiaceae, usually as an isolated tribe Pelliciereae. Tetrameristaceae may be a more appropriate choice. Current opinion tends to place it in a separate family, Pellicieraceae (Planch. and Tri.) Beauvisage, and to emphasize its affinities with more remote groups such as Marcgraviaceae. It is distinguished by its regularly pentamerous flowers enclosed by a large pair of bracteoles, mangal habitat, one-seeded indehiscent fruit with corky pericarp, and the large, naked embryo.

> *Pelliciera* Planchon and Triana 1862 [*Benth. Hook. Gen. Pl.* I: 186]
> *Pelliciera rhizophoreae* Triana and Planchon 1862 [*Ann.* Sci. *Nat.* Series 4, 17: 381]
> Including *P. rhizophoreae* var. *benthamii* Triana and Planchon 1862 [loc. cit.]
> For a detailed citation, see Kobuski (1951). (Figs. 4.1; B.43, B.44; Plate 19)

Small trees, 5–10 (but up to 18) m high with somewhat tiered, distally arcuate branches. Trunk swollen and markedly fluted below (Plate 19A), the ridges each originating as an acropetally developed series of short aerial roots. Bark dark, rough, fissured. Stems with conspicuous circular leaf scars and stubs of stalks of fallen fruits at intervals. Leaves spirally arranged, with regular two/five phyllotaxis, each sessile, asymmetric oblong-lanceolate and broadest at the middle, 10–12 cm by 2.5–4 cm; blade glabrous, leathery, dark glossy green; apex bluntly rounded, margin initially with a series of prominent but ephemeral glands (presumed salt glands) on the wider inrolled side, becoming entire with age. Leaf base abruptly narrowed to the insertion with a pair of glands (extrafloral nectaries). Young leaves involute in bud. Stipules and bud scales absent.

Flower (Plate 19B) large, showy, pentamerous up to 12 cm wide at anthesis, solitary in the one to three leaf axils immediately below the resting terminal bud, the leaves subtending the flowers broadened at the base. Flowers at first enclosed by a pair of narrow whitish or reddish bracteoles about 5 by 0.5 cm to 0.8 cm. Calyx of five short, free, imbricate, white, concave lobes; sometimes reddish without and with numerous small glands internally at the base (Fig. B.43c). Petals five, free, white, or rose pink with a white mid-vein, about 6 cm long, 1.5 cm wide at the base, tapered distally to a blunt point, initially erect but reflexed at anthesis, ephemeral and falling with the sepals and

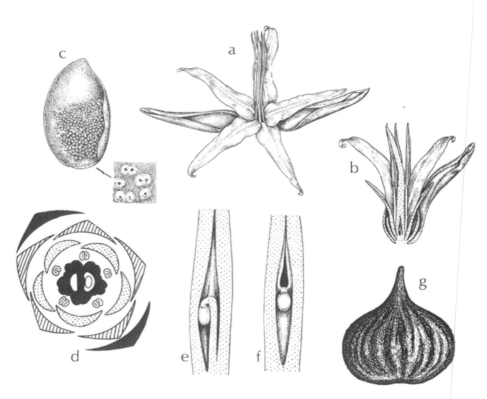

Figure B.43. *Pelliciera rhizophorae* (Pellicieraceae) flowers and fruit. (a) Flower from above (x1/2), stamens not yet reflexed. (b) Flower in L.S. (x1/2). (c) Sepal from within (x3/2), with inset detail showing adaxial glands (x8). (d) Floral diagram. (e) Ovary in L.S., transverse view (x3). (f) Ovary in L.S., dorsiventral view (x3). (g) Fruit (x1/4). (Material from Rio Tempisque near Puerto Humo, Costa Rica. [A.M.J.])

stamens. Stamens five free, 6 cm long, lying within the alternate grooves of the ovary, anthers narrow, 2.5 cm long, sagittate basally, pointed distally, dehiscence extrorse by longitudinal slits. Ovary narrow, cylindrical, woody (Fig. B.43d), about 6 cm long with 10 longitudinal grooves, tapering to a narrow style with a bifid stigma; locules two, each containing a single axile anatropous ovule, only one ovule developing or one locule sterile. Fruit top-shaped (Fig. B.43g), 8–12 cm long and about as wide, somewhat flattened, the style persisting as a pointed woody but brittle beak, at first green but (with age) becoming reddish brown and developing resinous pustules, the surface with many ridges. Fruit with the wall about 6 mm thick at the base, leathery externally but spongy within, splitting readily (Plate 19E). Seed single, without endosperm, and including a naked cordate embryo, testa at maturity represented by ribbon-like fragments, the hypocotyl curled in the seed, but extending into the stylar beak (Fig. B.44), plumule reddish, long, slender and sinuous, enclosed by the two fleshy cotyledons. Germination semi-epigeal, with rapid separation of the cotyledons and radicle and straightening of the plumule (Plate 19D).

Figure B.44. *Pelliciera rhizophoreae* (Pellicieraceae). Fruit dissected to show embryo morphology; two embryos each with a single cotyledon removed to show the characteristically curled plumule. (From a color transparency by A.M.J.)

Geographic and Ecological Distribution

The tree often forms extensive pure colonies, probably representing the limited dispersal ability of its seeds. Consequently, seedlings are commonly aggregated below a rather diffuse canopy (Plate 19C).

This species was once thought to have a limited distribution on the Pacific coast of central and northern South America, from Buenaventura Bay in western Colombia to Costa Rica. However, recent isolated collections on the Atlantic coast (Winograd, 1983; Jiménez, 1984; Roth and Grijalva, 1991) have been assumed to be relicts of its ancient, wider range, this reduction related to changes in salinity regimes. This situation is, in a sense, confounded by recent genetic analysis in which populations on the Atlantic and Pacific coasts of the CAI have been compared, suggesting a more recent reintroduction of Atlantic from Pacific populations (Castillo-Cardenas et al., 2015). These authors thus raise interesting new concepts about the prehistory of this species.

On the other hand, evidence from its occurrence as a microfossil (pollen) shows that it once had a circum-Caribbean distribution and extended to Mexico (Graham, 1977), and it is found in Eocene deposits in Panama (Fig. 4.1). Its progressive range restriction is attributed to climatic and sea-level changes and the increasing competition from *Rhizophora*. *Pelliciera* provides the best evidence for the mangrove "refugium" theory of Gentry (1982). It occupies the intertidal zone, commonly in association with *Rhizophora*, but with a much more localized distribution, typically in sheltered sites such as estuarine banks or protected beaches. Reasons for its limited distribution are summarized in Dangremond and Feller (2014).

Growth and Reproduction

Plants are evergreen and seem to develop resting terminal buds (which are not protected by bud scales) only in association with flowering, which occurs from the axils of one or sometimes more of the subterminal leaves so that the flowers, because of their size, superficially appear terminal and overtop the displaced apex. Flowers appear on trees only about 1 m high (Howe, 1911), a distinctive feature of pioneering species. Flowering is said to be over an extended season in winter and early spring (November to February), probably due to nonsynchronous flowering of different shoots, but there is a peak of heavy fruiting late in the year (October). The seasonality shown by individual shoots is reflected in the fairly regular series of stubs of fruit stalks on each shoot, separated by a long sequence of nonflowering nodes.

Each leaf subtends a single vegetative bud (except at flowering nodes), but most buds remain undeveloped and may have a limited life span because prolepsis and suckering are rare. Branching is therefore mainly by syllepsis and can result in a series of sylleptic shoots on vigorous specimens. Branching on the trunk is discontinuous (Attims' model) and thus results in tiers of plagiotropic axes that are themselves little branched but curved upward and bearing the leaves in terminal clusters (hence the superficial similarity with *Rhizophora* commented upon by several observers). Leaves are asymmetric, with the wider, glandular half rolled innermost in bud. Leaves on one shoot seem to retain the same symmetry, which may change on branches.

The floral parts surrounding the ovary are ephemeral and tend to fall together. There is no information about the floral mechanism or pollination biology, but Howe (1911) suggests that flowers are scented. The fruits fall and usually release the seeds, which are dispersed in the water, germinating immediately they become stranded. Howe suggests that the beak of the fruit is an adaptation for anchoring or even planting, implying that the point is forced into the mud when the fruit falls, a seeming extension of the "plunk" hypothesis applied to *Rhizophora*. This is contradicted by Johnston (1949, p. 207), who points out that the heavy end of the fruit is not the pointed end and that fallen fruits tend to land on their side or the blunt end. He suggests that broad dispersal by ocean currents is unlikely, as the protection of the seed by the fruit wall is not only poor but also of short duration (Plate 19E). Seeds in water swell after only a few hours and sink (Collins et al., 1977).

Root System

The short radicle may persist or abort, but its presence is not important in the development of the root system. The first adventitious roots arise as a series at the base of the hypocotyl; as in *Rhizophora*, they are differentiated in the embryo. They are replaced by wider aerial roots on the plumular axis. The anatomy of above-ground and subterranean roots is sharply contrasted. With age, several orthostichies of aerial roots are formed, each younger root above the next older root in the series. These roots do not immediately penetrate the stem surface but grow down inside the bark closely appressed to older roots. This raises the cortical tissue (which seems to be supplemented by some kind of secondary growth) so that the surface remains unbroken, except for numerous conspicuous lenticels. These vertical series of adventitious roots below a covering layer of bark form the fluted buttresses (Plate 19A). The roots become secondarily thickened and branch within the bark. Roots emerge below the substrate with the submerged type of anatomy; they may be seen exposed in specimens on the banks of creeks. A clear picture of the system can be obtained from dead specimens, from which the rotted bark is easily removed; the stem proper is shortly obconical. The whole buttress system in this species, with its numerous lenticels, seems to function as one giant pneumatophore.

Ant Protection

Collins et al. (1977) drew attention to the frequent presence of an aggressive species of *Azteca* that is suggested to provide protection against some insect predators, which the ants attack vigorously. The association may be mutualistic, as the ants may colonize *Pelliciera* and derive some nutrients from the pair of extrafloral nectaries at the base of each leaf blade and also from the insects they attack. The association seems nonobligatory, however, because healthy trees of *Pelliciera* may lack ants and trees with foraging ants may be extensively damaged by insects. In some parts of its range, *Pelliciera* lacks both nectaries and ants.

History and Systematics

There are relatively few illustrations of the plant that show diagnostic details because Triana (who first collected the plant) produced only scanty specimens, presumably used as the basis for Baillon's illustrations (1875, p. 245). The only extensive figures are provided by Hemsley (1879–1888) and reproduced by both Szyszylowicz (1893) and Beauvisage (1918); Howe's excellent account is illustrated by photographs. Beauvisage, in his description of the genus, repeats the earlier statement that the ovary is five-locular; this is incorrect, although it refers to the type collection (from Buenaventura, Colombia). Later collections have all shown two-celled ovaries, including specimens collected by Sutton-Hayes on which Hemsley's description and illustrations were based (Howe, 1911). The reportedly five-celled ovary of the more southern form is the basis for var. *benthamii* of Triana and Planchon, but the existence of such forms is doubtful and was admittedly based on incomplete specimens.

The place of this genus within the angiosperms is somewhat problematic. More recent study suggests the genus is closest to the small family Tetrameristaceae (Kubitzki, 2004).

Bentham and Hooker (1862) included it between the genera *Gordonia* and *Pyrenaria* in the tribe Gordonieae of their Temstroemiaceae. Baillon (1875) included it within the same family as an isolated tribe Pellicerieae, close to the Marcgravieae. Beauvisage

(1918), emphasizing anatomical features, drew attention to its similarities with certain Marcgraviaceae, especially the genus *Souroubea*, and effected the compromise of creating the family Pellicieracees intermediate between Ternstroemiaceae and Marcgraviaceae, largely in deference to Baillon. Beauvisage used the spelling *Pelliciera* throughout his account, and presumably on this basis, this spelling has been accepted by most subsequent authors. This is appropriate, as it commemorates the naturalist Bishop Pellicier of Montpellier; *Pelliceria*, as originally used by Triana and Planchon, seems to be an error (Kobuski, 1951).

Family: Plumbaginaceae

A cosmopolitan family, mainly of herbs or low shrubs, frequently in saline habitats, and including one mangrove representative (van Steenis, 1948). Genera are often distinguished by secretory epidermal glands, said to secrete either mucilage or chalk (Metcalfe and Chalk, 1950, p. 852). The genus *Aegialitis* (two species) has been segregated from the Plumbaginaceae as the family Aegialitidaceae because of its numerous distinctive features, including anomalous secondary thickening, abundant sclereids, and incipiently viviparous seeds. Molecular and morphological evidence (Lledo et al., 2001) now places the genus within the family as a well-separated clade (tribe Aegalitidae) sister to the subfamily Staticoideae. Pollen morphology is distinctive, but the greatest resemblance was said to be with the tribe Plumbagineae and particularly with *Plumbago* (Weber-El Ghobary, 1984).

> *Aegialitis* R. Brown 1810 [*Prodromus Florae Novae Hollandae*: 118: 426]
> (Figs. 4.3; B.45; Plate 9F)

A low-growing treelet or small shrub with perfect flowers, to 3 m high, with a basally swollen, fluted axis; typically occurring in exposed, often rocky or sandy sites. Anomalous secondary thickening producing included phloem. Bark dark, smooth, fissured or flaking with age. Buds mucilaginous. Twigs with conspicuous annular leaf scars. Leaves to 15 cm long, somewhat fleshy, spirally arranged, erect, and clustered terminally on the shoots. Petiole to 8 cm long, grooved adaxially and extended basally into a tubular leaf sheath with a completely encircling insertion. Blade broadly ovate, 6–8 cm by 2–5 cm, glabrous, margin entire, apex rounded or bluntly acuminate.

Flowers pentamerous, in many-flowered terminal, irregularly one-sided cymes (modified dichasia), with pairs of opposite linear bracteoles. Pedicel smooth, long (to 2 cm), up to three-fourths the length of the unexpanded flower bud. Calyx 7–13 mm long, tubular, somewhat inflated, fluted externally, five-lobed apically, the lobes about 3 mm high and minutely but bluntly apiculate by extension of median vein. Petals five, contorted, 8–10 mm long, white, imbricate, free above with bluntly rounded lobes, shortly fused basally to form a corolla tube 2–3 mm long. Stamens five, about 10 mm long, inserted on the corolla tube opposite the petals (cf. Fig. B.45e & f), filaments 6–7 mm long, slender but inflated close to their attachment, anthers 2–3 mm long, basifixed but extended horizontally at the time of pollen release; dehiscence extrorse by longitudinal slits; pollen sparse with large grains.

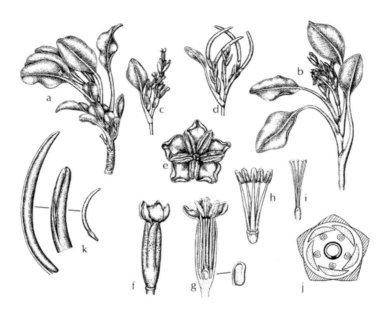

Figure B.45. *Aegialitis annulata* (Plumbaginaceae) vegetative, flowering, and fruiting shoots. (a) Vegetative shoot with congested internodes (x1/2). (b) Flowering shoot with extended internodes (x1/2). (c) Inflorescence (x1/2). (d) Infructescence (x1/2). (e) Flower from above (x6). (f) Flower from side (x3). (g) Flower in L.S. (x3); inset: single ovule and recurved funicle (x12). (h) Stamens (x3). (i) Ovary (x3). (j) Floral diagram, corolla contorted. (k) Fruit (x1/2); inset: details of apical part of capsule dehiscing to expose seed (center x3/2) and detached seed (left x3/2). (Material from Three Mile Creek, Townsville, Queensland.)

Ovary superior, grooved or angular below, narrowed above, with each lobe extending into one of the five free styles, each style 6–8 mm long with an extended oblique peltate stigma initially facing inward. Ovary unilocular with a single basally attached anatropous ovule (Fig. B.45g), the funicle recurved to orient the micropyle apically. Fruit an elongated, bluntly pointed, longitudinally dehiscent capsule, 4–5 cm long by 3–5 mm wide, enveloped basally by a persistent calyx. Pericarp thin, thickened somewhat distally. Single seed with a thick testa; endosperm absent. Embryo elongated, with an extended hypocotyl up to 4 cm long; cotyledons short, bluntly pointed, enclosing plumular leaves within a profuse mucilage (see earlier). Germination immediate, epigeal.

Two very similar species are known, distinguished as follows:

1A. Leaves dull above, calyx 7–8 mm long. Corolla with nine or 10 lobes. . . . .
. . . .*A. annulata*
1B. Leaves shining above, calyx 13 mm long. Corolla with 12 lobes. . . . .
. . . .*A. rotundifolia*
   The two species are said to be distinguishable in pollen morphology (Weber-El Ghobary, 1984).

Aegialitis annulata R. Brown 1810 [*Prodromus Flora Novae Hollandae*: 426]
This species is recorded for northern Australia and east Malesia.

Aegialitis rotundifolia Roxburgh 1824 [*Fl. Ind.* 2: 111]
This species is recorded in Myanmar, Bengal, and the Andaman Islands.

Geographic and Ecological Distribution

The genus is used by van Steenis (1949) as an example of vicarious distribution, as the two species together occupy a range somewhat comparable to that of many other Asiatic mangroves, but their individual ranges do not overlap (Fig. 4.3). Fossil pollen of *Aegialitis* is known from the intermediate area, however.

   Aegialitis is a characteristic mangrove associate but does not itself occur within closed mangrove communities, as it prefers or even requires exposed sites. At the same time, it withstands wave and tidal action. Consequently, it is most characteristic of open rocky shores or exposed beaches, sometimes as the outermost zone of a narrow mangrove belt. It is less common in back mangal, although it can become established in highly saline soils.

Anatomy

Aegialitis is unusual in the large number of trichosclereids (Plate 9F), which are produced in all parts except the flowers (Maury, 1886). These sclereids are absent from other Plumbaginaceae. The genus has anomalous secondary thickening from successive discontinuous cambia. Derivatives of these cambia may themselves differentiate as sclereids, an unusual, if not unique, condition. Epidermal glands are said to secrete mucilage (Metcalfe and Chalk, 1950, p. 852).

Family: Primulaceae (Myrsinoideae: Myrsinaceae)

A large assemblage of over 1000 species in about 50 genera distributed throughout the tropics and subtropics. In its present circumscription, several herbaceous genera formerly included in Primulaceae are now included here. The subfamily is characterized by free-central placentation and translucent, often elongated and resinous mesophyll cells.

The following taxa occur in, or in association with, mangroves:

1A. Fruit elongated, not fleshy, enclosing the single elongated seed; sunken epidermal glands present. . . . .*Aegiceras* (2 spp.)
1B. Fruit globose, fleshy, ripening black with one or more minute seeds; sunken epidermal glands absent. 2
2A. Anthers multilocellate, calyx lobes rounded. . . . .*Ardisia elliptica*
2B. Anthers not multilocellate, calyx lobes pointed. . . . .*Myrsine umbellulata*

> **Aegiceras** Gaertner 1788 [*Fruct. et Sem.* I: 216, pl. 46; based on *A. majus* Gaertn. (not *A. minus*)]

A genus of two species restricted to mangrove communities in the Asian tropics, with a distribution from India and Sri Lanka to South China and Hong Kong, through Malesia to the Philippines, New Guinea, and Australia, but not known from the Pacific islands.

The genus is distinctive within the family by virtue of its elongated (not rounded) capsular, dehiscent (not fleshy, indehiscent) fruit, and elongated seed without endosperm. It has distinctive multilocellate anthers, but these also occur in at least one species of *Ardisia* (*A. elliptica* Thunb.). A feature associated with the habitat is the sunken epidermal glands that excrete salt (Metcalfe and Chalk, 1950, p. 864). The discreteness of the genus has long been recognized, and de Candolle (1844, pp. 141–143) even included it as a separate "family," Aegiceraceae, in his *Prodromus*. However, these distinctive features largely relate to its unusual fruit and seed biology, which in turn may be adaptations to its habitat. In other characters, it resembles other Myrsinaceae closely (e.g., in the resinous cells), and it has the placenta with numerous embedded ovules in several series that characterizes the tribe Ardiseae, which is where it was placed earlier (Mez, 1902).

Two species are distinguished as follows (see Backer and Bakhuizen van den Brink, 1965):

1A.   Inflorescence consistently an umbel, with a short peduncle at most 5 mm long, flowers all on first-order branches. Pedicels 8–12 mm long, corolla tube 5–6 mm long, the mouth with a dense weft of fine hairs. Adult fruit strongly curved. Leaves large, up to 11 cm by 6 cm. . . . .*A. corniculatum*

1B.   Larger inflorescences racemose, with an extended peduncle up to 2 cm long, with the flowers on second-order branches. Pedicels 4–6 mm long, corolla tube about 4 mm long, with a thin weft of hairs at the mouth. Adult fruit only slightly curved. Leaves smaller, not more than 6 cm by 3 cm. . . . .*A. floridum*

Backer and Bakhuizen van den Brink record that the species differ in flower odor: *A. corniculatum* has sweet-scented flowers (hence the synonym *A. fragrans* König) and *A. floridum* has sour-smelling flowers. There is no information on floral biology or insect visitors that might account for these differences.

> **Aegiceras corniculatum** (L.) Blanco 1837 [*Fl. Filip.*: 79]
> *A. majus* Gaertn. 1788 [*Fruct. et Sem.* 1: 126. pl. 46, Fig. 1]
> *A. fragrans* König 1805, [in König and Sims. *Ann. Bot.* I: 131, Fig. 3]
> (Figs. B.46, B.47; Plates 8A, 9H)

Low evergreen tree or shrub growing to a height of 6 m. Bark smooth, dark gray. Shoots initially monopodial with articulate construction, forming resting terminal buds enclosed by bud scales, or later sympodial below terminal inflorescences. Leaves alternate, spirally arranged (rarely subopposite), stipules absent; petiole short (0.5–1.0 cm), terete but slightly two-keeled laterally; blade coriaceous, 4–8(–11) by 3–4(–6) cm, entire, elliptic to obovate, cuneate at the base, apex rounded to slightly emarginate. Surface often with salt crystals (Plate 8A), glabrous but with minute pustules turning black with age, and reddish punctate glands, especially on or near the slightly ribbed margin (Metcalfe and Chalk, 1950, p. 864). Midrib prominent below, often slightly red. Inflorescences as simple umbels, either terminating long shoots or on short leafy or leafless lateral shoots in the axils of foliage leaves, the older inflorescences resembling short shoots with numerous overlapping conspicuous flower scars. Bracts minute (1–3 mm), ephemeral; bracteoles absent. Flowers fragrant, perfect, pentamerous, pointed in bud with a slender pedicel 1–2 cm long. Calyx with five free imbricate contorted blunt and asymmetric lobes with surface glands like those of the leaves, remaining erect. Petals five, white, pointed, contorted, and always twisted to the left, fused basally to form a short tube 5–6 mm long, reflexed at maturity, with a dense series of hairs in the mouth of the corolla tube and shorter, capitate hairs at the base. Stamens five, opposite the corolla lobes; filaments about 3 mm long, united below into a short tube with a ring of internal and external hairs at the level of the mouth of the corolla tube; the facing surfaces of the stamen and corolla tube glandular basally. Anthers medifixed and versatile, at maturity extended horizontally above the reflexed petals, multilocellate, but dehiscing by longitudinal slits. Ovary 8 mm long, conical, with a single loculus, extended into a long solid, simple style beyond the

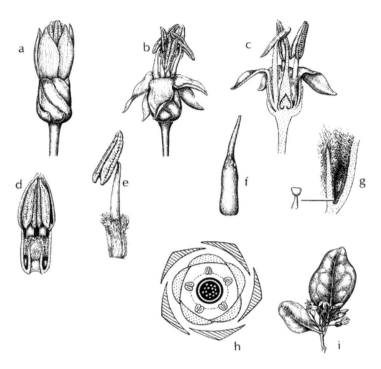

Figure B.46. *Aegiceras corniculatum* (Primulaceae) floral morphology. (a) Flower bud (x3) to show aestivation. (b) Open flower (x3), petals reflexed, stamens reflexing and dehiscing. (c) Flower in L.S. (x3), stamen to left completely reflexed. (d) Corolla from flower bud in L.S. (x4) to show corolla tube and attached stamen tube. (e) Single stamen (x4) dissected from stamen tube in bud position prior to becoming reflexed. (f) Ovary (x4). (g) Details of base of corolla tube (x8), with inset of single capitate hair from base of tube (x70). (h) Floral diagram. (i) Flowering axis (x1/2) with terminal umbel. (Material from Missionary Bay, Hinchinbrook Island, Queensland.)

mouth of the corolla tube, with pedicellate glands at the base of the style and a nectariferous area at the base of the ovary. Placentation free central, with numerous flattened ovules embedded in the rounded, but apically conical, somewhat fleshy, and shortly stalked placenta. Fruit 5–8 cm long, curved, with a persistent calyx, a one-seeded capsule pointed apically and filled at maturity by the embryo (i.e., without endosperm, as is usual in Myrsinaceae) and with an extended radicle, the whole attached laterally by a long funicle-like structure. Germination epigeal, immediately on release *(Rhizophora* type).

Distribution

This species has a wide distribution from Sri Lanka and India to South China, through the Malay Archipelago to the Solomon Islands and northern Australia, further south in New South Wales than in West Australia. It seems most characteristic of the outer

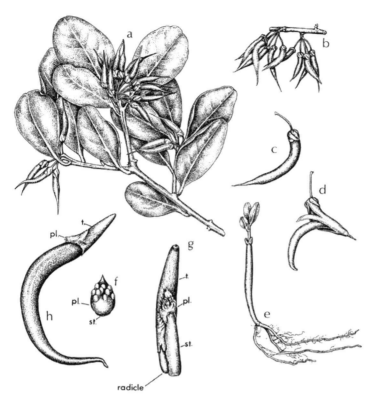

Figure B.47. *Aegiceras corniculatum* (Primulaceae) shoot, fruits, and seedling. (a) Habit with fruit (x1/2). (b) Older axis with fruiting clusters on short shoots (x1/2). (c) Single mature capsule (x3/4). (d) Dehiscent capsule extruding single seed (x3/8). (e) Germinating seedling (x1/2). (f) Placenta (pl.) (x9) showing numerous ovules and short placental stalk (st.). (g) Developing embryo (x3) with extended placental stalk (st.), flattened placenta (pl.), and embryonic radicle protruding from seed coat (t.). (h) Mature embryo (x3/2) with extruded hypocotyl and remains of seed coat (t.) attached to placental remains (pl.). (Material from various sources: (a) Semetan, Sarawak; (b–d) Botuma, Papua New Guinea; (e–h) Missionary Bay, Hinchinbrook Island, Queensland.)

(seaward) mangal fringe and occurs typically as an isolated low shrub, never forming a conspicuous part of the community.

Growth and Reproduction

The tree is initially monopodial in its construction, with articulate growth resulting from periodic rest of the terminal buds, which become enveloped by a series of reduced leaves as bud scales. Upon later regrowth, these leave a distinct series of scars when they fall. Branching is also rhythmic and seems loosely correlated with periodicity of extension growth so that tiers of branches may be formed. Each leaf subtends a single axillary meristem, which may grow out proleptically with a close series of basal bud-scale scars. Shoots may eventually terminate in umbels so that substitution growth occurs, with one branch eventually overtopping all others. Most inflorescences, however, terminate longer or shorter lateral shoots. Where these are leafless, they are recognized as axillary inflorescences.

Floral biology. Both the calyx and petals are contorted, always seemingly with the same direction of twist, which is usually described as "left-handed" (Fig. B.46a). Petals soon exceed the calyx lobes, expand, and become reflexed (Fig. B.46b, c). The anthers are more or less versatile (Fig. B. 46e) and become extruded as much by the extension of the stamen tube as by that of the free portion of the filament. The anthers become horizontal and bend toward the style, which they more or less enclose. Pollen grains are large. The hairs that close the corolla tube are short and stiff, some are gland-tipped, and there are additional sessile glands on the outside of the stamen tube and the inside of the corolla tube. One may speculate that these function as osmophores in view of the conspicuous floral fragrance. The flowers secrete nectar, apparently from the ring of tissue at the base of the ovary. Pollination experiments suggest that the species is predominantly pollinator dependent in fruit setting, even though flowers are self-compatible and capable of autogamy (Ge and Sun, 1999). Aluri (1990) has reported bees as visitors. There is considerable synchrony of flower development in one umbel, but an acropetal succession of flowers extends the flowering period.

Fruit and seed biology. *Aegiceras* is distinct in its fruit and seed morphology within the Myrsinoidae, although it has the same placentation as the related *Ardisia*. For example, seeds lack an oily endosperm. De Candolle (1841) first clearly described seed development and resolved the discrepancies in Gaertner's original description to which König first drew attention. Carey and Fraser (1932) provided a later account. It is clearly viviparous.

The placenta is central, shortly stalked, and somewhat conical apically. The sides of the placenta (Fig. B.47f) are covered by numerous semi-anatropous ovules, but in later development, the functional ovule is best described as campylotropous. In fertilized flowers after the corolla falls, one ovule enlarges and the others remain undeveloped. The developing ovule displaces the placenta laterally and the young seed elongates. An unusual feature is the elongation of the placental stalk, which has the same function as a funicle (Fig. B.47g, st.) but is topographically distinct from the ovular stalk. De Candolle makes an analogy between floral pedicel and peduncle in describing this structure. The structure had been referred to by Gaertner and several subsequent authors (e.g., König) as an "umbilical cord" or "umbilicle." The base of the fertilized ovule itself resembles a flattened aril. The fruit and seed develop further by intercalary growth, which is slightly unequal on opposite sides, resulting in the characteristic curvature of these organs (Fig. B.47h). Initially the seed coat covers it, but this is pierced by the elongating radicle in the lower (micropylar) region. This elongation of the embryo is parallel to the direction of extension of the umbilicle.

The seed coat remains complete around the distal (cotyledonary) end of the embryo (Fig. B.47h). Although de Candolle compared ovule development in

Aegiceras and *Rhizophora*, the major difference is in ovule orientation and the fact that the radicle of *Aegiceras* does not penetrate the fruit wall, as in *Rhizophora*. The propagule consists of the pedicellate fruit, which dehisces early to expose the green radicle, which curves away from the capsule wall. On the ground, the radicle penetrates the substrate and elongates to lift the plumule. The plumule itself extends through the fruit wall.

Genetic diversity. A study of genetic diversity in this species (Deng et al., 2009) has shown that this property is expressed most extensively among different populations, rather than within; a not surprising result and indicative of some outbreeding.

Aegiceras floridum Roemer and Schultes 1819 [*Syst. Veg.* 4: 512]
A. nigricans A. Rich 1834 [*Voy. Astrol. Bot.* 2: 57, Fig. 21]

Differing from *A. corniculatum* in the characters cited in the diagnostic key, this species tends to be somewhat diminutive in all its parts. The two species have not always been clearly distinguished, especially in the herbarium, but dried specimens of *A. floridum* can be recognized in the sterile state by the smaller leaves, which are characteristically dark beneath with a narrow light margin that is somewhat inrolled. The species has a much more restricted geographic distribution and is recorded from northern Borneo to Sulawesi and throughout the Philippines to Indochina. It is said to be uncommon, but because it may be overlooked, the range may be wider. Its ecological status in relation to the more common *A. corniculatum* has not been described, and it is not known whether the two either co-occur or exclude each other.

Ardisia SW. 1788 [*Prod. Veg. Ind. Occ.* 3: 48]

A more or less pantropical genus of some 250 species. The following is recorded in Southeast Asia as a mangrove associate in communities only occasionally inundated by the highest tides. Information about floral biology of *Ardisia* is discussed in Pascarella (1997a, b).

Ardisia elliptica Thunberg 1798 [*Nov. Gen. Pl.* 7: 119]
See also Trimen 1895 [*Fl. Ceylon* 3: 74]

This species is distributed widely from Sri Lanka and South India and Indochina to Java, Borneo, Sulawesi, and the Philippines.

A shrub growing to 5 m high, with spirally arranged entire elliptical leaves, about 10 cm by 4 cm, narrowed basally to a short (1 cm) petiole. The twigs are swollen at the base and easily detached. Flowers pentamerous, perfect, in axillary umbels or condensed racemes, about 1 cm in diameter. Calyx with five rounded, imbricate lobes; petals five, white or pink, pointed; stamens five, the anthers multilocellate. Ovary globose with a simple style, placentation free central. Fruit about 5 mm in diameter, with a globose few-seeded berry ripening black.

Myrsine L. 1753 [*Syst. Sp. Pl.*: 196]

A pantropical genus including some 10 species when considered in the strict sense, that is, distinct from *Rapanea*. The following species is recorded in Sarawak as a fringe mangal species, especially in regions transitional to heath forest; it has a limited distribution elsewhere in similar situations.

> **Myrsine umbellulata** A.DC. 1834 [*Trans. Linn. Soc.* 17: 135; see also 1844 *Pro-dromus* 8: 95]

A small tree growing to 7 m with spirally arranged ovate-elliptic leaves, about 8 cm by 3 cm, tapered to a short petiole (scarcely 5 mm long); apex rounded, flowers pentamerous, perfect, small, scarcely 3 mm in diameter, in sessile, condensed racemes, or umbels, on the older wood. Calyx cupulate, with five pointed lobes. Petals five, white, ephemeral; stamens five. Ovary globose with a simple style and free central placentation. Fruits globose, usually one-seeded berries, clustered on the older wood well below the leaf-bearing portion of the shoot, ripening black.

Family: Pteridaceae

A large and diverse family of the true ferns with about 35 genera and over 1000 species. It is divided into six tribes (Tryon and Tryon, 1982). The genus *Acrostichum* is often described as the "mangrove fern," although it is not restricted to mangal; it occurs in the back mangal but seems particularly suited to disturbed sites (Stokey and Atkinson, 1952). *Acrostichum* is included in the small subfamily Pteridoideae as a somewhat anomalous member with the sporangia covering the whole undersurface of the fertile pinna and not aggregated into sori, the "acrostichoid" condition (Plate 20D), shared with a number of unrelated ferns.

Fossil records indicate the occurrence of *Acrostichum* as early as the Eocene (Prado et al., 2007). However, these authors emphasize that extant ferns are of relatively recent origin and appear not to predate the angiosperms. Their analysis suggests that *Acrostichum* is sister to *Ceratopteris*, an essentially aquatic fern with unusual morphology. This clade is somewhat isolated and can be interpreted to be of relatively recent origin. The most significant feature of the sporophyte (and presumably the gametophyte) of *Acrostichum* is their significant tolerance of saline conditions.

> *Acrostichum* L. 1753 [*Sp. Pl.* 2: 1067, 73]
> (Fig. 5.11; Fig. B.48, 49; Plate 20)

Rhizomatous fern, common and often dominant in the understorey of the back mangal (sometimes inland in freshwater sites) and opportunistic in disturbed estuarine sites. Axis horizontal or erect, irregularly branched with a terminal cluster of erect, once-pinnate leaves up to 3 m long, with a terminal leaflet. Pinnae up to 40 cm by 6 cm, stalked (almost sessile distally), entire, narrowly oblong or lanceolate, the apex narrowly acute to abruptly acuminate or rounded. Basal pinnae sometimes reduced to distant spines along the petiole (Fig. B.48b). Venation reticulate, with uniform elongate areoles diverging from the thickened midrib, without free vein endings. Scales broad, restricted to base of frond, each growing to 1 cm long, with a thickened multicellular middle, margin entire. Fertile fronds with all or only a few distal pinnae fertile. Fertile pinnae without sori or indusia, the lower surface uniformly covered with sporangia mixed with capitate paraphyses, sporangia large; spores large, tetrahedral.

A pantropical genus; in the subsequent description, three species are recognized, as is traditional, but careful scrutiny of the genus throughout its range may reveal more. Individuals are very plastic in their development, and there are few reliable diagnostic characters. The species in the New World tropics seem much more clearly circumscribed than they are in the Asian tropics. Troll (1933b) provides a detailed comparison of features in which *A. aureum* and *A. speciosum* differ, especially in the sequence of leaf forms from juvenile to adult, which had caused earlier taxonomic confusion.

The following key outlines significant differences:

1A. Leaf usually at least 1 m (to 3 m long), apex of sterile pinnae rounded or truncate, at most abruptly acuminate; juvenile leaves with an oblong, blunt blade 2

1B. Leaf commonly less than 1 m long; apex of sterile pinnae narrowly acuminate; juvenile simple leaves with a lanceolate, pointed blade . . . .*A. speciosum*
      (Restricted to tropical Asia)

2A. Fertile fronds (Fig. B.48a) with only upper pinnae fertile (up to five pairs plus the terminal pinna: Plate 20B). Pinnae few, not more than 30, rather distant and often irregularly distributed, usually not overlapping, the lowest pinnae always distant, long stalked (up to 3 cm, Fig. B.48c). Areoles next to the midrib narrow, always three times longer than wide. Rachis rounded and smooth below, decidedly grooved above with the margin of the groove acute. Black spines (the midrib of aborted pinnae) frequent but distant on the lower part of the petiole (Fig. B.48b). Scales on petiole base not leaving a prominent scar, no scales up the leaf axis. Paraphyses (Fig. B.48f) with the terminal cell unextended, symmetric, the outline irregular and much lobed. Plant axis branched, horizontal . . . .*A. aureum*
      (Pantropical)

2B. Fertile fronds (Fig. B.49g) with all or most of the pinnae fertile (Plate 20C). Rarely only a few upper pinnae fertile. Pinnae many (40–60), closely set, overlapping and regularly arranged, often subopposite, the lowest pinnae relatively short stalked (less than 2 cm, Fig. B.49e). Areoles next to the midrib broad, never more than three times longer than wide. Rachis with several shallow grooves below, flat or scarcely grooved above, the margin of the groove blunt. Basal spines absent from petiole. Scales on petiole base leaving prominent scars. Scales also a little way up the petiole. Paraphyses (Fig. B.49i) with the terminal cell horizontally extended, often eccentric and with a smooth or little-lobed outline. Plant axis little branched, often erect . . . .*A. danaeifolium*
      (Restricted to tropical America and West Africa)

Tryon and Tryon (1982) indicate that the spores of *Acrostichum* represent a discrete type within the pteroid ferns; there is some variation in spore morphology within individual species and slight differences among species.

> ***Acrostichum aureum*** Linnaeus 1753 [*Sp. Pl.* 2: 1069, 73]
> (Fig. B.48)
> *Chrysodium aureum* Mett. 1856 [*Fil. Hart. Lips.*: 21]
> See Holttum 1954, *Flora of Malaya*, vol. 2, *Ferns of Malay*: 409.

This species is recognized by the few distal fertile pinnae on fertile fronds and by the shape of the paraphyses. In tropical America, it is readily distinguished from *A. danaeifolium*

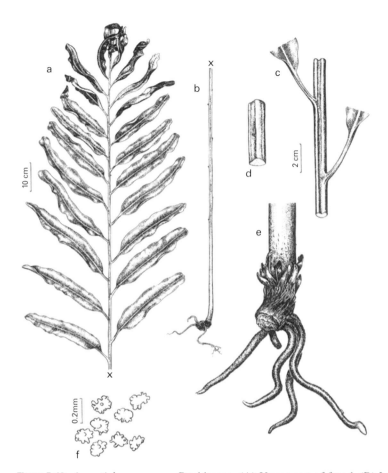

Figure B.48. *Acrostichum aureum.* Pteridaceae. (A) Upper part of frond. (B) Lower part of frond. (C) Middle of frond showing pinnule attachment. (D) Part of leaf axis with a marginal spine. (E) Leaf base with adventitious roots and scales. (F) Paraphyses. (After Adams and Tomlinson, 1979.)

(Adams and Tomlinson, 1979). The description in the key is largely based on American plants. In Southeast Asia, the young leaves have a characteristic crimson color.

Acrostichum danaeifolium Langsdorff and Fischer 1810 [*Ic. Fil.* 5 T1]
(Fig. B.49)

This species is restricted to the AEP region and readily recognized. Fertile fronds usually have all pinnae fertile; the pinnae are crowded, often overlapping, and relatively long stalked. It commonly occurs inland.

Acrostichum speciosum Willdenow 1810 [*Sp. Pl.* 5: 117]
See also Troll (1933b).

The distribution of this species is not well known, as it is imperfectly circumscribed from and commonly confused with *A. aureum.* It is described as being smaller in all its

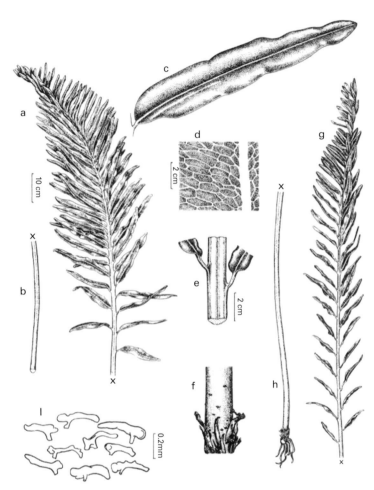

Figure B.49. *Acrostichum danaeifolium.* Pteridaceae (A) Upper part of sterile frond. (B) Unarmed axis of sterile frond. (C) Single pinna. (D) Areoles in midrib region. (E) Middle of frond showing pinnule attachment. (F) Base of leaf axis with persistent scale scars. (G) Upper part of fertile frond. (H) Lower part of fertile frond. (I) Paraphyses. (After Adams and Tomlinson, 1979.)

parts, with the sterile pinnae gradually tapering to a narrow point. Troll (1933b) emphasizes the difference between the early simple leaves of this species, which have a lanceolate, relatively short blade, and the long, oblong blade of *A. aureum*, as suggested in Plate 20E. It is the only species described for Australia (Duke, 2006).

Ecological and Geographic Distribution

The dimensions of individuals vary considerably according to the locality and there may be locally depauperate races. Jean-Marie Veillon (personal communication) refers to the fern on the tiny atoll of Ilot Matthew, southeast of New Caledonia, as being covered by a population (presumably of *A. aureum*) of plants all shorter than 25 cm. An autecological study of the genus, especially in relation to its taxonomy, needs to be done.

Although the fern is a characteristic element of back mangroves and associated tidally influenced estuarine communities, it is by no means restricted to them. Adams and Tomlinson (1979) comment that in South Florida, for example, *A. danaeifolium* extends into freshwater swamps and is commonly found inland in sink holes within hammocks (islands of forest). When *A. aureum* and *A. danaeifolium* grow together, the latter is more common and dominant (Plate 20A).

Acrostichum seems to have strong weedy tendencies and can be very aggressive in disturbed sites. In Vietnam, it is reported to suppress the regeneration of mangroves that were destroyed by military operations. These tendencies seem related to its high reproductive capacity and its light-tolerant or even light-demanding propensities. Watson (1928), Chai (1982), and others comment on its characteristic location on lobster mounds so that it grows beyond tidal influence although within tidal limits.

In the New World, *A. aureum* and *A. danaeifolium* have a somewhat similar range, although *A. danaeifolium* appears to have a wider distribution (e.g., it ranges farther north in South Florida, apparently in keeping with its larger stature and vigor). The two species often occur together, apparently without hybridization, although Garcia de Lopez (1978) reports intermediate forms between them in the Dominican Republic.

In tropical Asia, the situation is reversed. *A. aureum* is reported to be larger than *A. speciosum*, with fronds up to 4 m according to Holttum (1953), and has the wider geographic distribution and apparently greater ecological range. However, *A. speciosum* is apparently more restricted to saline environments; *A. aureum* can survive without regular tidal inundation.

Phenology

As presented by Mehltreter and Palcios-Ricos (2003), ferns have limited representation in phenological studies in the tropics. However, their study of *A. danaeifolium* in Mexico shows a distinctive seasonality in leaf production, which is correlated with rainfall; lowest in the dry winter months (October to March) and highest in the wet summer months (April to September). Leaf size and productivity overall is shortest in winter, with the maximum expansion of adult fertile leaves in summer. Leaf life span also varies, being longest (9.5 months) in sterile leaves and shortest (4.5 months) in fertile leaves. Individual leaves take 3–6 weeks to develop, with an average production of 14.5 sterile leaves per year. As a consequence, spore release is concentrated in the wet season, thus favoring spore germination.

Morphology and Reproduction

Acrostichum offers an unusual opportunity to study the population dynamics of plants in mangal, as it is rhizomatous and has a capacity for vegetative persistence and propagation different from that of all other members of the community, except *Nypa*, and it is sexually established from gametophytes via widely dispersed spores. The rhizome of *Acrostichum* has an extensive proliferative capacity because each leaf supports a branch meristem (Fig. 5.11); in fact, few of these develop so that rhizomes are rather infrequently branched and vegetative spread may be restricted. This restriction is clear in situations where the plant grows on a localized, specialized substrate, as mentioned on lobster mounds in the Malaysian mangroves. The extent of vegetative spread has not been investigated.

New individuals are readily established from spores via gametophytes, especially in disturbed sites (Plate 20E). This success is due, in part, to an appreciable salt tolerance by gametophytes (A. Juncosa, personal communication), although germination of spores seems most successful in fresh water. Sex-organ (antheridia and archegonia) ontogeny occurs in sequence so that outcrossing is promoted, but genetic analysis reveals frequent selfing (Lloyd and Gregg, 1975; Lloyd, 1980). This may account for the absence of hybrids.

Family: Rhizophoraceae

A small pantropical family of about 16 genera and 120 species of trees and shrubs, mainly in the Old World (Schwarzbach, 2014). For a tropical family, it has been surprisingly well studied partly because of the ecological importance of its mangrove genera (Raven and Tomlinson, 1988). Traditionally, it had been included in the Myrtales and a presumed relationship with the Anisophylleaceae, as reviewed by Dahlgren (1988). Critical evaluation, based primarily on molecular evidence, confirms the monophyly of Rhizophoraceae, but now places it in Malpighiales sister to Erythroxylaceae, perhaps somewhat surprisingly compared with earlier views (Schwarzbach and Ricklefs, 2000), whereas Anisophylleaceae is placed in Cucurbitales. Within the family, three or four tribes are recognized. Of them, the Rhizophoreae Blume, which includes only the four exclusively mangrove genera *Bruguiera*, *Ceriops*, *Kandelia*, and *Rhizophora*, is universally recognized as a natural and discrete taxon. It is convenient therefore to refer to the tribe as the "mangrove Rhizophoraceae." The uniformly viviparous condition is particularly distinctive. All reported chromosome counts are n=18 (Sidhu, 1968; Yoshioka et al., 1984).

The following account is based partly on that of Ding Hou (1958), but much up-dated from extensive subsequent fieldwork together with the analyses presented in a comprehensive series of papers in the *Annals of the Missouri Botanic Gardens* 75: 1988 1258–1425, and in more detail, of Behnke (1988), Juncosa and Tomlinson (1988a, b), and Keating and Randrianasolo (1988).

(Figs. B.50–61; Plates 14, 15, 21–24)

I. *Rhizophoraceae (Tribe Rhizophoreae): Key to Genera based on Characters of the Flowers*

1A.   Flowers tetramerous, calyx four-lobed; petals four, lanceolate, entire, lacking distal appendages, not enclosing pairs of stamens. Anthers sessile, multilocellate, dehiscing by an adaxial valve (Figs. B.53–56). *Rhizophora*

1B.   Flowers five- (or more, up to 16) merous, calyx with five or more lobes; petals deeply emarginate, with apical appendages, sometimes enclosing pairs of stamens. Anthers with long (short in *Ceriops decandra*) filaments, four-locular, dehiscing by slits. 2

2A.   Stamens numerous, more than twice as many as calyx lobes, petals bilobed with a long seta in the sinus, each lobe multifid, without marginal hairs. Petals not enclosing a stamen pair. Hypocotyl of seedling slender, gradually narrowed and distally pointed (Fig. B.61). . . . . *Kandelia*

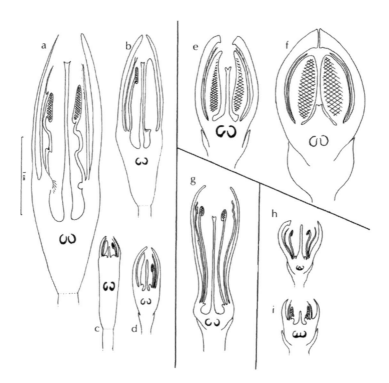

Figure B.50. Rhizophoraceae flower size. Longitudinal sections of flower buds, at a uniform scale, show the size range: (a–d) *Bruguiera*: (a) *B. gymnorrhiza*. (b) *B. exaristata*. (c) *B. parviflora*. (d) *B. cylindrica*. (e–f) *Rhizophora*: (e) *R. stylosa*. (f) *R. apiculata*. (g) *Kandelia candel*. (h–i) *Ceriops*: (h) *C. tagal*. (i) *C. decandra*. Extent of pollen-producing tissue indicated by crosshatching. (From Tomlinson et al. 1979.)

2B.   Stamens twice as many as calyx lobes. Petals with marginal hairs, each enclosing a pair of stamens (except *Ceriops decandra*). Hypocotyl of seedling blunt or apically pointed distally (Figs B.57–60)..3

3A.   Calyx eight to 16-lobed, lobes lanceolate, pointed, up to 15 mm long; petals 0.5–2 cm long. Stamens 16–32. Hypocotyl of seedling not or scarcely ridged (Figs. B.57, 58)... *Bruguiera*

3B.   Calyx five- or six-lobed, lobes blunt, 2–4 mm long; petals less than 0.5 cm long. Stamens 10 or 12. Hypocotyl of seedling more or less ridged (Fig. B.59)... *Ceriops*

II. *Rhizophoraceae (Tribe Rhizophoreae): Key to Genera based on Vegetative Characters*

1A.   Trees with extensively developed aerial trunk-borne stilt roots forming a looping or pendulous complex; leaves black dotted below (at least in dried material). (Trichosclereids present).. *Rhizophora*

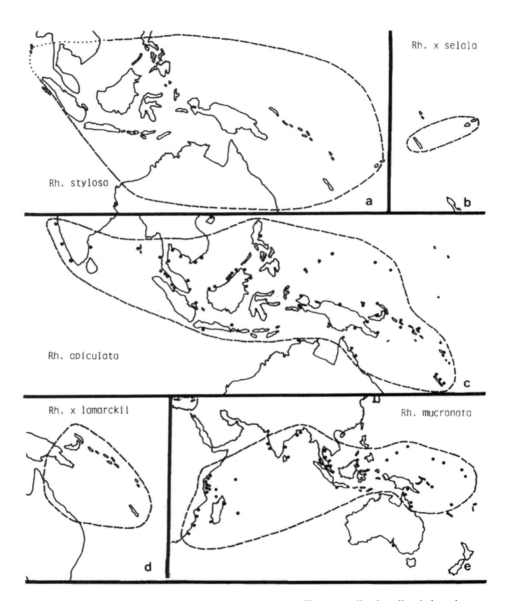

Figure B.51. Distribution of eastern species of *Rhizophora*. The generalized outline is based on field identification or herbarium records checked by the author. These are indicated by the dots in the distribution of *R. apiculata* and *R. mucronata*. (a) *R. stylosa*. (b) *R.* x *selala* in New Caledonia and Fiji. (c) *R. apiculata*. (d) *R.* x *lamarckii*. (e) *R. mucronata*.

1B.   Trees with extensively developed aerial roots (pneumatophores) of submersed origin, but not trunk-borne; leaves without black dots. (Trichosclereids absent). 2

2A.   Trees usually without pneumatophores, at most with inconspicuous, slender, negatively geotropic lateral roots. . . . . *Kandelia*

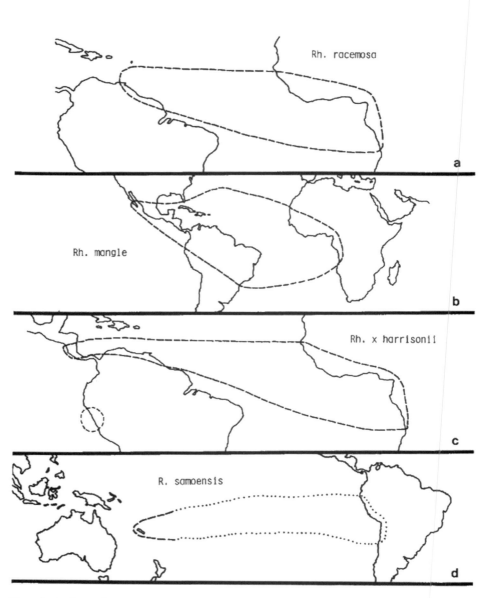

Figure B.52. Generalized distribution of western species of *Rhizophora*. (a) *R. racemosa*. (b) *R. mangle*. (c) *R.* x *harrisonii*. (d) *R. samoensis*, morphologically almost indistinguishable from *R. mangle*, known from New Caledonia and Fiji. The dotted extension of its range to South America is hypothetical.

2B.   Trees with knee-like (rounded and knobby) pneumatophores developed by looping of horizontal roots. 3

3A.   Leaves usually >10 cm long, acute at apex, terminal shoot buds not flattened. . . . . *Bruguiera*

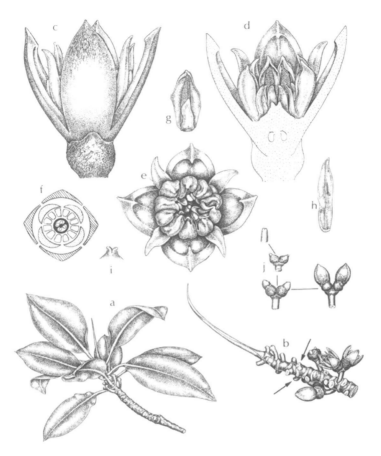

Figure B.53. *Rhizophora apiculata* (Rhizophoraceae) leaves and flowers. (a) Shoot with developing inflorescences (x1/4). (b) Flowering shoot with leaves removed (x1/4), stipules enclosing apex; open flowers only below level of insertion of oldest leaves (arrows). (c) Flowers from side (x3). (d) Flower in L.S. (x3). (e) Flower from above (x3). (f) Floral diagram, stamen number fairly constant. (g) Dehisced stamen (x3). (h) Petal (x3). (i) Stigma at receptive stage (x3). (j) Sequence of developmental stages (x1/2) of inflorescence (flower pair). (Material from Singaua, Lae, Papua New Guinea.)

3B. Leaves usually <10 cm long, rounded at apex, terminal shoot buds flattened. . . . . *Ceriops*

General Features of Rhizophoreae

Phyllotaxis. Although the phyllotaxis of this group is usually described as decussate (i.e., strictly with successive leaf pairs at 90° to each other), the successive pairs of leaves are not strictly at right angles to each other, as is evident from the arrangement of

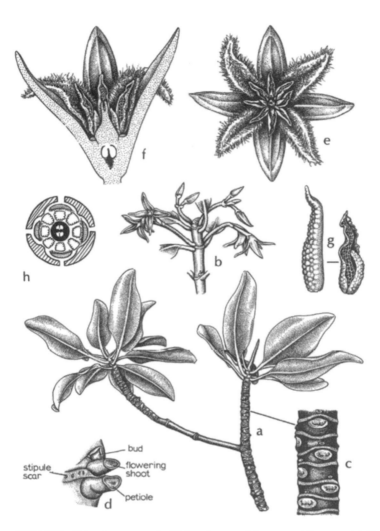

Figure B.54. *Rhizophora mangle* (Rhizophoraceae) leaves and flowers. (a) Shoot (x3/8) branching by apposition. (b) Part of flowering shoot with axillary inflorescences (x1/2). (c) Detail of scar pattern on terminal short shoot (x1) (cf. part [d]). (d) Node with leaf and flowering shoot cut off (x2), vegetative bud above inflorescence. (e) Flower from above (x3). (f) Flower in L.S. (x3). (g) undehisced and dehisced stamen (x4). (h) Floral diagram. (From Tomlinson, 2001.)

leaf scars on shoots with congested internodes. Tomlinson and Wheat (1979) have shown that the arrangement seen on mature shoots is primary (Fig. 5.9); that is, it represents the position of the initiation of leaf pairs by the apical meristem and is not the result of secondary twisting of the internode. The phyllotaxis is therefore strictly bijugate; that is, the pairs of leaves diverge from each other at an angle less than 90°. This arrangement has important architectural consequences, as mutual shading is

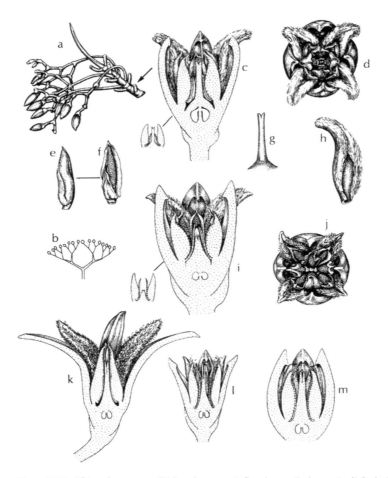

Figure B.55. *Rhizophora* spp. (Rhizophoraceae) floral morphology. (a–j) Indo-Pacific species. (a–h) *Rhizophora stylosa*: (a) shoot with leaves detached (x1/2), inflorescences borne within leafy shoot (cf. *R. apiculata*), lowest leaf on shoot at arrow. (b) Diagram of 16-flowered (2^4) inflorescence with branches shown in one plane. (c) L.S. flower (x3); inset shows relation of style to dehisced stamens. (d) Flower from above (x3). (e, f) Undehisced and dehisced stamen (x6). (g) Style x 3). (h) Single petal (x3). (i–j) *R. mucronata*: (i) Flower in L.S. (x3); inset shows relation of style to dehisced stamens. (j) Flower from above (x3). (k–m) Atlantic–Caribbean species, L.S. flower (x3): (k) *R. mangle*. (I) *R* x *harrisonii*. (m) *R. racemosa*. (Material sources: *R. stylosa*: Bootless Bay, Port Moresby, Papua New Guinea; *R. mucronata*: Semetan, Sarawak, P. Chai; *R. mangle*: Fairchild Tropical Garden, Miami, FL; *R.* x *harrisonii, R. racemosa*: Monrovia, Liberia, West Africa, David de May.)

reduced (Fig. 5.10) and branches diverge at varying angles. Bijugate phyllotaxis may, in fact, characterize most members of the family Rhizophoraceae.

Stipular and bud morphology. The arrangement of leaves in the buds in the Rhizophoreae (and many terrestrial Rhizophoraceae) is constant and characteristic. Each leaf pair has an associated pair of stipules that stand above the leaf insertion so

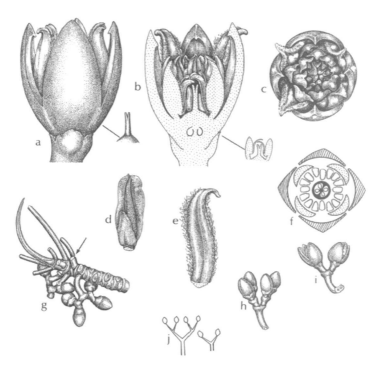

Figure B.56. *Rhizophora* x *lamarckii* (Rhizophoraceae) floral morphology. (a) Flower from the side (x3); inset: detail of style (x3). (b) Flower in L.S. (x3); inset: relation of style to dehisced stamens. (c) Flower from above (x3). (d) Dehisced stamen (x3). (e) Petal (x3). (f) Floral diagram, stamen number is not constant. (g) Shoot with leaves detached (x1/2), mature inflorescences just within or below leafy crown (arrow, position of oldest leaf in crown). (h) Four-flowered inflorescence. (i) Two-flowered inflorescence. (j) Diagram of two- and four-flowered inflorescences with branches shown in one plane. (Material from Barune, Port Moresby, Papua New Guinea; in Tomlinson and Womersley, 1976.)

that the stipules do not enclose the associated leaf but rather the younger appendages, including axillary appendages at the same node (inflorescences, branches, buds). Each stipule is a folded structure open on the ventral side, but encircling the node at its insertion. Adjacent margins of opposite stipules overlap the same way, so that the left side of one is overlapped by the right side of the other; that is, only one side of each stipule is exposed (Fig. 5.8). The leaf blade, which is more tightly rolled than the stipule, overwraps in the same way. Within the base of each stipule is a series of glandular colleters (Fig. 5.8; Figs. B.59e, B.61f).

A feature of the buds of Rhizophoreae is the secretion of fluid (the viscosity varies with the taxa), apparently emanating from this palisade of colleters. Bud components are usually loosely packed and the secretion can fill space that otherwise would be unoccupied, notably in *Ceriops*. The color and consistency of the exudate are variable and somewhat diagnostic for different species. It can be a rather milky fluid in *Rhizophora*, but a more resinous material that dries to a varnish in *Ceriops*. Primack

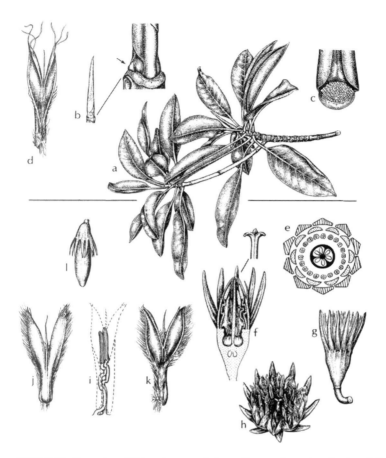

Figure B.57. *Bruguiera* (Rhizophoraceae): large-flowered species shoot and flowers. (a–d) *Bruguiera gymnorrhiza*: (a) Leafy shoot (x1/4) showing apposition growth. (b) Lower: terminal bud (x1/2) with expanded leaves removed to show stipule pair; upper: detail (x3) with stipules removed to show axillary bud (arrow) and palisade of colleters. (c) Detail of detached stipule from inside (x3) to show colleters. (d) Petal from inside, partly unfolded (x3) to show diagnostic features of lobe tips. (e–l) *Bruguiera sexangula*: (e) Floral diagram. (f) Flower in L.S. (x3/2), the petals enclose the stamens; inset: detail of stigmas (x3). (g) Flower from side (x3/2). (h) Flower from above (x3/2), all petals "exploded." (i) Stamen pair (x3/2) with outline of enclosing petal. (j) Petal from abaxial (outer) side (x3/2). (k) Petal from adaxial (inner) side (x3/2). (l) Young fruit (x1/2). (*B. gymnorrhiza* cultivated at "The Kampong," Douglas Road, Miami, FL; *B. sexangula* from Bootless Bay, Port Moresby, Papua New Guinea, in Tomlinson et al., 1979.)

and Tomlinson (1978) have observed honeyeaters licking this fluid exudate from the buds of *Rhizophora stylosa*, and suggest that this encourages the visits of the birds, which also eat predatory insects. In a detailed systematic study of the whole of the Rhizophoraceae, Sheue et al. (2013) discuss stipule form and colleter structure in great detail. They conclude that those of the Rhizophoreae are not only the largest, but also the most highly derived in a phylogenetic sense. *Kandelia* is distinctive in the presence

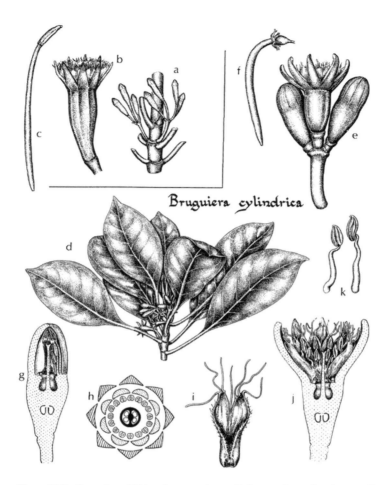

Figure B.58. *Bruguiera* (Rhizophoraceae): small-flowered species, leaves, flowers, and seedlings. (a–c) *Bruguiera parviflora*: (a) Part of shoot with leaves removed (x1). (b) Flower from side (x3), petals all "exploded." (c) Mature detached seedling (x1/2). (d–k) *Bruguiera cylindrica*: (d) Flowering shoot (x1/2). (e) Three-flowered cymose inflorescence (x3). (f) Mature detached seedling (x1/2). (g) Unopened flower in L.S. (x4). (h) Floral diagram. (i) Petal (x4) after pollen release. (j) Flower in L.S. (x4) after petals have "exploded." (k) Antepetalous (left) and antesepalous (right) stamens (x9). (*B. parviflora* from Labu Lagoon, Lae, Papua New Guinea; *B. cylindrica* from Tarauma Link Road, Port Moresby, Papua New Guinea, in Tomlinson et al., 1979.)

of some vascular tissues in the colleters. In terrestrial members, which occupy different tribes, the tribe Macarisieae is regarded as plesiomorphic in its stipular simplicity.

The stipule color is somewhat variable, even within a species; it may be red in *Rhizophora* (e.g., *R. apiculata*) and especially in *Bruguiera gymnorrhiza*. The stipules are always ephemeral and abscise cleanly as soon as the enclosed organs begin to expand. This leaves an annular stipule scar, which in older shoots stands above the petiole scar but below the scar of branch structures or dormant buds (Fig. B.54d).

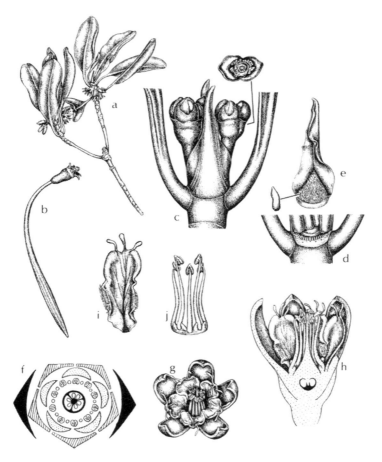

Figure B.59. *Ceriops tagal* (Rhizophoraceae) flower, fruit, and seedling. (a) Flowering shoot (x1/2). (b) Mature detached seedling (x1/2). (c) Details of terminal bud with expanding inflorescences (x3); inset: cupular bracteoles with flower removed (x4) to show colleters. (d) Detail of terminal bud (x3) as in part (a) but with stipules removed to show palisade of colleters. (e) Stipules from within (x3); inset: single colleter (x20). (f) Floral diagram. (g) Flower from above (x6), petals all "exploded." (h) Flower in L.S. (x6). (i) Petal (x10). (j) Stamens and presumed nectaries (x10). (Material from Barune, Port Moresby, Papua New Guinea, in Tomlinson et al., 1979.)

In shoots with short internodes, the helical arrangement of the petiole scars, which results from the bijugate phyllotaxis, is very clear. Ding Hou (1960) and Kenneally et al. (1978) have commented independently on the following diagnostic differences between *Rhizophora* and *Bruguiera*, based on the different arrangement of the vascular bundles that appear on the petiole scar:

1A. Leaf   scars   with   several   vascular   bundles,   arranged   in   two rows. . . . . *Rhizophora*
1B. Leaf scars with one series of three bundles. . . . . *Bruguiera*

Figure B.60 *Ceriops* cf. *tagal* (Rhizophoraceae). Mature root system. Dead plant on the left shows the flanged, flattened aerial roots. Townsville, Queensland. (From a color transparency by A.M.G.)

This difference is the result of the earlier dispersion of the leaf traces in *Rhizophora* because the nodal anatomy is otherwise similar in all members of the Rhizophoraceae. The configuration of nodal scars has been used to estimate seedling age (Duke and Pinzon, 1992).

Terminal bud construction in the Rhizophoreae is further characterized by the small and constant number of appendages within the exposed pair of enveloping stipules, to a maximum of three nodal sets (Gill and Tomlinson, 1971b). As a pair of leaves expands out of the terminal bud, they are replaced by a new pair, initiated by the shoot apex. The rate of expansion fluctuates considerably, however, and buds can remain relatively inactive for extended periods, neither initiating new primordia nor expanding older ones. The bud retains the capability to develop and expand new appendages at any time, depending on exogenous and endogenous factors. There is neither periodic accumulation of appendage primordia to form a resting bud nor their subsequent rapid expansion in a pronounced "flush" of growth, as occurs in many tropical trees. The size and shape of buds vary in different species, but are somewhat diagnostic and largely dependent on stipule size. Stipules may be as long as 8 cm in *Rhizophora apiculata*, but less than 3 cm long in *Ceriops* and the small-flowered species of *Bruguiera*. *Ceriops* has characteristically flattened buds with the dorsal sides almost angular. As mentioned, in *Ceriops*, the developing primordia are not close-packed and there is ample free space, to a large extent occupied by colleter secretions.

Figure B.61. *Kandelia candel* (Rhizophoraceae) leaves, flowers, and seedling. (a) Flowering shoot (x1/3). (b) Inflorescence (x1/2). (c) Mature detached seedling (x1/3). (d) Floral diagram. (e) Terminal bud represented by enveloping stipule pair (x3/2); inset: vernation of vegetative parts at one node; stipules cross-hatched, leaves and branches solid. (f) Bud in part (e) with stipules removed (x3); inset: detail of detached stipule from within to show colleters. (g) Flower from side (x3/2). (h) Flower in L.S. (x3); inset: detail of stigma (x6). (i) Petal (x4). (Material from Semetan, Sarawak, P. Chai; from Tomlinson et al., 1979.)

Architecture and Crown Shape

Although the Rhizophoreae have a basically uniform shoot construction, they are somewhat diverse architecturally, have some differences in reiterative responses, and consequently show different crown shapes. Constant features include continuous growth (in the broad sense), as there are no obvious resting terminal buds and unbranched shoots lack obvious discontinuities. Continuous growth can result in continuous branching, but more usually branching is discontinuous.

Sapling growth is characterized by an initial unbranched phase followed by diffuse branching. The extent of the unbranched phase varies and seems dependent on general vigor. In dense shade, plants up to 3 m tall with short internodes can develop. In the branched phase, intermittent tiers of branches develop, sometimes rather regularly. Branching is always by syllepsis (Fig. 5. 10); reserve buds are produced but they seem never to contribute to crown development. In *Rhizophora mangle*, for example, reserve buds are known to last for three years at a maximum, hence the tree has no capability for sprouting from older wood. This seems a general property of *Rhizophora*, which has a rather uniform physiognomy with no specific architectural variation. The inherent deterministic features of crown shape nevertheless allow the development of a consider- able plasticity in crown shapes in which branches may revert from a plagiotropic to an orthotropic orientation.

In the tribe Rhizophoreae, there are two architectural models (that of Attims and Aubréville, following the system of Hallé et al., 1978), but the physiognomy of the crown has three fairly distinct types.

1. *Attims' model (Rhizophora).* In the definition of this model, the lateral axes are, at least initially, orthotropic, whereas growth and branching are continuous or diffuse. Sylleptic branches are produced, usually in tiers of one to three pairs at successive nodes, separated by about the same number of branch-free nodes. The orthotropic nature of the branches is at first apparent and they tend to repeat the branching pattern of the parent tree, but as the branch ages and extends, it reclines, with the branches becoming restricted to the lower side and plagiotropy is gradually imposed. Progressively, in axes of two or three successive orders, the distal portion of the axis develops short internodes and eventually functions as a terminal short shoot, which is more or less abruptly erect. In this way, plagiotropy by apposition is gradually achieved, leading to a branch complex that has much the same physiognomy as that characteristic of Aubréville's model.

2. *Attims' model (Kandelia, Ceriops, small-flowered Bruguiera species).* These correspond to the preceding growth pattern, but the axes remain much more strongly orthotropic, and plagiotropy by apposition is little developed.

3. *Aubréville's model (large-flowered Bruguiera species).* In the definition of this model, the trunk axis shows continuous or rhythmic growth, but always with the production of regular tiers of branches that show plagiotropy by apposition, with regular and pronounced sympodial development (Fig. 5.3). The plagiotropy in *Bruguiera* is not as precise as the clear pattern described for *Terminalia*, but the same physiognomy results, with erect terminal short shoots substituted by extended lateral axes, each of which in turn repeats the process. Proliferation of the system results when two, instead of one, lateral renewal shoots develop. The greater precision of this pattern in *Bruguiera* compared with *Rhizophora* is best appreciated when the young leader shoot complexes of the two types are com- pared; the abrupt plagiotropy of the former is then contrasted with the gradual plagiotropy of the latter.

Reiteration. *Bruguiera* more regularly conforms to its architectural model because reiteration seems difficult and results in shoots in which a direct substitution of the leader is most common. Consequently, the trunk remains single and the crown shape narrow and conical. In *Rhizophora*, because of the strong branch orthotropy, the leader is readily replaced by one or more existing laterals. Several leader complexes can be developed; the crown becomes open, broad, and irregular (Fig. 5.1C). The reiterative ability of *Rhizophora* is best revealed in low scrubby plants in marginal habitats, with numerous erect axes and asymmetry of the compact crown. It is important to emphasize that prolepsis (development of shoots from dormant buds) plays no role in this plasticity of crown development.

Differences between the two major contrasted tree forms are inherently subtle, but the resultant crown shapes are readily distinguished where the trees grow together. Quantification of the differences has not been attempted, but one can speculate that the more plastic method of organization in *Rhizophora* accounts, in part, for its greater success. The difference in the root system of *Rhizophora* compared with other members of the tribe may aid in this differential success.

Floral Biology

The floral biology of mangrove Rhizophoraceae is complex and best understood in a comparative context (Juncosa and Tomlinson, 1987). The diversity in flower size, reflecting a diversity of mechanisms, is encapsulated in Fig. B.50. The subject has been dealt with in some detail by Tomlinson et al. (1979), which includes only an account of the different kinds of floral mechanisms and the kinds of flower visitors, with some reasonable inference about the likely method of pollen transfer. There is very little information about incompatibility and other isolating mechanisms because no experimental work has been carried out.

The contrasted pollination mechanisms are summarized in the following key:

1A. **Pollination predominantly by wind** (anthers dehiscent in the flower bud). 2
2A. Mature flowers below the leafy cluster. Petals without hairs. . . . . *R. apiculata*
2B. Mature flowers within the leafy cluster. Petals hairy. . . . .
. . . . *Rhizophora* (other species)
1B. **Pollination by animals** (only the usual and presumed most effective pollinator is included; flowers may receive a diversity of visitors). 3
3A. Pollination mechanism explosive (anthers dehiscent in the flower bud). 4
3B. Pollination mechanism not explosive (anthers not dehiscent in the flower bud). 7
4A. Pollination by day-time visitors. 5
4B. Pollination by night-time visitors. 6
5A. Pollination by birds, large-flowered *Bruguiera* spp. (*B. exarista, B. gymnorrhiza, B. sexangula*)
5B. Pollination by butterflies, small-flowered *Bruguiera* spp. (*B. cylindrical, B. hainesii*,* *B. parviflora*)

6.   Pollination by moths. . . . . *Ceriops tagal*
7A.   Pollination presumably by small, short-tongued insects. . . . . . *Ceriops decandra*
7B.   Pollination presumably by fairly large, long-tongued insects. . . . . .*Kandelia candel*

*, The record by Noske (1993) of this as bird pollinated seems anomalous.

Wind pollination in Rhizophora. Evidence for this mechanism comes from the floral syndrome and from the observation that *Rhizophora* has an appreciably larger pollen/ovule ratio (the ratio of the total number of pollen grains to the total number of ovules per flower). When this value is high, it has been used as an indication that a plant is pollinated by wind, as the method of pollen transfer is very wasteful of pollen (Cruden, 1977). High pollen/ovule ratios on a relative scale are characteristic of familiar wind-pollinated taxa. For example, *Rhizophora* has a ratio as much as an order of magnitude higher than the ratio in other genera. Kondo et al. (1987) report a minute amount of nectar (compared with *Bruguiera*) but emphasize the similarity of the dry pollen in both genera. The stigma is also papillose in both and secretes mucilage, which retains pollen. Both are said to be self-fertile.

Other characters that are indicative of wind pollination are the brief pollen presentation time and the short functional life of the flower (Plate 22C, D). The anthers dehisce before the flower bud opens and, in all species except *R. apiculata*, are drawn out with the marginal hairs of the petals as they expand. The petals fall within a day. Flowers are also typically pendulous. In *R. apiculata*, which lacks petal hairs (Fig. B.53h), the flowers point downward at maturity and also mature below the leafy crown, so that the foliage does not hinder pollen dispersal. There is no copious secretion of nectar in any species of *Rhizophora*. In addition, the style of *Rhizophora* is not elaborated in the manner typical of wind-pollinated flowers; nevertheless, flowers are visited by insects (e.g., bees), although these may only seek pollen, whereas thrips are common in even freshly opened flowers. It is possible that the pollination mechanism is being transformed from an original animal-pollinated ancestral condition (which persists in all other genera of the tribe) to a wind-pollinated condition. In this way, *Rhizophora* escapes the competition for animal pollination that otherwise predominates in mangrove communities, as we have discussed (Chapter 9).

The abundance of *Rhizophora* pollen in marine sediments indicates the productivity of the tree in this respect, and does not conflict with the idea of the plant as a wind-pollinated species.

Unspecified pollination by insects. This is characteristic of *Kandelia* and *Ceriops decandra*. *Kandelia* has a generalized flower but a well-developed calyx cup into which nectar is secreted. *C. decandra* differs from *C. tagal* because it lacks the elaborate pollen-discharge mechanism of that species. It is unusual to have such contrasted pollination devices in so small a genus.

Explosive pollen discharge and animal pollination. The original description of the explosive mechanism was by Gehrmann (1911) and described in *Bruguiera sexangula* (as *B. eriopetala*).

This group includes *C. tagal* and all six species of *Bruguiera*. The petals enclose pairs of stamens; the antisepalous set has a characteristic basal twist of the filament to allow this. Opposite margins of the petal are held together by a set of interlocking hairs. The anthers dehisce precociously so that, at the time the flower opens, each unexploded petal pouch includes loose pollen. The petals are under tension, generated during floral expansion; the tension is released suddenly by a slight touch to the base of each petal, such as is provided by the probing mouth or beak of a flower visitor searching for nectar in the floral cup. This explodes the petal, which unzips instantly, scattering a cloud of pollen on the head of the visitor. Each complex explodes independently so that unexploded and exploded petals may occur in a single flower (Plate 23B). Consequently, pollen discharge can involve more than one visitor; the presentation time, in the absence of a visitor, is several days. It could be argued that the mechanism is designed simply to disperse pollen freely for wind pollination, but this seems unlikely: first, in view of the nectar, which is copious in larger flowers; second, in the elaborateness of the mechanism; third, in the usual inability of petals to explode in the absence of precise triggering by a visitor. Scented flowers also attract visitors. These conclusions are verified by Kondo et al. (1987).

Although the mechanism applies to all seven species that have this structure, individual species are designed to attract a given class of visitor because of their flower size and orientation and time of anthesis. The large-flowered species of *Bruguiera* (*B. exaristata*, *B. gymnorrhiza*, and *B. sexangula*) are pollinated by birds, and have downward-pointed flowers. Flowers open in the early morning, coincident with the maximum time of bird visitation. Typical visitors are migrant honeyeaters (*Microscelis* and *Zosterops*), which feed from below from pendulous flowers on the nectar retained in the hypanthium cavity (capacity 61.3 µl). Insect visitors are too small to trigger the mechanism. Nectar can be produced for up to a month, although pollination (presumably pollen discharge) may stop nectar secretion.The smaller-flowered species *(B. cylindrica*, *B. hainesii*, and *B. parviflora)* are pollinated mainly by day-flying insects (e.g., butterflies) and have erect or divergent inflorescences. *C. tagal* is pollinated by small night-flying moths. The clear impression is that competition is minimized by this division of labor.

Isolating mechanism. Some degree of dichogamy, involving weak protandry, seems characteristic of most species, although the time of stigma receptivity has not been established precisely. There is evidence that *Rhizophora* is self-compatible from the ability of cultivated plants to set fruit, but no careful and extensive experimental work has been done. On theoretical grounds, Primack and Tomlinson (1980) suggest that mangrove plants should be self-compatible, as this is a characteristic of pioneering species that may become established as isolated individuals in new environments remote from parental sources (Table 2.2). The available evidence does not conflict with this view.

Seedling Anatomy

In studying the structure of fossilized propagules of *Ceriops cantiensis* Chand. in the London Clay flora (Eocene, Lower Tertiary), Wilkinson (1981) contrasted the hypocotyl anatomy of several species in the tribe Rhizophoreae and summarized older literature. She notes that *Rhizophora* can be distinguished immediately by the spherical masses of stone cells in the outer cortex and the multiradiate sclereids in the cortex and pith (the latter corresponding closely to those found abundantly in other vegetative parts). Differences in epidermal features and cuticle texture are said to distinguish other taxa. That *Ceriops* can be recognized unequivocally from deposits of this age is a further clear indication of the ancestry of mangrove Rhizophoraceae. Wilkinson also comments on the absence of pollen of Rhizophoraceae from the London Clay flora. This could be evidence that the mangrove propagules were delivered to the site as flotsam rather than for the existence of mangal *in situ*. Further information on the systemic anatomy of seedlings of Rhizophoraceae is summarized by Tomlinson and Cox (2000) in a diagnostic key, but then extended to a suggested explanation for the viviparous condition (Chapter 10).

Generic Accounts

Rhizophora L. 1753 [*Sp. Pl.*: 443]
Ding Hou (1958) includes an extended synonymy.
(Figs. B.53–B.56; Plates 21, 22)

A pantropical genus of seven species (and at least four named hybrids), all very uniform in vegetative construction. Recent extensive fieldwork and the addition of molecular and genetic evidence has greatly illuminated the study of this important genus, necessitating a detailed account. A limited account is provided in Graham (1964).

Evergreen trees growing to 30 m tall, with characteristic aerial stilt roots, perfect flowers; often depauperate, scrubby, and much branched in marginal habitats (Fig. 5.4). Leaves with bijugate phyllotaxis, each pair associated with a pair of lanceolate, interlocked stipules to form a distinctive pointed terminal bud, the stipules with a two-ranked palisade of glandular colleters internally at their insertion; stipules early caducous to leave an annular scar, the petiole scar itself elliptical and evident below the associated stipule scar. Leaves simple, entire, blade elliptical, margin often somewhat recurved, 10–20 cm by 4–10 cm, but often somewhat larger in saplings, the range of size and shape fairly characteristic for each species but much influenced by the habitat and position of the tree. Petiole short, 2.0–4.0 cm by 3–5 mm, cylindrical. Leaves and sometimes stipules abruptly yellowing with senescence. Leaf texture coriaceous, glabrous but with numerous microscopic cork warts (Plate 9C) on the lower surface, visible on older leaves as fine black dots in dried, if not always in fresh specimens; veins evident but not prominent below. Apex with a short persistent and erect mucro 2–3 mm long (one group of species) or apex blunt and recurved (a second group of species); base usually acute to attenuate.

Floral morphology is highly diagnostic, especially when seen in L.S., as in Fig. B.55 & B.56). Inflorescence axillary either within the leafy crown, or maturing below the leafy crown in *R. apiculata*, with a shorter (1–3 cm) or longer (8–12 cm) peduncle, the length and thickness fairly diagnostic for each species. Inflorescence cymose, often pendulous in species with longer peduncles. Flowers two to four or more (up to 128) per inflorescence, depending on the number of bifurcations, but with the frequent abortion of one branch; however, in *R. mangle*, often with a trifurcation at the first (rarely subsequent) node and with a small flower number. The range of flower number usually diagnostic for each species (e.g., Plate 24). Each node on inflorescence with a pair of short cupular connate bracts, the bracts either fleshy or even corky (*R. apiculata*), with a pair of similar bracteoles below each flower. Pedicel above bracteoles absent or very short. Flowers perfect, tetramerous; calyx green, yellow, white, or brown, as a shallow tube with four thick, fleshy, valvate, acute lobes, persistent in fruit. Petals four, ephemeral, delicate, white or greenish white, lanceolate and somewhat longer than the calyx lobes, glabrous (*R. apiculata*) or otherwise with a slightly to markedly hairy margin. Stamens eight, 12, or 16 (or some number close to these) inserted on margin or receptacular disc each sessile (*R. apiculata*) or at most with a very short (to 1 mm) filament, acute, multilocellar, dehiscing introrsely by a short adaxial valve. Ovary semi-inferior, the apex obscurely to distinctly conical and extended into a single style 1–5 mm long or stigmas almost sessile (*R. mucronata*). Stigma obscurely to evidently bilobed. Ovary two-locular, with two anatropous ovules on an axile placenta in each loculus. Fruit ovoid, somewhat extended apically, with a brown or olive green, somewhat leathery pericarp and including one (rarely more) developing seed. Seeds viviparous (Plates 14, 21), germinating by extension of hypocotyl through the stylar canal, hypocotyl maturing on the tree to a length of 15–30 cm (70 cm or more in *R. mucronata*). Cotyledons without stipules and fused to form a tubular collar that remains on the tree after seedling is lost (Plate 14C). Cotyledonary collar extended and becoming exposed before abscission of seedling. Propagule a seedling with extending hypocotyl, the plumule protected by a stipule pair of the first (but aborted) plumular leaves.

The genus is distinguished anatomically from other members of the tribe by the cork warts on the leaves and the abundant sclereids (Plate 9A), including characteristic H-shaped trichosclereids, which are especially abundant in the aerial roots and seedling hypocotyl. They can be seen with a hand lens, protruding from any broken surface, and are a useful diagnostic field character.

Nomenclatural history of Rhizophora. *Rhizophora* as used by Linnaeus was a generic name for all mangrove plants, much as Rumphius (1743, 3:102–120) used the name *Mangium* (Chapter 1). The characteristics of the genus were based largely on vivipary and habitat, and Linnaeus listed seven species (including two more in the second edition of *Species plantarum*, 1762). Of Linnaeus' original seven species, six are now assigned elsewhere; four to other genera of the mangrove Rhizophoraceae (*Bruguiera*, *Ceriops*, and *Kandelia*), whereas *Aegiceras* (Myrsinaceae) and *Sonneratia* (Sonneratiaceae) were appropriately recognized and segregated by subsequent workers. The only remaining Linnaean species in the genus is *R. mangle*, based on the name *Mangle*, which goes back at least as far as Oviedo (1526, *Hist. Gen. Nat. Ind.*) and was used for the genus by pre-Linnaean authors. A complete history of the origin and use of the name *Rhizophora* is provided by Salvoza (1936).

Brief Diagnosis of Rhizophora Species

The taxonomic subdivision into sect. Aerope (= *R. apiculata*) and sect. Mangle (the remaining species) suggested by Blume (1849) does emphasize morphological differences but is little used.

> **Rhizophora apiculata** BL. 1827 [*En. Pl. Jav.* 1: 91]
> *R. conjugata* Am. (non L.) 1838 [*Ann. Mag. Nat. Hist.* 1:363]
> A more complete citation is found in Ding Hou (1958, 1960); see also Salvoza (1936).
> (Fig. B.53)

This species is perhaps the most distinctive in the genus, with large, dark glossy green leaves and flowers, usually in pairs on a stubby axis, maturing below the rosette of leaves (Fig. B.53b). In parts of its range it can form pure stands, the trees straight trunked and with limited development of basal aerial roots. As a consequence, it is easily harvested and finds much use as a source of poles and charcoal (Plate 16D).

> **Rhizophora mangle** L. 1753 (Fig. B.67) [*Sp. Pl.* ed. 1: 443]
> (Fig. B.54)

This species is distinguished by its usually rather small leaves, with the blade 10–12 cm by 5 cm, the petiole about 2 cm long, pale and extending into the midrib, which is pale green to almost white below, and particularly by the blunt apex. Stipules are 6–8 cm long. Older shoots are light brown. Fruits are elongated, 2–3 cm long, often asymmetric and tapered to the apex, which is attenuated to the persistent style 2–3 mm long. The mature fruit is characteristic scurfy brown, but germinating rapidly (Plate 5F). Peduncles in more robust shoots may be 6 cm long but are usually shorter, the pedicels about one-third this length. Seedlings are 15–20 cm long, blunt, and without conspicuous lenticels.

> **Rhizophora samoensis** (Hochr.) Salvoza 1936 [*Bull. Nat. Appl. Sci. Univ. Philipp.* 5: 220, pl. 6]
> *Rhizophora mangle* Guppy 1906 (non L.) [*Obs. Nat. Pacif.* 2: 441, frontisp., as "*mangle chico*"]
> *Rhizophora mangle* var. *samoensis* Hochr. 1925 [*Gand.* 2: 447]

This species is scarcely distinguishable in its morphology from *R. mangle*, but it has blunt (not pointed) flower buds and obscure (not obvious) bracteoles. Geographic isolation alone is not sufficient reason for maintaining a separate species, but it is useful to retain the name for populations in New Caledonia, Fiji, Samoa, and Tonga.

> **Rhizophora racemosa** Meyer 1818 [*Prim. Fl. Esseq.*: 185]
> *R. mangle* var. *racemosa* Engler 1872 [*Fl. Bras.* 12(2): 427]
> (Fig. B.55m)

This species is distinguished most readily·by the much-branched inflorescence, with up to six orders of bifurcation, leading to a potential flower number of 128 on any one axillary flowering branch, although numbers are fewer than this because of abortion of some axes. The flower buds are characteristically pointed and the flowers rather small, the sepals being 8–10 mm long.

> **Rhizophora x harrisonii** Leechman 1918 [*Kew Bull.* 1918: 8, Fig. A]
> *R. brevistyla* Salvoza 1936 [*Bull. Nat. Appl. Sci. Univ. Philipp.* 5: 211]
> The synonymy follows the interpretations implied in Ding Hou (1960)
> (Fig. B.55l)

This species is, on circumstantial evidence, considered to be a hybrid between *R. mangle* and *R. racemosa* (intermediate in its morphology, high degree of sterility, and overlapping geographic range). It remains morphologically distinct and is recognized by its many-flowered inflorescences with blunt or rounded flower buds, the sepals being 10 mm or longer. It can achieve a spectacular appearance, suggesting hybrid vigor (Plate 13A). Descriptions of its reproductive capacity also suggest that it sets abundant seed, unlike the other *Rhizophora* hybrids of the IWP species. *R. brevistyla* refers to specimens from the Pacific coast of Central and South America, which Salvoza (1936) considered specifically distinct from *R. x harrisonii*. Further field study is needed to see whether his opinion can be substantiated. Differentiation between Pacific and Atlantic populations of *Rhizophora* is to be expected in view of their presumed long separation by formation of the Isthmus of Panama about 3×10^6 years BP.

> **Rhizophora mucronata** Lamk. 1804 [*Encycl.* 6: 189]
> *R. mucronata* Lamk. var. *typica* Schimp. 1891 [*Bot. Mitt. Trop.* 3: 92]
> (Fig. B.55i–j)

The flowers of this species differ from those of *R. stylosa* most consistently in the sessile stigma (Fig. B.55i); inflorescences in the two species are very similar. The leaves of *R. mucronata* are typically much larger than those of *R. stylosa*. Another conspicuous character is the very long seedling hypocotyl (Plate 14B), often with a rough, warty surface.

> **Rhizophora stylosa** Griff. 1854 (Fig.B.68a–h) [*Nat. Pl. As.* 4: 665]
> *R.mucronata* Lamk. var. *stylosa* Schimp. 1891 [*Bot. Mitt. Trop.* 3: 92]
> (Fig. 55a–h)

It is interesting that the type specimen of this species is an apparent mixture, only some of which is *R. stylosa*. In the field, *R. stylosa* can be distinguished from *R. mucronata* by the characters cited in the following key.

> **Rhizophora x lamarckii** Montr. 1860 [*Mem. Acad. Sci. Lyon* 10: 201]
> *R. conjugata* var. *lamarckii* (Montr.) Guillaumin 1914 [*Notul. Syst.* 3: 56]
> (Fig. B.56; Plate 24)

This taxon is interpreted as a sterile F1 hybrid between *R. apiculata* and *R. stylosa* by Tomlinson and Womersley (1976). Inflorescence and floral bud morphology shows features intermediate between it and its parents (Plate 24, and in detail by comparison with Fig. B.56 and with Figs. B.53, B.55). It was originally considered to be endemic to the northeast coast of New Caledonia, and at its type locality at Canala it grows not only with its putative parents but also with *R.* x *selala* and *R. samoensis*, so that here we have the largest concentration of *Rhizophora* anywhere (five out of a possible eight taxa). Its wider distribution is now well established (Tomlinson, 1978; e.g., Queensland, Papua New Guinea, and the Solomon Islands). It is always consistently intermediate in its morphology between its putative parents (Tomlinson and Womersley, 1976; Tomlinson, 1978; Duke and Bunt, 1979) and occurs only where the two grow together. The pollen appears to be sterile and seedlings are rarely produced.

Duke and Bunt (1979) note that in Queensland, *R.* x *lamarckii* is multitrunked and rambling, forming communities of gnarled and stunted individuals in higher tidal contours and lower to middle tidal reaches in areas of moderate to high rainfall. The frequently rather battered appearance of trees suggests that once established, this taxon is very tenacious and plants may be considerably old; unfortunately, precise age cannot be estimated.

> **_Rhizophora_ x _selala_** (Salvoza) Tomlinson 1978 [*J. Arnold Arbor.* 59: 159]
> "*Selala*" Guppy 1906 [*Obs. Nat. Pacif.* 2: 445, 487, frontisp.]
> *R. mucronata* Lamie var. *selala* Salvoza 1936 [*Bull. Nat. Appl. Sci. Univ. Philipp.* 5: 219]
> *R. selala* Tomlinson and Womersley 1976 [*Contrib. Herb. Aust.* 19: 9]

Interpreted as a sterile F1 hybrid between *R. stylosa* and *R. samoensis*. This taxon was first recognized by Guppy in Fiji. It is characterized by its distinct but recurved or deciduous leaf tip, the inflorescence with two to nine flowers commonly branched beyond two orders, the white and scarcely angled flower buds, and its sterile condition. In my experience, it always co-occurs with its putative parents from which it is readily distinguished, at least in the field, in its consistent intermediate morphology.

Rhizophora: Key to Species and Some Hybrids

1A. Leaf apex with an erect mucro 2–3 mm long. 2
     (IWP "Old World" species, ranging from East Africa to the Pacific in the Tonga-Samoan region)

1B. Leaf apex without an erect mucro, usually blunt with a recurved margin, at most (*R.* x *selala*) with an irregular recurved (but not prominent) mucro. 5
     (CEP = "New World" species of the Atlantic coasts of West Africa, tropical America, the Caribbean, and Pacific coasts of central and southern tropical America, with an extension into the Pacific as far west as New Caledonia and Vanuatu)

     * IWP (Eastern or "Old World" Species)

2A. Flowers in twos (rarely in fours) on stout peduncles (about 10 mm long) shorter than petiole, borne below the leafy crown, that is, in the axil of a leaf scar; each flower sessile with a pair of massive cupular corky bracteoles. Petals glabrous. Stamens usually 12 (sometimes fewer or more), sessile. Stigmas almost sessile.

Fruits and seedlings always maturing well below the leaf rosette (Fig. B.53).. . . .
. . . . *R. apiculate*
(Indo-Malaya)

2B. Flowers in fours or higher numbers (rarely in twos) on stout to somewhat slender
peduncles (15–40 mm long), longer than the petiole, borne within the leafy
crown, that is, in the axil of an attached foliage leaf; each flower with an evident
(even if short) pedicel, bracteoles green, fleshy but not corky. Petals slightly to
conspicuously hairy on the margins. Stamens either eight or about 16 with a
short (1–2 mm) filament. Stigmas sessile or ovary extended into a slender
style.   Fruits   and   seeds   borne   within   or   just   below   leafy
rosette. 3
(Remaining eastern species)

3A. Flowers per inflorescence usually in fours (sometimes twos), peduncle short (about
15 mm long) and rather stout. Stamens variable in number (eight to) 16 (up to 22)
and some often distorted, aborted, or represented by a filamentous staminode.
Petals with inconspicuous marginal hairs. Styles 2–3 mm long. Trees usually sterile
and not producing seedlings. Pollen sterile.. . . . .*R. x lamarckii*
(An assumed hybrid *R. apiculata* x *R. stylosa* in New Caledonia, Vanuatu,
Queensland, New Guinea, Solomon Islands)

3B. Flowers four, eight, or more per inflorescence (rarely four or fewer by reduction
in depauperate specimens), peduncle usually long (25–40 mm), slender. Stamens
almost always eight, rarely aborted. Petals with densely woolly, conspicuous
marginal hairs. Style either long or absent. Trees fertile (able to develop numer-
ous seedlings). Pollen fertile. 4

4A. Stigmas sessile, seedlings warted, 50–70 cm long. Leaf blade broad (to 10 cm)
and long (to 20 cm). . . . .*R. mucronata*
(East Africa to the Western Pacific, possibly as far as Vanuatu and Kiribati)

4B. Stigmas on a slender style 4–5 (to 6) mm long, seedlings smooth, not exceeding
30 cm. Leaf blade narrow (to 7 cm) and short (to 12 cm). . . . . *R. stylosa*
(India to Tonga and Samoa)

AEP (Western or "New World" Species)

5A. Peduncle commonly trifurcate at first node; flowers per inflorescence two to nine.
Flower   buds   yellow   or   yellowish   white,   somewhat   to   distinctly
angular. 6

5B. Peduncle bifurcate at all nodes, flowers per inflorescence usually more than 10
(up   to   64).   Flower   buds   white   or   greenish   white,   not
angular. 8

6A. Leaf apex with an irregular curved, sometimes deciduous mucro. Peduncle
2.5–3.0 cm long, but sometimes longer, 2.6–3.2 mm wide, often with more than
two orders of branching. Flowers two to nine per inflorescence. Mature flower
buds white, neither sharply angular in cross-section nor abruptly narrowed to a
distinct shoulder at the base, 12–14 mm long. Apex of ovary extended into a
distinct style 1–2 mm long. (Plants sterile, lacking fruits and viviparous
seedlings). . . . . *R.x selala*

(An assumed hybrid *R. samoensis* x *R. stylosa*) in New Caledonia, Fiji, and Vanuatu)

6B.   Leaf apex blunt, recurved, without a mucro. Peduncle 2.0–2.5 cm long (or longer), 2.0–2.4 mm wide, rarely with more than two orders of branching. Flowers usually two to five per inflorescence. Mature flower buds yellow, sharply angular in cross-section, and distinctly shouldered at the base, 10–12 mm long. Apex of ovary steeply conical and without a discrete slender style. (Plants fertile, capable of developing fruits and seedlings). 7

7A.   Bracteoles conspicuous, 1 mm or longer, flower buds pointed at apex. . . . .
. . . . *R. mangle*
(West Africa to tropical America on the Atlantic and Pacific coasts)

7B.   Bracteoles inconspicuous, less than 1 mm long, flower buds rounded at apex. . .
. . . . *R. samoensis*
(New Caledonia, New Hebrides, Fiji, and probably elsewhere in the Western Pacific; possibly occurring on the Pacific coast of tropical America)

8A.   Bracteoles thick, short, rounded; flower buds rounded, flowers often 32–64 per inflorescence. . . . . *R. racemosa*
(West Africa and Atlantic coast of tropical America)

8B.   Bracteoles thin, long, pointed; flower buds pointed apically, flowers usually eight to 32 per inflorescence. . . . . *R.* x *harrisonii*
(Regarded as a hybrid between *R. mangle* and *R. racemosa* but reported as occurring outside the range of one of its putative parents. West Africa and Atlantic and Pacific coasts of tropical America)

Further notes on the taxonomy of *Rhizophora* are found in the discussion on the variation in morphological features that are used diagnostically.

Species Aggregates in *Rhizophora*

The species of *Rhizophora* may be aggregated in various ways. The most morphologically distinctive species is *R. apiculata* in terms of flower number per inflorescence, inflorescence position, bracteole structure, petal indumentum, and stamen number; this is the basis for its segregation as a monotypic section *Aerope* Blume, in contrast to the remaining species forming section *Rhizophora* (sect. *Mangle* Blume). However, representatives of the two sections are demonstrably capable of crossing (*R.* x *lamarckii*, *R.* x *tomlinsonii*).

Another method of aggregation could recognize two groups distinguished by the morphology of the leaf apex; one group with a prominent mucro (*R. apiculata, stylosa, mucronata,* and x *lamarckii*) and the other without, but the leaf apex recurved (*R. samoensis, mangle,* x *harrisonii,* and *racemosa*). This arrangement is strongly correlated with geographic distribution, the first group constituting the IWP species and the second the AEP species. Again, however, members of the two groups appear capable of crossing where they coincide, as in the production of *R.* x *selala*, which is morphologically intermediate in its leaf apex and occurs in the area of geographic overlap between the two groups.

A third method of segregation could contrast species with the capability of trifurcating at the first node of the inflorescence (*R. mangle* and *R. samoensis*) with the remaining species that do not. Once again, the two groups are interfertile, in the hybrids *R.* x *selala* and *R.* x *harrisonii.*

Hybrid Taxa in *Rhizophora*

Evidence for the existence of hybrid populations in this genus has long been appreciated, based on morphological intermediacy and hybrid sterility (e.g., *R.* x *harrisonii*, *R.* x *lamarckii*, and *R.* x *selala*). Plate 24 indicates the kind of morphological evidence used in recognizing such taxa and they are included in the previous key. The hybrid status of *R.* x *lamarckii* and *R.* x *annamalayana* has been confirmed by genetic methods in Parani et al. (1997) and Ng et al. (2014), and they suggest that *R. mucronata* x *R. stylosa* could exist. The result has been to demonstrate an almost complete level of interspecies fertility for the genus as a whole in Cerón-Souza et al. (2010) and Duke (2010). In particular, New Caledonia has emerged as a center for taxonomic diversity in *Rhizophora*, where seven taxa occur; for instance, four of the widely distributed IWP species and three of their hybrids, two newly described (Duke, 2010). No less than six taxa have been collected within a single estuary at Canala.

The following listing includes both earlier and later described taxa.

> ***Rhizophora annamalayana*** Kathiresan 1995 [*Envir. Ecol.* 13: 240–241
> = *apiculata* x *mucronata*]

This taxon has all the characters of a hybrid (Seetharaman and Kandasamy, 2011). Seed set is rare because pollen is rarely viable; there is a suggestion of cleistogamy because of the precocious dehiscence of anthers, although this is a feature common in *Rhizophora*.

> ***Rhizophora* x *harrisonii*** Leechman 1918 [*Kew Bull.* 1918: 8, Fig. A] 1918
> = *mangle* x *racemosa*
> *R. brevistyla* Salvoza 1936 [*Bull. Nat. Appl. Sci. Univ. Philipp.* 5: 211]

There is circumstantial evidence that the species is a hybrid between *R. mangle* and *R. racemosa*, but if so, its geographic distribution is peculiar, being greater than one of its putative parents. It co-occurs with its two parents in West Africa (Keay, 1953) and the Atlantic coast of tropical America. However, Breteler (1969, 1977) claims that it occurs on the Pacific coast of tropical America in the absence of *R. racemosa*. Since the Panama Isthmus was last open over three million years ago, it seems that the situation needs re-examining, because it is unlikely that a species could exist in a vegetative condition for this length of time if *R. harrisonii* is infertile. However, Prance (1975) has recorded *R. racemosa* on the Pacific coast of Colombia.

> ***Rhizophora* x *lamarckii*** Montr. 1860 [*Mem. Acad. Sci. Lyon* 10: 201]
> = *apiculata* x *stylosa*
> *R. conjugata* var. *lamarckii* (Montr.) Guillaumin 1914 [*Notul. Syst.* 3: 56]

This species was first described from, and considered to be endemic to, New Caledonia. Its existence in isolated localities in Queensland, New Guinea, the Solomon Islands, and Vanuatu has been established, based on field observation and herbarium records. On abundant circumstantial evidence, it is recognized as a hybrid of *R. apiculata* with *R. stylosa*, and always coexists with its putative parents. In New Caledonia, for example, it is restricted, like *R. apiculata*, to the east coast (Duke, 2010).

> ***Rhizophora*** x ***selala*** (Salvoza) Tomlinson 1978 [*Arnold Arbor.* 59: 159]
> = *samoensis* x *stylosa*
> "*Selala*" Guppy 1906 [*Obs. Nat. Pacif.* 2: 445, 487, frontisp.]
> *R. mucronata* Lamk. var. *selala* Salvoza 1936 [*Bull. Nat. Appl. Sci. Univ. Philipp.*
>   5: 219]
> *R. selala* Tomlinson and Womersley 1976 [*Contrib. Herb. Aust.* 19: 9]

This taxon was first recognized by Guppy (1906) in Fiji, who suggested that it was a hybrid between *R. stylosa* (which he incorrectly called *R. mucronata)* and *R. samoensis* (which he called, more appropriately, *R. mangle).* Salvoza accepted the taxon but did not provide a formal diagnosis, which was first done by Tomlinson (1978), who subsequently demonstrated its existence in New Caledonia It is also recorded for Vanuatu. It is likely to be found elsewhere in places where its parents coexist.

> ***Rhizophora*** x ***tomlinsonii*** N.C. Duke 2010 [*Blumea* 55: 185]
> = samoensis x apiculata
> Named and recorded by Duke (2010) with a restricted distribution in New Caledonia.

In general, *Rhizophora* species and their hybrids are very similar, and diagnostic features are often quantitative, referring to a range of sizes or numbers that are readily appreciated in the field where large samples are available, but not so in the herbarium. Examples of measurements that show the ranges of some of these features are found in Tomlinson (1978) and Duke and Bunt (1979). Normally where more than one species of *Rhizophora* grow together, the fieldworker soon learns that they can be distinguished easily and completely by the combination of characters in the key, together with more elusive "Gestalt" features, such as leaf color and size, which are unavailable to the herbarium worker. Some field characters, however, may not be permanent; Duke and Bunt (1979) record the absence of leaf dots in their key to distinguish *R. apiculata.* This species does have the cork warts, which are a generic diagnostic feature; they are simply obscure in fresh leaves, but appear in dried or fluid-preserved leaves. It would be misleading to apply this character elsewhere than in the field. Furthermore, the ranges of variation so far recorded in any detail refer to local populations and may not be applicable throughout the entire range of one species, although in my experience, *Rhizophora* species are remarkably uniform in their morphology at the extremes of their ranges. Field study so far demonstrates that even though the measured ranges of given characters overlap among different species, there is no continuum and species (and even hybrids) remain as clearly recognizable entities.

Distribution of *Rhizophora* Species

Rhizophora is a pantropical genus in climatically rather uniform coastal environments, but with limited extension into the subtropics, always much less than that of *Avicennia*. Although the range of most species is appreciable, there are considerable differences in longitudinal ranges that are difficult to account for ecologically, especially because some species co-occur (Jiménez, 1987). Latitudinal limits are known with some precision in many parts of the world because botanists have been interested in the limits of the climatic tolerance of mangrove species at the margins of their range. The chief restriction is cold; *Rhizophora* is killed by frost and cannot survive extended periods of near-freezing temperatures. Some species (e.g., *R. apiculata* and *R. mucronata*) seem restricted to wetter climates; *R. stylosa* seems tolerant of drier climates. Plants carried inland as propagules by extreme storms can persist in essentially freshwater environments, as with *R. mangle* in the Florida Everglades (Plate 1B). The limits of the latitudinal range are not precise because populations may be killed in exceptional winters but are restored subsequently by seedlings brought some distance via ocean currents.

Recent experience shows that despite the frequency and wide distribution of *Rhizophora*, significant new records are easily made; *R. stylosa* was first collected on Singapore, a botanically accessible island, by H. M. Burkill in 1960.

Maps that outline the distribution of AEP species can be found in Cerón-Souza et al. (2010) and of the IWP species in Duke (2010), both in relation to their studies of hybridization. A more complete expression, useful politically, is the listing country by country in Spalding et al. (2010). The maps in Figs. B.51, B.52 are crude, but originally based on herbarium specimens.

The species are discussed further in decreasing order of their range. In summary, the broad facts of *Rhizophora* phytogeography are known, and have increasingly been subjected to phylogenetic discussion, but genetic analysis is specially needed.

1.
Eastern (IWP) Taxa (Fig. B.51)

Rhizophora mucronata (Fig. B.51e)
Ranges from East Africa to the Western Pacific. It has a southern limit in Africa at about the latitude of Durban; its limit in the Persian Gulf, where it is poorly developed by virtue of the dry climate, is not precisely known. In much of the Indian Ocean areas, it is the sole representative of the genus but does coexist with *Bruguiera* and *Ceriops*. Further eastward in the Indo-Malayan region it grows with *R. apiculata* and *R. stylosa*, but becomes progressively a less conspicuous element of mangrove floras as one moves eastward. It is thus not known in Western Australia (possibly because of the dry climate) and has a localized distribution in northern Australia, Queensland, and into eastern New Guinea and the Solomon Islands. There is a record for Kiribati. Older records for Fiji and Tonga are based on misidentification and confusion with *R. stylosa*. In the northern hemisphere it occurs in the northern Philippines and Indochina, but without precisely ascertained limits.

Rhizophora stylosa (Fig. B.51a)

Ranges from southern India to the Eastern Pacific (Samoa), southward to New South Wales, and westward to Indochina. This species becomes a progressively more conspicuous element of mangrove vegetation in an easterly direction (cf. *R. mucronata*). In the Western Pacific, this species seems to be particularly vigorous and will survive in extreme habitats (coral reefs, rocky coastlines) although often depauperate. Probably for this reason, it appears to be the sole representative of the genus in Western Australia. Its precise limit in the North Pacific is not certain; here it is largely supplanted by *R. mucronata.*

Rhizophora apiculata (Fig. B.51c)

This species has a more restricted range than *R. stylosa*; it is not known west of the Andaman Islands. Otherwise, it is a common and even dominant constituent of mangroves in the Malesian region as far west as Queensland and Papua New Guinea in the south and the Philippines in the north. It occurs on New Caledonia (its easternmost limit) but restricted to the east (north) coast of this island, probably because of temperature and rainfall. In the central Pacific, its easternmost limit is Ponape (Fosberg, 1975).

Rhizophora samoensis (Fig. B.52d)

This species is distinguished from *R. mangle* with difficulty by morphology alone and may be regarded as a geographic outlier of that species in the Western Pacific, occurring from New Caledonia and the New Hebrides to Tonga and Samoa. Its range certainly overlaps that of *R. stylosa* by virtue of the hybrid *R.* x *selala* (Fig. B.51b). Tentatively it may occur in eastern South America (Fig. B.52d).

2.   ## Western (AEP) Taxa (Fig. B.52)

Rhizophora mangle (Fig. B.52b)

Ranges from western Africa to the Pacific coast of tropical America. In Africa, its precise latitudinal limits are not certain, but it is recorded as far south as Angola and as far north as Mauritania. In the Americas it has, on the Atlantic side, a wide distribution to about latitude 25°N in Florida and to eastern Brazil, and on the Pacific side from Mexico to northern Chile, where its southern range is limited by the cold, dry climate. A better understanding of its possible co-occurrence with *R. samoensis* in Pacific South America is desired.

Rhizophora x *harrisonii* (Fig. B.52c)

Although there is evidence that the taxon is a hybrid between *R. mangle* and *R. racemosa*, its geographic distribution is peculiar, being greater than one of its putative parents. It co-occurs with its two parents in West Africa (Keay, 1953) and the Atlantic coast of tropical America. However, Breteler (1977) claims that it occurs on the Pacific coast of tropical America in the absence of *R. racemosa*. Since the Panama Isthmus was last open over three million years ago, it seems that the situation needs re-examining, because it is unlikely that a species could exist in a vegetative condition for this length of time if *R.* x *harrisonii* is relatively infertile. However, Prance (1975) has recorded *R. racemosa* on the Pacific coast of Colombia.

Rhizophora racemosa (Fig. B.52a)

Ranges from western Africa to northern South America, but the precise limits of its distribution are uncertain, especially its occurrence on the Pacific coast.

> **Bruguiera** Lamarck 1793–1797 [*Tabl. Enc. Meth.* t. 397; see also *Encycl. Meth. Bot.* 4: 696, 1797–1978]
>
> For a full synonymy, see Ding Hou (1957, 1958).
> (Fig. B.57, B.58; Plate 8F, L, 12C, D, 22E–G, 23)

Trees growing to 40 m high, evergreen, with continuous growth, hermaphroditic; short aerial roots of sapling developing into stout buttresses of adult trees; pneumatophores knee-like (Plate 12C, D). Leaves opposite, bijugate; petiolate, entire; blade glabrous, usually ovate to oblong-ovate, with a cuneate to acuminate base, apex usually bluntly pointed. Stipules in pairs, 1–4 cm long, sometimes reddish; orientation as in *Rhizophora*. Flowers either solitary or in shortly stalked, bifurcate, two- to six-flowered cymes. Bracteoles never enclosing base of flower. Flowers either large (2–4 cm long, Fig. B.57) or smaller (1–1.5 cm long, Fig. B.58) with either about 13 (large-flowered) or about eight (small-flowered) pointed, initially valvate, fleshy calyx lobes. Petals delicate, as many as and alternating with the calyx lobes, bilobed and with marginal, interlocking hairs, the lobes commonly with appendages distally and usually with a rigid filament between them. Stamens with slender, somewhat contorted filaments, twice as many as the sepals and enclosed in pairs by the pouched petals (Plate 23A, E), dehiscing precociously, the pouches exploding when triggered (Plate 23B, C, I). Ovary inferior, obscurely two- to four-locular, with two ovules in each loculus. Floral disk cup-shaped; style slender with an obscurely three-lobed stigma. Fruit one-seeded with persistent, erect, or reflexed lobes. Hypocotyl terete or obscurely ribbed, cigar-shaped, blunt apically. Fruit falling with the seedling, the cotyledonary collar not extending.

This genus is distinguished from other mangrove Rhizophoraceae in its numerous (more than six) calyx lobes and the propagule, which consists of both seedling and fruit (Plate 22F), abscissing independently after the seedling is detached (cf., *B. parviflora*), unlike the other genera in which the seedling on the trees is detached by abscission at the cotyledonary node so that cotyledons and fruit remain on the tree. *Bruguiera* is also consistent in its explosive mechanism of pollen release, which is triggered by a visiting pollinator. Only *Ceriops tagal* of the other Rhizophoreae resembles it.

Nomenclature

The genus is somewhat imperfectly segregated into two groups of species; the large, solitary-flowered group (e.g., *B. gymnorrhiza)* and the small, many-flowered group (e.g., *B. parviflora*). The difference in flower size is correlated with differences in leaf size and fruit size. This distinction was the basis for the recognition of two genera by Blume: *Bruguiera* (*sensu stricto*) representing the first and *Kanilia* the second group. Subsequently, Miquel [1855, *Fl. Ned. Ind.* I (1): 585] reduced *Kanilia* to a section of *Bruguiera* and applied the name *Mangium* (from Rumphius) to the other section. However, Ding Hou (1958) pointed out that the distinction between the two sections was obscured by the later discovery of the two species *B. hainesii* and *B. exaristata*, which in some features are morphologically intermediate. The first species sometimes has solitary flowers; the second sometimes has flowers in pairs and the size difference between flowers is not great. On the other hand, there is a fairly sharp distinction between species on the basis of flower orientation in relation to contrasted classes of pollinators (birds versus insects). The morphological continuum is thus transcended by a functional discontinuity, but this is insufficient justification to revive the sectional names.

The following key provides diagnostic information about the six species recognized in Bruguiera:

1A. Inflorescence one (to two)-flowered; flowers recurved, 2–3 cm long, calyx lobes about 13; sinus between lobes of petals empty or at most occupied by a short bristle not exceeding the lobes (Fig. B.57j, k). Fruit stout, 0.5–2 cm in diameter. Pollinated mainly by birds. 2

1B. Inflorescences (one to) two- to six-flowered; flowers oblique or erect, 1–1.5 (to 2) cm long, calyx lobes about eight (to 10); sinus between lobes at petals always occupied by a long bristle exceeding the length of the lobes (Fig. B.58i). Fruit slender, 0.4–0.8 cm in diameter. Pollinated by insects. 4

2.   *Large-flowered species* (Fig. B.57)

2A. Leaves obovate, 5–9 cm by 3–4 cm. Flowers sometimes in twos, 2–2.5 cm long, yellowish green. Petals 9–11 mm long, usually without a bristle in the sinus between the petal lobes. . . . .*B. exaristata*

2B. Leaves elliptic-oblong, 10–20 cm by 5–7 cm, usually longer than 10 cm. Flowers 3–4 cm long, reddish. Petals 12–15 mm long, with a distinct bristle in the sinus between the petal lobes. 3

3A. Tips of petal lobes blunt, without filamentous appendages. . . . . *B. sexangula*

3B. Tips of petal lobes acute, each extended into three filamentous appendages. . . . . *B. gymnorrhiza*

4.   *Small-flowered species* (Fig. B.58)

4A. Calyx lobes slender, short (less than 3 mm), only one-fourth to one-fifth the length of the calyx, erect or at most slightly spreading in fruit. Petals 1.5–2 mm. long. . . . . *B. parviflora*

4B. Calyx lobes stout, long (more than 3 mm), half the length of the calyx, horizontal or reflexed in fruit. Petals 4–9 mm long. 5

5A. Mature flowers 10–12 mm long. Petals 3–4 mm long. Calyx tube at mouth 2 mm in diameter; calyx lobes eight and completely reflexed in fruit. . . . . .*B. cylindrica*

5B.   Mature flowers 18–22 mm long. Petals 7–9 mm long. Calyx tube at mouth 5 mm
       in diameter; calyx lobes 10 and extended at right angles in fruit......*B. hainesii*

Detailed descriptions are provided for one representative of each group. A complete
synonymy is provided by Ding Hou (1958); see also Ding Hou (1957).

Large-flowered *Bruguiera* species (Fig. B.57)

The large-flowered species are beautifully and completely illustrated in Duke and Ge
(2011). The detail is important where the diagnostic characters used in the previous
keys, although precise, may otherwise seem miniscule.

> ***Bruguiera gymnorrhiza*** (L.) Lamk. 1797–1798 [*Encycl. Meth. Bot.* 4: 696; see also
>   *Tabl. Encycl. Meth.* 2, 1819: 517, t. 397, as *Gymnorhiza*]
> *B. rheedii* Blume 1827 [*En. Pl. Jav.*: 92]
> *B. conjugata* Merr. 1914 [*Philipp . J. Sci. Bot.* 9: 118]
> (Figs. 5.3, B.57a–d; Plate 4F, 8F, 12C ,D, 22B, G, 23A–E)

Trees to 30 m or more high with short buttresses. Bark rough, black, fissured in a regular
checkered pattern (Plate 4F). Shoots plagiotropic by apposition, developing terminal
short shoots (Aubréville's model). Leaves elliptic-oblong, coriaceous, 8–22 cm by 5–10 cm,
apex bluntly pointed, base cuneate. Petiole up to 4 cm long, often glaucous with a white
wax, as on the young stems, leaves often reddish beneath. Stipules to 4 cm long, often
reddish. Flowers solitary, up to 3.5 cm long, reflexed at anthesis, pedicel naked, up to
2.5 cm long. Calyx usually reddish with 12 to 14 lobes that remain erect in fruit. Petals
to 15 mm long, two-lobed with the deep median sinus occupied by a slender bristle,
additional three to four bristles on the acute petal lobes. Stamens enclosed in pairs by
the petals. Calyx cup deep, style slender, filiform with three or four stigmatic lobes,
calyx tube scarcely elongating in fruit. Hypocotyl to 25 cm long, cigar-shaped, blunt
apically, slightly angular.

Ecological and geographic distribution. This tree is a characteristic element of the
middle mangrove community throughout its range, extending into the transitional
landward communities. Ding Hou (1958) describes it as the "largest and probably the
longest-lived of the mangrove community." It may grow on exposed shores where the
mangrove belt is thin or depauperate. It has the broadest range of the genus, and indeed,
of all mangroves, extending from East Africa (including Madagascar) through Sri
Lanka and the Malay Archipelago into Micronesia and Polynesia (Samoa) northward
to the Ryukyu Islands and southward in tropical Australia (Western Australia and
Queensland).

Variation. The red to almost scarlet color of the flowers, which is repeated in the
reddish stipules and leaf undersurfaces, is not a constant feature of this species. I have
seen scattered individuals without this color among populations of the plant in New
Caledonia; the contrast renders them very striking but they do not seem to differ in any
morphological feature.

Architecture. Bruguiera gymnorrhiza offers a good example of Aubréville's model and may be contrasted with *Rhizophora* (Attims' model). The difference between these two architectural models leads to important differences in crown shape, and particularly in crown plasticity. The sapling axis is orthotropic and with radial symmetry as a result of the bijugate phyllotaxis. It goes through a period during which it remains unbranched, with the length of time depending on the vigor of the sapling. This vigor is reflected by the spacing of the leaves; axes with numerous, close-set leaf scars suggest slow growth; axes with few distant leaf scars suggest rapid growth.

Branching is by syllepsis, as is usual in the family. The branches are markedly plagiotropic and sympodial by apposition (Fig. 5.3). After a limited time of horizontal growth, the axis turns erect and continues its activity as an erect short shoot, producing a rosette of leaves as a consequence of the bijugate phyllotaxis. A renewal shoot arises in the axil of a leaf at the base of the erect shoot; this extends horizontally to repeat the developmental process. The system is proliferated when two instead of one renewal shoots develop, the two renewal shoots usually diverging at an angle of about 60° to form an irregular hexagonal pattern.

The arrangement of branches is not as regular as in *Terminalia* (q.v.), which has the same architecture, but one can appreciate that the overall pattern (as in *Terminalia*) minimizes the path length and maximizes the effective photosynthetic area. The regularity is not achieved in *Bruguiera* because the phyllotaxis does not permit a precise repetition of leaf positions and the units of extension are not of such uniform length, so that the number of internodes between successive branches and their consequent spacing is rather variable.

The net result, however, is a young tree with regular tiers of precisely oriented, almost pagoda-like branches. This regularity is lost in older trees as units are damaged or broken off. In the process of reiteration, the trunk axis is replaced by an existing orthotropic branch axis, so that the mature tree characteristically retains a single trunk. This should be contrasted with that of *Rhizophora* (q.v.), which because of imposed instead of obligate plagiotropic branches, responds to damage by "dedifferentiation" in the sense of Hallé et al. (1978). Consequently, the crown readily loses its regular shape, and in stressed individuals the process results in shrubby or deformed trees, a condition not usual in *Bruguiera*.

The obligate versus imposed plagiotropy in these two genera is best appreciated by examining the leader of a vigorous sapling. The pronounced orthotropy of young branches found in *Rhizophora* contrasts with the plagiotropy of *Bruguiera*. This simple difference, reflecting contrasted architectures, leads to a limited plasticity of crown development in *Bruguiera* but a considerable plasticity in *Rhizophora*.

Bruguiera sexangula (Lour.) Poir. 1816 [*Lamk. Encycl.* 4: 262]
B. eriopetala Wight and Arnold 1838 ex Arn. [*Ann. Mag. Nat. Hist.* l: 368]
B. malabarica F.-Vill. (non Arn.) 1880 [*Nov. App.*: 79]

(Fig. B.57e–l; Plate 22F)

This species resembles *B. gymnorrhiza* closely, but differs in the somewhat smaller, thinner leaves and most clearly in the absence of long appendages on the blunt petal lobes. Wyatt-Smith (1953a) suggests that the calyx is smaller and more distinctly ribbed. The twigs and petioles lack the white waxy covering that is often characteristic of *B. gymnorrhiza*: unlike that species, the calyx is never a conspicuous scarlet, although it is occasionally reddish. The aerial roots at the base of the trunk, which appear in the sapling, may be quite well developed.

> ***Bruguiera* x *rhynchopetala*** (W.C. Ko) X.-J. Ge & N.C. Duke 2006 [*Austr. Mangr.* 122]
>
> = *gymnorrhiza* x *sexangula*
>
> [Ko, W. C. 1978 *Acta Phytotax. Sinica* 16: 109]

Described in Duke (2006), this hybrid is said to provide an example of unidirectional hybridization (Zhou et al., 2008). Genomic markers can also reveal the extent of introgressive hybridization (Sun and Lo, 2011). It has a peculiar disjunct distribution. It was recognized first in Queensland together with Papua New Guinea, and subsequently in Hainan, where its hybrid status has been confirmed. It is distinguished by large solitary-flowered inflorescences, the petals with a short spine between the petal lobes, the lobes rounded with one to two short bristles. It grows as a tall straight-trunked emergent with prominent flared buttresses and with fissured and checkered bark. It produces apparently viable seedlings.

> ***Bruguiera exaristata*** Ding Hou 1957 [*Nova Guinea n.s.* 8: 166, f. 1–21]
>
> (Plate 23F–G)

It seems remarkable that this distinctive species should have remained overlooked for some time. It is recorded from the Malay Peninsula to Papua New Guinea and northern Australia (Spalding et al., 2010), possibly in areas that were poorly explored botanically.

It is distinguished from the two other species of the large, solitary-flower group in its appreciably smaller, yellowish green flowers (never reddish), with an eight (to 10)-lobed calyx, smaller leaves, and limited stature (not exceeding 20 m). In Queensland, it is a common constituent of the landward part of mangal. The most frequent flower visitors are honey-eaters.

Small-flowered *Bruguiera* species (Fig. B.58)

In their description of hybridization, Duke and Ge (2011) include a tentative phylogeny in which the small-flowered species (with floral aggregates) are presented as basal to the large-flowered species (with largely solitary flowers).

> ***Bruguiera parviflora*** Wight and Arnold ex Griffith 1936 [*Trans. Med. Phys. Soc. Cal.* 8: 10]
>
> (Fig. B.58a–c; Plate 23I)

Tree growing to 24 m with smooth, gray, obscurely lenticellate bark. Leaves 7–13 cm by 2–4 cm, elliptic, apex bluntly pointed, base narrowly cuneate. Stipules 4–6 cm long.

Inflorescence three- to four-flowered, peduncle 2 cm long with obscure bracteoles. Flowers yellowish green, erect at anthesis, pedicels to 13 mm long. Calyx tube ridged, to 9 mm long; calyx lobes usually eight, short, that is, less than one-fourth the length of the calyx tube. Petals to 2 mm long, each petal lobe with bristles; each petal enclosing two stamens. Style 1–1.5 mm long, stigmas two or three. Fruit with calyx tube enlarged two- to three-fold, calyx lobes remaining erect. Hypocotyl smooth, to 15 cm long, truncate at the apex, scarcely 5 mm in diameter, at first green and erect, becoming brown and pendulous.

This species is representative of the small, many-flowered group. It is distributed throughout Southeast Asia, from the Malay Peninsula to tropical Australia, the Solomon Islands and Vanuatu, but it is not recorded for New Caledonia. It occurs on the inner mangrove fringe and riverbanks and has, to some extent, the characteristics of a pioneering species. Good field characters are the light green stipules that form a slender tube and the delicate light green foliage with slender petioles scarcely 2 mm wide.

Floral biology. In this species, the flowers remain erect and thus point distally, that is, outward in the crown. They are visited by small insects including butterflies, which seem capable of the delicate triggering of the explosive pollen release and are probably effective pollinators.

Seedling establishment. The seedling is released with the fruit attached, and on establishment, the plumule pushes its way through the basal end of the fruit (Ng, 2014). The hypocotyl becomes arched a few days after falling to the ground so that its radicular end usually remains in contact with the mud. The cortex at this end becomes stellately expanded and aids in fixing the hypocotyl, facilitating the penetration of the extending radicle, which becomes a taproot.

Architecture. This species conforms more closely to Attims' s model, in contrast to the clear-cut Aubréville's model shown in *B. gymnorrhiza*. Shoots have a strong tendency to remain orthotropic, and the crown of saplings is fairly narrow and conical.

Root system. The looping aerial roots that produce pneumatophores are rather poorly or at most inconspicuously developed in this species.

Bruguiera cylindrica (L.) Bl. 1827 [*En. Pl. Jav.* 1: 93]
Rhizophora caryophylloides Burm.f. 1768 [*Fl. Ind.*: 109 (excl. var. B)]
B. malabarica Arnold 1838 [*Ann. Mag. Nat. Hist.* 1: 369]

(Fig. B.58d–k; Plate 22E)

Tree to about 20 m high with short buttresses, bark grayish. Leaves elliptic, 7–17 cm by 2–8 cm, with a bluntly pointed apex and cuneate base, petiole to 4 cm long, often reddish. Stipules to 3 cm long. Inflorescence usually three-flowered, peduncle about 1 cm long. Flowers greenish, erect at anthesis. Calyx tube 4–6 mm long, 2 mm in diameter, with eight lobes about as long as the tube. Petals shortly bilobed with two or three bristles at the apex of each lobe (Fig. B.58i); each petal enclosing a pair of stamens. Calyx tube somewhat enlarged in fruit, the lobes reflexed, detached with the seedling. Hypocotyl up to 15 cm long at maturity, grooved or angled.

Ecological and geographic distribution. The plant usually grows like a small tree in the inner mangroves but is described as specifically characteristic of newly established substrates, being also somewhat susceptible to prolonged submersion. Because of its pioneering propensities, it may form pure stands, but it is also said to be very slow growing. It is distributed from India through the Malay Archipelago as far as New Guinea and north Queensland.

Growth and floral biology. This species has the same habit as *B. parviflora*, with which it may occasionally grow. The floral mechanism is also the same and involves explosive pollen discharge after triggering by small insect visitors.

Bruguiera hainesii C. G. Rogers 1919 [*Kew Bull.*1919: 225]
B. eriopetala Wight, of Watson 1928 [*Malay. For. Rec.* 6: 109]

This species resembles the other small-flowered *Bruguieras* but is distinguished by its larger flowers and the usually 10-lobed calyx, which stands out at right angles to the calyx tube in the fruit. It is somewhat intermediate between the two flower types in the genus. The inflorescences are usually two- or three-flowered, but solitary flowers occur. In its erect flower position it suggests that it is suited to small insects that are the putative pollinators of *B. parviflora* and *B. cylindrica*, but there are no records of flower visitors. It has a wide distribution from south Myanmar and Thailand through the Malay Archipelago to the Port Moresby region of Papua New Guinea, with a northern limit in the Philippines, but has not been recorded in Queensland. It is rather infrequent in the inland side of mangroves in regions that are not regularly inundated by normal tides.

Ceriops Arn. 1838 [*Ann. Mag. Nat. Hist.* 1: 363]
For a full synonymy and citation, see Ding Hou (1958).

(Figs. B.59, B.60; Plates 10E, 11A, 12A, B, 13F)

Evergreen, small to moderately tall (up to 20 m high) tree with perfect flowers, typically with a distinct trunk and short basal buttresses (Fig. B.60) originating as short basal stilt roots (Plate 13F). Pneumatophores sometimes developed as looped surface roots (Plate 12B), as in *Bruguiera*. Sclereids absent from all parts. Bark smooth or slightly fissured, pale grayish or white, often with a red tinge, peeling in thick strips from the buttressed portion. Twigs jointed with swollen nodes. Leaves opposite, blade of the order of 7–12 cm by 3–5 cm, with a well-developed terete petiole 1–2 cm long; blade ovate to slightly obovate or elliptic-oblong, glabrous, without prominent veins, apex rounded or slightly emarginate, never apiculate; primordia including minute protruding hydathodes on lower surface. Cork warts lacking. Buds 1.5–4 cm long, characteristically spear-shaped, flattened, the individual stipules thickened dorsally and including copious mucilage drying to a varnish-like consistency, both within the free space of the bud and covering young expanded portions to make them shiny. Inflorescence solitary, axillary with a short (5 mm) or moderately long (1.5 cm) axis and few (two to eight), rarely many, flowers in compact bifurcating cymes, the flowers and axes protected by thick enveloping bracts and bracteoles. Flowers small (5 mm long) each enclosed immediately in a cup-shaped pair of partially fused bracteoles. Calyx with five (to six) pointed valvate lobes. Petals five (to six) free, keeled, 3 mm long with apical appendages and marginal hairs, at first white but becoming brown with age. Stamens twice as many as the petals, either with long filaments and, at first, enclosed in pairs by the petals (*C. tagal*), or with short filaments and uniformly spaced (*C. decandra*). Disk within the stamen ring well developed, lobes enclosing the base of the stamens. Ovary semi-inferior, calyx tube short; locules three with two ovules per loculus; style slender and usually with three obscure stigmas. Fruits conical mainly by extrusion of the upper part of the ovary, surface brown and roughened, calyx tube warty, lobes persistent. Seedling and ultimately the cotyledonary tube penetrating the fruit. Hypocotyl up to 20 cm long, 5 mm at the widest part, terete or ridged, detached from the cotyledon at the moment of release. In *C. tagal*, n=18 (*Taxon* 27:29, Kilwa Rd, Dar-es-Salaam, Tanzania).

Geographic and ecological distribution. The genus is widely distributed from East Africa and Madagascar throughout tropical Asia and Queensland, to Melanesia (New Caledonia and the Solomon Islands) and through Micronesia north to Hong Kong. The precise eastern limit is not known. In Melanesia, *C. tagal* is recorded only for the very north of New Caledonia (Diohot, e.g., Schmid 1594 21.9.66, as *C. timoriensis*) and only from Winton Bay, Malekula (Gillison RSNH 3556, 9.10.71) Vanuatu. The species are typical constituents of the inner mangrove, often forming pure stands on better-drained sites, becoming stunted in exposed and highly saline sites, but rarely losing their simple trunk. They are easily distinguished from *Bruguiera* and *Rhizophora* by their glossy green, rounded (not pointed or apiculate) leaves, flattened buds, slender and rather short hypocotyls, and pale-barked trunk with restricted flaky-barked buttresses. The clusters of small shiny flower buds or short peduncles encased in varnish-like material are also characteristic. Wilkinson (1981) lists anatomical features of seedlings in which the species may differ.

Species. Subsequent descriptions of the genus are based largely on the two better-know and longer-recognized species. Clearly *C. australis* and *C. tagal* are sibling species, most notably in the explosive pollination mechanism.

The three species can be distinguished in technical details as follows:

1A. Inflorescence axis short and thick, expanded distally (10 mm or less by 3–4 mm). Petals not enclosing stamens at anthesis, the stamens in a single series not opening explosively, the apex of the slightly keeled petals with a fringe of filamentous appendages. Stamens with a short filament, equalled or exceeded by the anther, which ends in a longer or shorter appendage. Calyx tube swollen in fruit. Hypocotyl sharply ridged. . . . . *C. decandra*

1B. Inflorescence axis relatively long and uniformly slender (10–20 cm by 2 mm). Petals enclosing the stamens in pairs at anthesis and opening explosively, the apex of the petal with usually three clavate appendages (Fig. B.59i). Stamens with long, slender filaments, much exceeding the blunt anthers (Fig. B.59j). Calyx tube not swollen in fruit. Hypocotyl terete or ridged. 2

2A. Leaves yellow-green, blade usually longer than 6.5 cm, stipules <1.2 cm long; base of calyx lobes 12–15 mm wide; petals 5 mm long with three to five slender long appendages; style 2.7–4 mm long, fruit 9–14 mm long, hypocotyl terete, not exceeding 12 cm in length. . . . . *C. australis*

2B.   Leaves dark green, blade usually shorter than 6.5 cm, stipules >1.4 cm long; base
of calyx lobes usually 18–25 mm wide; petals 3 mm long with short appendages;
style 1.5 mm long, fruit 18–25 mm long, hypocotyl up to 35 cm long and
prominently ridged (Fig. B.59b). . . . . *C. tagal*

Ceriops tagal (Perr.) C. B. Robinson 1908 [*Aust. Syst. Bot.* 50: 391]
(Fig. B.59)
[*Aust. Syst. Bot.* 50: 391. *C. candolleana* Arn. 1838 [*Ann. Mag. Nat. Hist.* 1: 364]
C. timoriensis Domin 1928 [*Bibl. Bot.* 89: 444]
C. boiviniana Tulasne 1856 [*Ann. Sci. Nat.* 6: 112, see *Fl. Veg. Madagas.*:
589, 1974]

From East Africa and Madagascar throughout the Indo-Malayan region to Australia and
Vanuatu, northward to Hong Kong. The specific name comes from the Philippine
vernacular.

Although several specific names had been recognized, subsequent opinion relegated
them all to one rather variable entity. In Queensland, two varieties were recognized by
White (1926): *C. tagal* (Perr.) C. B. Rob. var. *tagal* for plants with distinctly fluted
seedling hypocotyls and *C. tagal* (Perr.) C. B. Rob. var. *australis* for plants with terete,
not fluted, seedling hypocotyls. Jones (1971) indicated that var. *australis* has a more
southerly range than var. *tagal*. Ballment et al. (1988) described them as distinct
species, creating the name *C. australis*, an approach subsequently justified by the more
complete analysis and discussion by Sheue et al. (2009).

Ceriops australis (White) Ballment, Smith, and Stoddard 1988 [*Austr. Syst. Bot.*
50: 391]
C. tagal Perr. C.B. Rob. var. *australis* C.T. White1926 [*J. Bot. Lond.* 64: 220]

This species is recognized by the characters used in the previous key based on Duke
(2006), with confirmation found in Sheue et al. (2009), where there is an added
thorough morphological analysis and molecular evidence. They also demonstrate a
wider distribution outside Australia in Indonesia and Papua New Guinea.

Ceriops decandra (Griff.) Ding Hou 1958 [*Fl. Males. Series 1*, vol. 5(4): 471]
C. roxburghiana Arn. 1838 [*Ann. Mus. Nat. Hist.* 1: 364]

A similar distribution to *C. tagal*, but absent from Africa and a more limited northern
distribution.

Ecological distribution. This species has a more limited distribution and is typically a plant of the inner mangroves, apparently on better-drained soils; flooding with
salt water is infrequent but produces high soil salinity by evaporation. Under these
circumstances, it will form pure stands. *C. decandra* more often grows within the tidal
zone mixed with other Rhizophoraceae. The two species differ in the major features of
floral biology, as is discussed later; otherwise, they are sufficiently similar that the
general account of vegetative features illustrates both. In Queensland, *C. decandra*
seems to have larger and darker leaves, but this may be a consequence of habitat
differences.

Seedling and sapling. Dispersal and establishment of seedlings follow the pattern usual for the Rhizophoreae. The sapling is initially unbranched, but once branching begins, it may be continuous or intermittent, depending on the vigor of the plant. Branching is by syllepsis, with the lateral axes developing in pairs within the terminal bud. This pattern is repeated on successive orders of branches but without any marked tendency to develop plagiotropy by apposition, as in *Rhizophora* and *Bruguiera*. Consequently, the crown remains relatively narrow. Vegetative branching becomes less frequent on distal axes. This minimizes the congestion of axes.

The terminal buds of *Ceriops* are distinctive in their construction. The individual stipules, which are shorter than those in the other Rhizophoreae, have a marked dorsal thickening so that the bud as a whole is flattened. There is much free space even within this flattened bud, because only a single younger pair of leaf primordia or stipules is present at any one time. In addition, the fluid, presumably secreted by the stipular colleters, dries to form a semi-solid plug of material that occupies the free space within the bud. The same material coats the surface of the newly extended internodes and young appendages in the manner of a varnish that flakes and falls off as the organs expand.

With the onset of flowering, which occurs in quite small saplings, the full complement of lateral appendages developed by the terminal bud is present so that each leaf may subtend the following:

1.   A single resting bud (the most usual condition).
2.   A single sylleptic branch, which is commonly associated with a supernumerary bud between branch and petiole.
3.   An inflorescence, which also may be associated with a supernumerary bud.

Internodes on distal shoots are shortened, characteristically in relation to flowering.

Architecture. *Ceriops* corresponds closely to Attims' model, as axes, at least in early stages of development, show the potential for continuous growth. Rauh's model is suggested by the peripheral axes, however, which branch infrequently and intermittently rather than rhythmically because the growth of shoots is essentially continuous. In early stages of crown development, there is much replacement of damaged axes by the direct substitution of an existing (sylleptic) lateral shoot for the damaged leader, but prolepsis can occur, although seemingly as a minor mechanism of repair. There is no basal sprouting.

Phenology. In seasonal climates, as in north Queensland, there is a clear restriction of flowering to summer (Duke, 2006), the flower buds developing during the previous winter. Fruiting follows from these flowers in the subsequent summer, with seedlings on the tree in late summer (February to March). This suggests that seedling development takes about a year, but some fruits formed in late summer overwinter. This extends the seedling season, as embryos of two summers mature at the same time.

Less evident periodicity of leaf development occurs simply by fluctuations in the time of leaf production according to climate (temperature). Otherwise, leaf expansion can occur at any time of the year.

Inflorescence structure. The lateral inflorescence is a modified cyme with regular dichotomy of axes; the plane of orientation changes by 180° at each successive fork. Axes are enclosed by thick bracts, and each flower, which is the ultimate product of branching, is at first enclosed by a pair of bracteoles. Irregularities occur in the system by the abortion of axes or flowers. Flower number is low (two to eight); *C. decandra* usually has fewer flowers than other species; its slightly larger flowers are congested on a short inflorescence axis. However, populations of *C. decandra* in north Queensland may have larger flower clusters, with up to 40 flowers per cluster.

Floral biology. The two common species differ strikingly in their floral mechanism. Most information about pollination is available for *C. tagal.*

In *C. tagal*, the flower buds appear to open mainly in the evening, emitting a faint but fragrant odor, and they may remain open the following day. Anther dehiscence occurs in bud, and at anthesis, the petals are closed, enveloping the stamens in pairs exactly as in *Bruguiera.* Pollination is probably mainly by night-flying insects; moths have been observed visiting the flowers, presumably for the small quantity of nectar secreted by the disk, but bees may be daytime visitors. Pollen release is explosive and is triggered by a delicate touch of the petals. This would presumably throw pollen onto a suitably positioned flower visitor. The tension that sets this mechanism is generated by the enclosed stamen pair held back by the pouched petal.

In *C. decandra*, the stamens are not enclosed at anthesis, and they retain their antesepalous or antepetalous position without displacement. Antepetalous stamens are appressed to the subtending petal, but there is no explosive release of pollen; the short filament is not suitable for developing tension. Details of flower visitors are lacking for this species, but the floral mechanism is apparently much less specialized than that of *C. tagal.*

Root architecture. Development of an above-ground root system is less conspicuous in *Ceriops* than in *Bruguiera* but follows the same pattern. Buttresses and aerial roots are restricted to the base of the trunk; the aerial roots appear first on seedlings, becoming the buttresses of the mature tree (Plate 12A). These in turn may continue to develop short descending aerial roots, so that the base of the trunk comes to be supported by a series of narrow flanges that form a conical structure enclosing a central hollow, into which further aerial roots develop (Fig. B.60). In more exposed and presumably better-drained sites, this may be the sole elaboration of above-ground roots, and the relative instability of the system is demonstrated because dead trees are easily pushed over. On wetter sites, where trees are presumably more vigorous, pneumatophores of the *Bruguiera* type are developed and well-developed projecting knobs can be seen, initiated as loops in the horizontal root system. This plasticity of the *Ceriops* root system has not been appreciated by several workers, so that there are reports of it lacking pneumatophores.

Kandelia Wight and Arnold 1834 [*Prodromus*: 310]
Fig. B.61; Plate 22A)

A genus of two species, ranging from India, the Ganges Delta, Myanmar, through Southeast Asia to south China, the Ryukyu Islands, and south Japan. The status of the genus in the Philippines is uncertain. Originally described as a single species, more recent research has recognized a second species in the northern part of its range, but without overlap in their distribution. In the southern part of its range, it is confined to northeast Sumatra and northern Borneo, an unusually restricted southern limit for mangrove Rhizophoreae. It is distinguished from related genera in the Rhizophoreae by the slender calyx lobes, the numerous, indefinite number of stamens with long slender filaments, the multifid, glabrous petals, the essentially unilocular ovaries, and the absence of either buttresses or specialized aerial roots. The hypocotyl of the seedling is characteristically slender and tapered at each end.

Kandelia occupies a narrow niche in the mangrove forest, as it is nowhere abundant and typically occurs in the back-mangrove communities or on the banks of tidal rivers farther inland. It still has some pioneering propensity, as recorded by Steup (in Ding Hou, 1958). The effect of plant density and aging on the mating system of *Kandelia candel* has been explored by Chen (2000).

Growth and reproduction. The architecture in the sapling stage corresponds precisely to Attims' model, with continuous growth and continuous to diffuse branching. The floral biology is not well known but seems to be the least specialized of all mangrove Rhizophoraceae, as the flowers apparently attract a diversity of small flying insects. Petals and stamens are ephemeral. In its floral biology, it most resembles *C. decandra*, although its flowers are much larger. The generalized floral construction of this species may represent the ancestral condition in the tribe. Sun et al. (1998) studied genetic diversity of 13 populations of *Kandelia* in Hong Kong (probably *K. obovata*), and described a nonspecialized floral mechanism such that seed set is still pollinator dependent. They suggest that the populations have a recent co-ancestry that limits genetic diversity.

The two species are distinguished in a preliminary way, as follows:

1A. Leaves elliptic-oblong; involucel of bracteoles V-shaped; sepals light green abaxially; each half petal with three to five even, long threads; hypocotyl 20–40 cm long.. *Kandelia candel*
    (More easterly and southerly in its distribution, from India to Borneo)

1B. Leaves obovate to obovate-elliptic; involucel of bracteoles U-shaped; sepals becoming white upon blooming (Plate 22A); each half petal with (six to) eight to 12 uneven, long threads; hypocotyl 15–23 cm long.........*Kandelia obovata*
    (More northerly and westerly in its distribution, from South China to Ryukyus)

Other distinctive features are included in the species descriptions.

Kandelia candel (L.) Druce 1914 [*Rep. Bot. Exch. Club Brit. Isles* 1913 (3): 420] *K. rheedii* Wight and Arnold 1834 [*Prodromus*: 311] (Fig. B.61)

Small understorey tree growing to 7 m. Buttresses and pneumatophores absent. Bark smooth, grayish or reddish brown. Leaves opposite, bijugate, 6–13 cm by 2.5–6 cm, oblong-elliptic to narrowly elliptic-lanceolate, apex obtuse, base cuneate, margin entire, somewhat reflexed. Petiole 1–1.5 cm long, terete. Stipules to 2 cm long, overlapping as in *Rhizophora.* Inflorescence axillary, bifurcating with four or more flowers, peduncles 2 or 3 cm but up to 5 cm long, slender. Bracteoles cup-shaped, the pair below the flower either enclosing the calyx cup or separated from it by a pedicel 2–3 mm long. Flowers (Fig. B.61g, h) 1.5–2 cm long, white; calyx lobes five (to six) linear, 15 mm by 2 mm, reflexed at anthesis. Petals 5 (to 6) –15 mm long, scarcely 1 mm wide, with a filamentous appendage between the lobes, the lobes themselves with three or four further linear appendages 5–6 mm long. Stamens numerous (30 to 40) inserted on the rim of the calyx cup, filaments uneven in length, 8–15 mm long, anthers minute. Floral disk cup-shaped, enclosing the shallow nectar cup. Ovary inferior, essentially unilocular but with evidence of three carpels; ovules six, anatropous, attached to basal projecting placenta. Style filiform, 1 cm long, stigma minutely three-lobed. Fruit 1.5–2 cm long with persistent reflexed sepals (Fig. B.61c), peduncle elongating. Seedling hypocotyl up to 40 cm long at maturity, slender, pointed apically, and tapered at each end.

Kandelia obovata Sheue, Liu & Yong 2003 [*Taxon* 52: 287]

In addition to the characters used in the key, this species differs from *K. candel* in chromosome number (2n 38 versus 2n 36), leaf anatomy, hypocotyl shape, stamen and style length, and pollen size. The clue to the recognition of this species was provided by a study of ecogeographic variation in *Kandelia* that showed some of the morphological differences, but most significantly, the greater cold tolerance of northern populations (Maxwell, 1995). In addition, genetic discontinuities have been recognized (Huang and Chen, 2000).

Family: Rubiaceae

A very large, cosmopolitan, and natural family with about 600 genera and over 10 000 species, represented in the tropics mainly by woody plants. Interpetiolar stipules (Fig. B.62b) with associated glandular structures (colleters) are a diagnostic feature; they protect terminal buds and usually persist on mature stems. The family includes several commercially important species, notably coffee.

Scyphiphora is a frequent but minor mangrove constituent in the Old World; a mangrove associate from the New World is also described.

Scyphiphora Gaertn. f. 1805 [*Fruct. Sem. Pl.* 3: 91, t. 196, Fig. 2]

A monotypic genus with a range from South India and Sri Lanka, Indochina and Hainan, through the Malay Archipelago and Philippines to tropical Australia and New Caledonia, northward to the Solomon Islands and Palau. It is an uncommon constituent of mangroves and more exposed coastal sites including beaches. It has been extensively studied because of its relative rarity, and even suggested to have a threatened status (Raju and Rajesh, 2014). Tan and Rao (1988), in a study of sporogenesis and gametogenesis, note a number of characters in this species that are anomalous for the family Rubiaceae, thus indicative of relative systematic isolation. *Scyphiphora* is one of a number of monotypic genera allied in the family (Ixoroideae-Vanguerieae), reflecting their early evolutionary divergence (Razafimandimbison et al., 2011).

Scyphiphora hydrophylacea Gaertn.f. 1805 [*Fruct. Sem. Pl.* 3: 91]
[*Ixora manila* Blanco 1837 *Fl. Filip:* 635]
(Figs. B.62, 63)
Gaertner used the spelling *hydrophylacea*, which is correct as it is derived from *Hydrophylax* L.f., for a plant collected by Sir Joseph Banks. The frequent use of *hydrophyllacea* is thus incorrect.

An erect shrub, rarely exceeding 2 m, but thicker stems develop a flaky bark. Leaves opposite, decussate, branching mainly by syllepsis but discontinuous (Attims' model, Hallé et al., 1978). Leaves glabrous, somewhat fleshy or coriaceous, oblong-ovate, 4–9 cm by 2–5 cm, acute at base, apex rounded to bluntly pointed or slightly emarginate, margin entire; petiole 1–2 cm long. Interpetiolar stipules short (to 3 mm), rounded, minutely hairy on the margin, at first enclosing younger parts, with well-developed colleters internally at the base. Flowers perfect, protandrous, usually tetramerous, in axillary regular, usually condensed cymes with three to seven (up to 13) flowers,

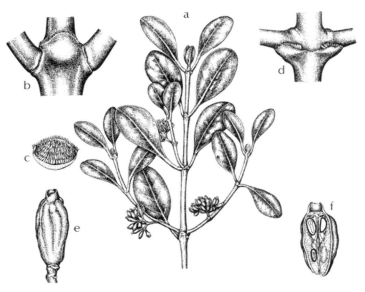

Figure B.62. *Scyphiphora hydrophylacea.* (Rubiaceae) shoot and fruit morphology. (a) Distal shoot (x1/2) with flowers and young fruits. (b) Detail of node (x4) with interpetiolar stipules. (c) Stipule (x4) detached to show colleters on inner face. (d) First node on inflorescence (x4) to show bracts with internal colleters. (e) Fruit. (f) Fruit in L.S. (x3) with four seeds (cf. Fig. B.63c). (Material from Barune, Port Moresby, Papua New Guinea.)

bracteoles obscure. Peduncles 2–15 mm long; flowers sessile or shortly but obscurely pedicellate. Calyx tube glabrous, 3–5 mm long, free portion of calyx scarcely 2 mm long entire with four(to five) obscure teeth. Corolla tube 3–4 mm long with four (to five) at first contorted white (or slightly red), bluntly pointed petal lobes, reflexed at anthesis; throat of corolla tube occluded with dense hairs. Stamens four (to five) inserted on the mouth of the corolla tube with a short (1 mm) filament, anthers 2 mm long, medifixed and dehiscing introrsely before the flower opens. Ovary inferior, indistinctly separate from calyx tube and pedicel; two-locular, with two ovules in each cell, one above the other, the upper ovules erect, the lower pendulous. Style slender with a club-shaped bilobed stigma, the lobes eventually spreading. Fruit (Fig. B.62e) green, turning brown, glabrous, up to 1 cm long, crowned with persistent calyx, with eight (to 10) longitudinal ridges. Outer layers of fruit fleshy, inner corky. Seeds four (or fewer) separated horizontally by an incomplete septum; each with a straight embryo, endosperm present, testa thin. Germination hypogeal.

Growth and Reproduction

Vegetative development. The terminal buds are protected by the hood-like stipules that secrete a copious varnish-like substance (presumably originating in the palisade of colleters internally at the base of each stipule pair), forming a plug enclosing the bud.

This material also coats the exposed surfaces of younger organs, especially inflorescences. The stipules are persistent. Branches are wide-angled and sylleptic, inserted a little above the node so they characteristically extend from above the mouth of the stipules. Growth is seemingly intermittent but irregular. There are apparently no supernumerary buds so that reserve buds occur only at nodes that do not develop sylleptic branches. This restricts the reiterative capacity of the plant.

Floral biology. Pollination biology of this species is among the best studied of all mangrove taxa (Raju and Rajesh, 2014). The flowers are in fairly regular compound but congested dichasia, but the order of flowering does not follow their age very precisely. Flowers in one head are often synchronous in development and protandrous. The anthers dehisce introrsely before the flower opens in a form of secondary pollen presentation (Fig. B.63e). The large and sticky pollen grains form a mass that is retained between the club-shaped stigmas and the stamens, being prevented from falling into the corolla tube by the weft of stiff hairs at the mouth of the corolla. The petals reflex, exposing the still-closed, club-shaped stigmas; pollen is further exposed by elongation of the style (Fig. B.63c). The shriveled anthers fall back with the petals and thus do not

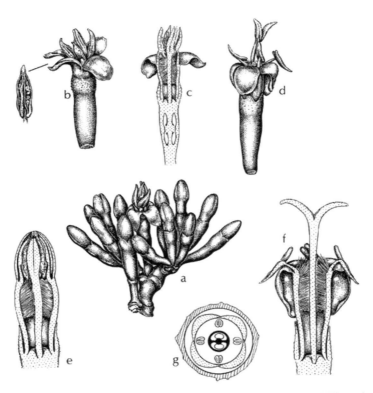

Figure B.63. *Scyphiphora hydrophylacea* (Rubiaceae) flowers. (a) Flowering branch (x4). (b) Flower just opening (x4); inset: detail of anther. (c) Flower in L.S. (x4). (d) Flower at late stage (x4), stamens reflexing stigma lobes diverged. (e) Flower bud in L.S. (x9). (f) Flower at late stage (x9); stigmas diverged, stamens reflexed. (g) Floral diagram. (Material from Barune, Port Moresby, Papua New Guinea.)

impede access to the corolla tube. Nectar is secreted by the glandular disk at the base of the corolla tube; it is accessible to short-tongued insects. Bees and other small insects are recorded visitors. Pollen is picked up either from the stamens themselves or from the back of the stigmas (much as in the related genus *Ixora*). The stigma lobes now diverge (Fig. B.63f) and present their receptive inner surface to a visiting insect transporting pollen from another flower. Eventually, the corolla lobes close again and the style falls to one side of the tube. Pollen adherent to the outside of the style can promote selfing. This can be important because of competition for pollinators from other mangroves.

Fruit biology. Incidence of fruit set can be high (almost 100 percent), but depends on the method of pollination that can be due to autogamy or xenogamy, the latter dependent on insect visitation whereas selfing has been demonstrated. There are reports of a low percentage of seed germination, however. The fruit turns brown at maturity as or before it is shed; it floats because of the spongy layers of the inner pericarp.

Nomenclature and Taxonomy

The genus has been regarded as isolated since it was erected by Gaertner. It is included by Schumann (1891) in Cinchonoideae–Gardeniinae–Gardenieae, but is distinguished by its evergreen habit, cymose inflorescence, two ovules per loculus with their characteristic orientation, few-seeded, drupe-like fruit with smooth (not hairy) seeds, and marine habitat. A more recent placement is in Rubioideae–Spermacocceae, which seems anomalous.

Merrill (1918) suggests that the city of Manila in the Philippines takes its name from this plant, the vernacular Tagalog name of which is reported by Blanco as *nilad* or *nilar*, from which he took his specific name *manila*, literally "the place where *nilar* grows." Ironically, the species has been essentially extirpated from this locality.

Rustia Klotsch in Hayne 1846 [*Arzn. Gew.* 14, sub. t. 14 and 15]

A genus of 15 species in tropical America and the West Indies.

Rustia occidentalis (Benth.) Hemsl. 1879 [*Biol. Cent. Am. Bot.* 2: 14]

This species has a coastal distribution and may be found as a mangrove associate. It is a shrub or small tree growing to 15 m, with terminal panicles of pale violet flowers that produce two-valved capsules. The petiolate leaves are large (15–25 cm by 5–15 cm), ovate-lanceolate, in opposite pairs, but with inconspicuous interpetiolar stipules.

Family: Rutaceae

A large, cosmopolitan family with over 1800 species in about 160 genera. Many species are the source of aromatic oils. The family includes the genus *Citrus*.

Merope Roem. 1846 [*Syn. Monogr.* 1: 144]

A single species, sometimes referred to as "mangrove lime."

Merope angulata (Willd.) Swingle 1915 [*Bull. Wash. Bioi. Soc.* 7: 22–23]

A shrubby, low tree or even scrambling plant with axillary, often paired spines. Leaves alternate, simple, coriaceous, aromatic with pellucid dots; blade 7–16 cm by 3–7 cm; petiole short, slender, unwinged. Flowers white, axillary, pentamerous, with 10 stamens; ovary three-locular, with four ovules in each loculus. Fruits triangular in section, 2–3 cm long with large long, flattened seeds, dispersed by tides.

This species seems restricted to back mangal and riverbanks, but is rather scattered. It occurs in India and throughout Malesia. Interest had been expressed in it as a possible salt-tolerant root stock for *Citrus* (Swingle, 1943), but it is placed some distance from that genus in recent analysis (Kubitzki, 2011). Genetic analysis shows that plants are predominantly outcrossing and with no local differentiation. In India, there is concern for it as a threatened species (Jena et al., 2015).

Family: Sapindaceae

A large tropical family of about 140 genera and 1900 species of trees, shrubs, and vines, usually with compound, exstipulate leaves. The following species is recorded as a mangrove associate in the Asian tropics.

Allophylus L. 1747 [*Fl. Zeyl.*: 58; *Sp. Pl.* 1753: 348.]

A large genus and of some complexity (Leenhouts, 1967).

Allophylus cobbe (L.) Bl. 1847 [*Rumphia* 3: 131]
See also Corner (1939) *Gardens Bull. Straits Settlements* 10: 38.

A species with a wide range from India, Sri Lanka, and eastward to New Guinea, but not the Philippines. It is variably recorded as a small tree, but more usually as an understorey shrub and frequently as a climber, a common habit in the family. It is recognized by its trifoliate leaves with stout woody petioles about 4 cm long, the leaflets ovate, up to 15 cm long, with incised margins, the apex blunt to pointed. Flowers of varying gender are small, about 2 mm long, white, in slender axillary spikes 8 mm long. Fruits are 5 mm or more in diameter, in hanging bunches, green turning red and fleshy; they are described as edible in some reports, but used as a fish poison in others. Germination epigeal.

The species also has a wide range, both ecological and altitudinal, in inland communities, and is somewhat variable in its vegetative parts. Radlkofer (1932) divided it into numerous species, but Corner (1939) prefers to treat it as one variable species, but still creates a number of varieties, which do not necessarily correspond to Radlkofer's species: Leenhouts (1967) is of the same opinion. Four of Corner's varieties have a coastal distribution, at least in part: var. *limosus* (with lanceolate leaflets 2–6.5 cm wide, becoming glabrous with age) is restricted to back mangal; var. *marinus* (with larger ovate glabrous leaflets, 4–10 cm wide, and whitish twigs) is said to be restricted to rocky and sandy shores. The flowers arise as condensed cushions on the spike and have greenish white sepals; stamens protrude at anthesis and render the flower relatively conspicuous. In a study of the floral structure of the related *Allophylus edulis* (A. St. Hil.) Niederl. (tribe Thouineae), Gonzalez et al. (2014) suggest that it is a model for all species for the family. Flowers are unisexual, the male with a reduced gynoecium, the female with indehiscent stamens.

Family: Sapotaceae

A mainly tropical, woody family with about 40 not well circumscribed genera and 600 species, commonly with a milky latex. The family is said to have affinities that "remain obscure" (Pennington, 2004).

> ***Pouteria*** Aubl. 1775 [*Pl. Guian.* 1: 85]

A large genus of over 200 species, not clearly distinguished from a number of other genera that are often included and its status as a monophyletic group is not supported (Triono et al., 2007). Several species provide edible fruit. The following, with a wide distribution, is sometimes recorded in back-mangal communities.

> ***Pouteria obovata*** (R. Br.) Baehni 1942 [*Candollea* 9: 423]
> [*Planchonella obovata* H. J. Lam 1925 [*Sap.* 209 etc., Fig. 58]
> See Baehni for a full synonymy.

The species may be divisible into two varieties based on leaf indumentum (Triono et al., 2007).

Recorded as a tree to 30 m but usually smaller, exuding white latex but not copiously. Leaves simple entire, obovate, to oblong-lanceolate, 8–12 cm by 3–5 cm, attenuate basally to a short petiole (2–3 cm long), the apex bluntly pointed or rounded. Surface of young twigs, petiole and lower leaf with a reddish brown tomentum. Flowers in axillary clusters, often on older wood; scent foetid. Fruit rounded, about 1 cm long, a one- (two- to three-) seeded drupe with a brown tomentum.

Family: Sonneratiaceae (Lythraceae)

Traditionally a tropical family of two small genera in the Indo-Malayan region. *Sonneratia* is restricted to mangrove communities; it has essentially solitary flowers, numerous stamens, vestigial or no petals, a fruit that does not dehisce regularly, and seed without extended tails. *Duabanga* includes two species in lowland tropical rain forest and has a more restricted range. It is distinguished from *Sonneratia* by its several-flowered inflorescence, conspicuous petals, 12 (or more) stamens, and especially the capsular fruit, dehiscing by four or more valves to release the numerous tailed seeds (Graham, 2014); vegetatively it lacks the conspicuous pneumatophores found in *Sonneratia*. In modern analyses, the two genera are included in an expanded Lythraceae where they remain closely related as an early divergent subclade but not forming a subfamily in a taxonomic sense (Graham et al., 2005).

Similarities between seedling *Sonneratia* and species of *Lythrum* are said to be particularly striking. The following account is based initially on that by Backer and van Steenis (1954), where a complete synonymy is given.

Sonneratia L. f. 1781 [*Suppl. So. Pl.* 38, a conserved name]
(Figs. 7.3, 9.3, B.64, 65; Plates 5J, 6D, E, 9D, 11B, 12F)

A genus of five to six species, ranging from East Africa through Indo-Malaya to tropical Australia and into Micronesia and Melanesia. Its pharmaceutical properties have been much investigated.

Sonneratia is a typical constituent of mangrove communities throughout its range, often forming a seaward fringe, and it is recognized by its tall conical pneumatophores arising from horizontal roots (Plates 11B, 12F), its opposite, simple, orbicular leaves, large flowers, and large, globose fruits with persistent calyx (Fig. B.64h), the seeds embedded in an often fleshy pulp. It is sometimes referred to as the "firefly mangrove," because these insects congregate on the trees at night, providing a tourist attraction. Trees are relatively uniform in their vegetative features; the following account describes these for one representative species. Where vegetative differences are known, they are included in the key (cf. Wyatt-Smith, 1953b). The root system is well illustrated by Troll and Dragendorff (1931) reproduced here as Fig. 7.3. Duke and Jacques (1987) provided an extensive commentary on *Sonneratia* in Australasia, amplified later with the recognition of hybrids, two of which are rare. In the description of the taxa occurring in Australia, Duke (2006) uses a range of characters that includes texture and shape of calyx cup, petal color, leaf size together with outline and shape (Fig. B.65).

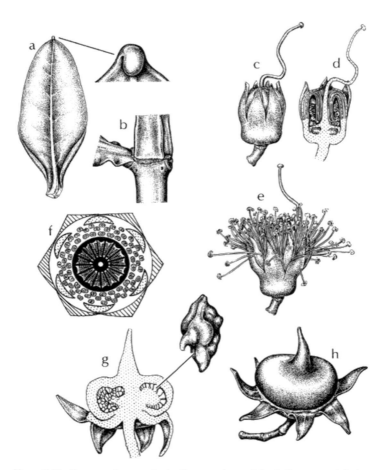

Figure B.64. *Sonneratia caseolaris* (Sonneratiaceae) leaf, flower, and fruit. (a) Leaf incompletely unrolled to show vernation (x1/2); inset: detail of recurved leaf tip (x9). (b) Node (x3/2) with one leaf removed. (c) Flower bud (x1/2). (d) Flower bud in L.S. (x1/2). (e) Flower at early anthesis (x1/2). (f) Floral diagram. (g) Fruit in L.S. (x1/2); inset: detail of seed (x9). (h) Fruit (x1/2). (Material from Semetan, Sarawak, together with P. Chai.)

The following descriptions suggest other characters that might be useful. However, accounts of this genus in different parts of its range can be very conflicting, and the present treatment is no exception. A thorough revision of the genus with genetic input is much needed, but a considerable task.

Suggested Key to Sonneratia Species

1A.   Calyx four- (to six-) lobed, ovary and calyx not exceeding 2 cm long, ovary five- to eight-celled, stigma expanded, mushroom-like; fruit 1.5–2 cm in diameter. Leaves narrow (less than 5 cm wide), but sometimes wider, gradually tapering toward the apex (Fig B.65C). . . . .*S. apetala*

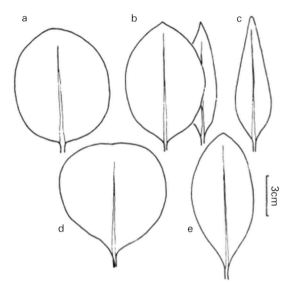

Figure B.65. *Sonneratia* spp. Outline of leaves. (A) *S. ovata.* (B) *S. caseolaris*, adult and "juvenile" or stenophyllous form behind. (C) *S. apetala.* (D, E) *S. alba* from two collections. Identification of *Sonneratia* species from leaf morphology is sometimes possible, but there is considerable overlap.

1B.   Calyx usually six- to eight-lobed, ovary and calyx 2.5 cm or longer, ovary 14- to 21-celled, stigma capitate but not expanded; fruit 3–7 cm in diameter. Leaves usually 5 cm wide or wider, abruptly narrowed to the rounded or even emarginate apex...2

2A.   Calyx flat, extended horizontally, not enclosing the ripe fruit, at most obscurely ribbed..3

2B.   Calyx cup-shaped, enclosing the base of the fruit, prominently ribbed..4

3A.   Twigs not pendulous. Leaves obovate to suborbicular, petiole scarcely developed, midrib green throughout; veins conspicuous, prominent on the upper blade surface. Petals absent; filaments white......................*S. griffithii*

3B.   Twigs slender, pendulous. Leaves elliptic-oblong or oval-obovate to narrowly elliptic, petiole short, midrib often red at base; veins inconspicuous, not prominent. Petals usually present, filaments red below, white distally..................*S. caseolaris*

4A.   Apex of fruit depressed at base of style. Tube of the fruiting calyx finely warted, lobes ascending, petals absent. Leaves broadly ovate, as broad as long, with a distinct, narrow petiole...*S. ovata*

4B.   Apex of fruit not depressed at base of style. Tube of the fruiting calyx smooth, lobes reflexed, petals white if present, stamens white. Leaves ovate to oblong-ovate, with a short thick petiole...*S. alba*

Wyatt-Smith (1953b) gives field characters for the three common species in Malaya. Characters of the calyx are much emphasized in some accounts (Duke, 2006), but as they refer to progressive changes from an erect position in the flower bud to an expanded and even recurved position in the mature fruit, they are much in need of quantitative analysis. Wang and Chen (2002) suggest a division within the genus of two sections determined by capitate versus peltate stigmas.

> ***Sonneratia caseolaris*** (L.) Engler 1897 [In Engler and Prantl, *Nachtr.*: 261]
> *S. acida* L.f. 1781 [*Suppl. Sp. Pl.*: 252]
> *S. lanceolata* Bl. 1851. [*Herb. Lugd. Batav.*: 567] (Figs. B.64; Plate 5J)

Trees to 15 m with continuous growth but diffuse branching (Attims' model), branches horizontal or drooping. Leaves glabrous, opposite, without stipules, shortly petiolate to almost sessile. Adult leaves (Fig. B.65B, left) broadly ovate or obovate, usually with a blunt apex, somewhat fleshy with an entire margin, 3–5 cm by 4–7 cm; juvenile leaves (Fig. B.65B, right) lanceolate, less fleshy, often with extended, red petioles. Apex in most leaves with a minute recurved spiculus serving apparently as a domatium rather than a hydathode. Nodes (including bracteolar nodes) with two lateral pairs of circular glands, one pair on each side of the stem. Axis angled, usually almost square in transverse section. Flowers ephemeral, opening in late evening and lasting one night. Flowers solitary or in few-flowered dichasia, terminal at the ends of outer twigs or in the axils of distal leaves, with at least one pair of subtending bracteoles. Flowers with a shallow, green calyx tube with six to eight somewhat longer valvate teeth; petals slender, reddish, alternate with calyx teeth (or absent), inconspicuous, and best seen in flower bud. Stamens numerous, inserted on the inner rim of the calyx tube, filaments slender, 2–3 cm long, red at least below, anther bilocular, medifixed. Ovary globose but flattened or even depressed above, style simple, about twice the length of the stamens, with a capitate stigma; ovary with numerous (up to 20) locules, each with numerous ovules on essentially axile placentas. Fruit a green, somewhat leathery "berry" with persistent subtending calyx (Plate 5J); seeds numerous in the fleshy pulp of the placenta, the individual seeds irregular and angular. Germination epigeal. Root system including an extended series of cable roots giving rise to narrow, shallowly descending lateral roots and erect pneumatophores. Pneumatophores at first greenish gray with a flaky back, extending to as much as 2 m at maturity, tapering, woody via secondary thickening and with numerous narrow second-order roots developing horizontally in the substrate. The characteristic habitat of the species is riverbanks and tidal areas with mud banks. It ascends the mouth of the river farther than other species.

This description may serve as an outline for the other species.

Distribution and Variation

S. caseolaris is common in the inner mangroves and extends inland along tidal creeks as far as the influence of salinity extends. It has a distribution from India and Sri Lanka throughout Southeast Asia, the Malay Archipelago and Philippines, north to Hainan, east to Vanuatu and into tropical Australia. It is distinguished by the large fruits in which the calyx lobes expand horizontally (not erect) in fruit and the red petals, which

Figure B.66. *Heritiera littoralis* (Sterculiaceae). Buttresses and bark characteristics. (Hinchinbrook Island, Queensland.)

contrast with the greenish or yellowish white of the surface of the sepals. The leaves do not have prominent veins and are usually pointed at the apex. The seeds, which tend to float and not sink, are smaller than those of *S. alba* and *S. ovata.*

Sonneratia lanceolata Bl. 1851 Bl. [*Herb. Lugd. Batav.*: 567]

Cited earlier as a synonym of *S. caseolaris*, it is clearly recognized in Australia by Duke (2006) and is listed as a separate species in Spalding et al. (2010). It is distinguished in its narrower leaves with pointed apices, narrow red petals, white stamen filaments and uniformly round (not constricted) flower buds, and a flat calyx whorl beneath the globose fruit. The pneumatophores are illustrated as slender and conical. It is referred to as a "sibling species" to *S. caseolaris*. Their ranges overlap in Queensland but the two do not co-occur in the same estuary. Furthermore, *S. lanceolata* is known from The Northern Territory whereas *S. caseolaris* is restricted to Queensland. It is recorded outside Australia.

Sonneratia alba J. Smith 1819 [*Rees Cycl.* 33(no. 2)]
S. griffithii Watson 1928 [*Malay. For. Rec.* 6: 120, 121, f. 24] (Plate 6D & E)

This is said to be the most widely distributed species; it ranges from east Africa and Madagascar to Southeast Asia, the Malay Archipelago to the Philippines, and tropical Australia to Micronesia, Vanuatu, and New Caledonia. According to Chai (1982), it is a pioneer species, colonizing newly formed sandy mud flats in sheltered situations. Flowers (Plate 6D & E) are distinguished by the numerous white petals (if present) that are tinged red basally, and the cupular calyx under the fruit with smooth (not hairy) lobes that are usually reflexed distally in a characteristic manner. The leaf is obovate to oval (Fig. B.66D & E), with the apex broadly rounded and even emarginate. The seeds are falcate and smooth. Duke (2006) suggests that petal-less flowers occur most commonly at higher latitudes, or may be segregated ecologically.

Sonneratia apetala Buch.-Ham. 1800 [*Symes Embassy Ava* 3: 477]

This species is perhaps the most distinctive in the genus but is the least common, being restricted to southern India and Myanmar. It is recorded as very rare in Sri Lanka, with a population of only six trees near Muttur in the estuary of the Koddiyar River (MacNae and Fosberg, 1981). It is a small to medium-sized tree, growing to 15 (–20) m high with narrow leaves (Fig. B.65C) gradually tapering toward the apex. The flowers (1.5–2 cm) are consistently smaller than those of the other species, and the usually four calyx lobes are twice as long as the tube. The ovary is five- to eight-celled, and the stigma is broad and mushroom-shaped at anthesis. The fruit (1–2 cm by 2 cm) is proportionately smaller than in the other species; the calyx lobes remain flat and do not enclose the fruit.

The specific name is not entirely appropriate, because other species of *Sonneratia* are apetalous, in flowers of populations that may elsewhere retain petals.

Sonneratia griffithii Kurz 1871 [*J. Asiat. Soc. Bengal* 2: 56 in the key]

This species has a restricted distribution along the shores of the Andaman Sea, northward to Bengal and southward to the upper Malay Peninsula and Sumatra. Although described as locally common, it is little collected. It is close to *S. caseolaris,* from which it is said to differ in the prominent leaf veins and absence of petals.

Sonneratia ovata Backer 1929 [*Bull. Jard. Bot. Buitenz.* 3(2): 329]
S. alba of several authors [e.g., Watson 1928, *Malay. For. Rec.* 6]
(Fig. B.65A)

This species has a distribution from Thailand through the Malay Peninsula and Malay Archipelago (but not Borneo) to eastern New Guinea but not Queensland. It is recorded by Chai (1982) as an inland riverbank species but still within salt-water influence. It lacks petals and is distinguished by the finely warty calyx, the lobes of which are red on the inner side, and by the erect calyx segments in fruit.

Hybrids in *Sonneratia*

Sonneratia provides a well-documented example of hybridization within mangroves (Muller and Hou-Liu, 1966; see also Muller and van Steenis, 1968). Hybrids were first recognized on the basis of reduced pollen fertility, and subsequently, with a mixture of parental characters. Populations in Brunei, northwest Borneo, are aggregates of *S. alba*, *S. caseolaris*, and *S. ovata*, together with intermediate forms suggested to be the hybrids *S. alba* x *ovata* and *S. alba* x *caseolaris*. Flowering and fruiting were poorest in the latter. Meiosis is normal in the parental species (n=11) but abnormal in the putative hybrids. The origin of these hybrids is not discussed. The pollen structure of *Sonneratia* (Fig. 9.3) is probably the best known of any mangrove genus, described in detail by Muller (1969). Field-work in Australia has extended the recognition of hybrids considerably, with the naming of two new taxa (Duke, 1984, 1994), in part based on the above-described characters. Unidirectional hybridisation, where one gender dominates the exchange, is described in *S. alba* by Zhou et al. (2008).

> **Sonneratia x gulngai** N.C. Duke 1984. (= *S. alba* x *S. caseolaris*) [*Austrobaileya* 2: 103–105]

The taxon is suggested to occur extensively elsewhere in southeast Asia, where it may therefore correspond to hybrids described by earlier authors. In Australia, it occurs most commonly where the ecological conditions of the two parents overlap.

> **Sonneratia x urama** N.C. Duke 1994 (= *S. alba* x *S. lanceolata*) [*Austral. Syst. Bot* 7: 521–526]

This is described as uncommon, but always in the presence of its putative parents. It is also recorded for adjacent Indonesia and New Guinea. It is named for the location in Papua New Guinea where it was first clearly described. In discussing its distribution, Duke emphasizes the difference in upstream and downstream occurrence of its putative parents.

S. haihanensis (= *S. apetala* x *S.ovata*) is supported by genetic analysis as is *S.* x *gulngai* (Li and Chen, 2008). *S. paracaseolaris* (= *S. alba* x *S. caseolaris*) was proposed for populations in China (Wang et al., 1999). The last name seems redundant as it is preceded by *S.* x *gulngai* (see earlier).

General Notes on *Sonneratia*

Heteroblasty. Leaf shape in *Sonneratia* is variable but diagnostically useful; the leaf margin is inrolled at the base, the apex likewise abaxially. Figure B.66 gives an indication of the range of variation in the four most common species. The phenomenon of leaf dimorphism between seedling and adult plants is pronounced in *Sonneratia* species, especially in *S. caseolaris* (Fig. B.65B), and to a lesser extent in S. *alba*. Leaves on seedlings (and sometimes on larger plants below tidal levels) are narrowly lanceolate (12–13 cm by 1–5 cm). Chai (1982) refers to the condition as "heterophylly," but perhaps "stenophylly" (adaptation of leaf shape to current flow in rheophytes – plants of moving waters) is the best term.

Pollination. The floral mechanism and floral biology in relation to pollinators are distinctive and have been much discussed (Ashton, 2014). *S. caseolaris* flowers continuously, but *S. alba* intermittently. Flower buds enlarge rapidly and open in the early evening, the style at first projecting. The stamens are recurved but expand abruptly as the calyx segments diverge. Collectively, they render the open flower conspicuous. At the same time, a sour, buttery odor is emitted. The most frequently noted visitors are bats (van der Pijl, 1936), which drink the copious nectar that fills the calyx cup. The contrast between flowers before- and after-bat visitation is shown in Plate 6D & E). In west Malaysia, several species of nectarivorous bats are responsible for the pollination of *Sonneratia*. The bat *Macroglossus minimus* Geoffroy is said to be dependent on *Sonneratia* as its major food source. It has not been recorded away from mangroves. Where bats are uncommon, hawk moths are an alternative visitor (Primack et al., 1981). Both kinds of pollinators touch stamens and stigmas in their search for nectar, the bats in particular being clumsy in their approach to the flowers. It is not known whether flowers are self-compatible, but the seed set is always high in developed fruits. Stamens and petals fall from the flowers within 12 hours of opening.

The relationship between bats and flowers in west Malaysia was investigated extensively by Start and Marshall (1976), and it has also been shown to involve fruiting of the commercially important durian (*Durio zibethinus* Mull.). Durian is strongly seasonal in its fruiting; the main season is July following an extensive flowering in April. Durian flowers are pollinated almost entirely by a single species of bat, *Eonycterus spelaea* Dobson, a widely distributed nocturnal bat that roosts primarily in large limestone caves, for example, Batu caves in Selangor. The bats are extremely fast flyers and may range up to 50 km from their roosts each night in search of food in the form of pollen and nectar from a diversity of plants. This range includes the mangroves, and species of *Sonneratia*, especially *S. alba*, are important food sources for the bats at certain times of the year. The mangroves therefore may sustain bat populations when durians are not in flower. This interdependence among bat, mangrove, and durian is a demonstration of the complexity of the processes that underlie conservation measures. The destruction of both caves and mangroves can adversely affect the production of durian, a fruit that plays an important role in the rural economy of Malaysia because it is a favorite of the city dweller (Lee, 1980).

Start and Marshall (1976) comment on the contrasted floral phenologies of *Sonneratia* species; *S. caseolaris* is described as aseasonal, which implies that flowers are available continuously, even though in small numbers, whereas *S. alba* and *S. ovata* are seasonal in their flowering and flowers are not continuously available in one locality. Further detail by A .N. Start includes a demonstration of the presence of *Sonneratia* pollen (Fig. 9.3), found in bat droppings in their roosting places, coinciding precisely with the periodicity of flowering of the different but distant mangrove species (Ashton, 2014).

Vegetative spread. *S. alba* is described by Holbrook and Putz (1982) as having a capacity for clonal development by reclining lower horizontal branches that root distally where they touch the mud. Repeated rooting produces extensive linear clones up to 37 m long, primarily oriented toward the sea. It is implied that this vegetative spread occurs only on rocky coastlines where seedlings do not become established.

Family: Sterculiaceae (Malvaceae: Sterculioideae)

A moderate-sized group (70 genera and 150 species) of tropical and subtropical trees and shrubs with alternate and simple leaves. Originally delimited as a distinct family, it is now subsumed within a more comprehensive Malvaceae (Alverson et al., 1999; Wilkie et al., 2006). It includes several commercially important species, notably *Theobroma cacao* (cocoa). One genus is represented in the Indo-Malayan mangroves.

> *Heritiera* Aiton. 1789 [*Hort. Kew.* 3: 546]
> [Including *Argyrodendron* F. v. Mueller and *Tarrietia* Bl.]

A genus of 29 species (Kostermans, 1959) mainly of rain forest trees with a distribution from eastern tropical Africa (two species) and the remainder from India to the Pacific (Tonga and Niue), with *H. littoralis* introduced farther east in Polynesia to Hawaii. The genus is characterized by its leaves with a scaly indumentum on the lower surface, the leaves either digitately compound or more usually simple. The fruit is usually a wind-dispersed samara, but in water-dispersed species the wing is reduced to a vestigial "keel."

A number of species has a coastal distribution. The genus is found predominantly in inland forests, but *H. littoralis* is a characteristic constituent of the back mangal, occupying almost the entire range of the genus; two others may be described as mangrove associates, both with a limited distribution. Several of the inland species are important timber species, for example, *H. sylvatica* Vid. in the Philippines.

Key to Mangrove or Salt-tolerant Species of Heritiera

1A. Petiole exceeding 2 cm (to 4 cm), fruit globose (Fig. B.67j), with a distal slightly recurved beak; plants of back-mangal fringe and lowland forests in western Borneo. . . . .*H. globosa*

1B. Petiole not exceeding 2 cm (1–2 cm), fruit ovoid, extended, flattened on dorsal side, without a distal beak; plants of back mangal, distribution wider. 2

2A. Fruit knobby with a ventral ridge together with a transverse, circular ridge; plants with a limited distribution in tidal swamps in the Ganges and Ayerarawady Deltas; pneumatophores sometimes developed. . . . .*H. fomes*

2B. Fruit smooth with a rudder-like crest (Fig. B.67b), but without a transverse circular ridge; plants with a wide distribution in back mangal, from East Africa to Melanesia; pneumatophores absent. . . . .*H. littoralis*

Figure B.67. *Heritiera* spp. (Sterculiaceae) leaf, fruit, and seedling. (a–i) *Heritiera littoralis*: (a) young fruits (x1/2) on portion of inflorescence that is still erect. (b) Detached mature fruit (x1/2). (c) Seed in T.S. (x1/2). (d) Fruit in L.S. (x1/2). (e) Fruit at early stage of germination (x1/2), radicle protruding. (f) Established fruit (x1/2) before emergence of plumule. (g) Young seedling (x1/3). (h) Leaf outline (x1/4); insets: details of abaxial epidermis (x35) and single epidermal scale (x90). (i) Detail of leaf insertion (x1) with bipulvinate petiole and one of the stipule pair. (j) *Heritiera globosa* fruit (x2); cf. part (b). (*H. littoralis* from Hinchinbrook Island, North Queensland; seedlings cultivated at Fairchild Tropical Garden; *H. globosa* from Kuching, Sarawak; P. Chai.)

The existence of these two local species, which resemble the widely distributed *H. littoralis* in their morphology and to some extent in ecology, indicates that the former species is less isolated from terrestrial relatives than is usual among mangrove taxa.

> **Heritiera littoralis** Dryand. in Aiton 1789 [*Hort . Kew.* ed. 1, 3: 546]
> *H. minor Lam.* 1797 [*Encycl.* 3: 299] (Figs. B.66, 67, 68)

This species is the type of the genus. A complete synonymy and citation are given in the monograph by Kostermans (1959). This tree is most easily identified by the silvery under surface of the large leaves with their short bipulvinate petioles. The woody, keeled fruits are highly distinctive, both on and off the tree.

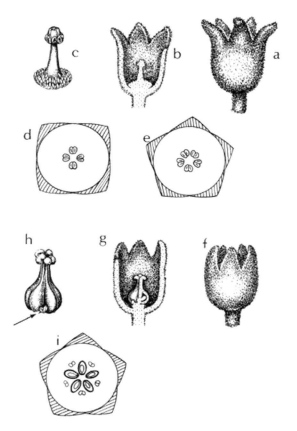

Figure B.68. *Heritiera littoralis* (Sterculiaceae). Floral morphology. (a–e) Male flower: (a) male flower (x4). (b) Male flower (x4) in L.S. (c) Stamen column and disk (x9). (d) Floral diagram of four-merous flower. (e) Floral diagram of five-merous flower. (f–i) Female flower: (f) female flower (x4). (g) Female flower (x4) in L.S. (h) Carpels (x9) with staminodes (arrow). (i) Floral diagram. (Material from Labu Lagoon, Lae, Papua New Guinea; J. S. Womersley s. n.)

Trees up to 25 m with a buttressed trunk (Fig. B.66) to 60 cm in diameter, evergreen, monoecious. Bark fissured, dark or gray. Leaves spirally arranged, simple (often regarded as a compound leaf with one leaflet, by reduction); stiffly coriaceous, oblong or ovate-elliptic, usually 10–20 cm by 5–10 cm (but up to 30 cm by 15–18 cm), with a usually obtusely pointed apex and somewhat cordate at the base. Margin entire. Veins prominent below. Leaves (Fig. B.67h) dark green above, lower leaf surface silvery beneath because of indumentum of overlapping, stellate scales (similar scales on young twigs). Petiole 1–2 cm long and 1 cm wide, shortly bipulvinate (Fig. B67i). Stipules in pairs at each node, short, subulate up to 1 cm long, caducous but conspicuous in resting buds and forming bud scales.

Flowers unisexual in complex tomentose panicles in axils mainly of distal foliage leaves on current growth, often subterminal, that is, arising immediately below resting terminal bud. Flowers (Fig. B.68; male [a–e], female [f–i] 3–4 mm in diameter, 4–5 mm long, with a slender short pedicel, the male somewhat smaller; both sexes with a cup-shaped calyx, the cup shallower in male flowers, reddish hairy inside, green and hairy outside with four or five (to six) short pointed lobes. Receptacle within the cup with a flattened tuberculate disk, the tubercules represented by minute fleshy structures that resemble short-stalked colleters. Petals absent. Male flowers more numerous, each with an androgynophore arising from the center of the disk and consisting of stamens fused together around a central column representing the pistillode; stamens four or five (to six), anthers minute and forming a ring around the central column, dehiscent extrorsely by two longitudinal slits to release the small amount of pollen. Female flowers less numerous and usually terminating distal panicles of branches; disc scarcely developed, staminodes minute below the carpels. Carpels four or five (to six) sessile, united loosely, each laterally compressed and with one (to two) basal ovule, styles as a short extension of the carpel, stigmas minute and recurved. Fruits maturing in pendulous clusters, usually only one developing from a single carpel in each flower; always one-seeded, 6–8 cm long, 5–6 cm wide, ellipsoidal, flattened on one side (ventrally) and with a raised (dorsal) keel 5 mm (but sometimes up to 2 cm) high on the opposite side, epicarp woody, mesocarp fibrous, endocarp hard. The keel represents the vestige of the wing found in terrestrial species. Embryo represented by fused cotyledons, radicle directed ventrally.

Fruits float with the "keel" upward (i.e., the ridge is not a functional keel). In contrast, Duke (2006) sees it as a sail. Seeds germinate readily in muddy substrates at the upper limits of tidal influences.

Floral morphology. The terminology applied to the flower is not clear and some accounts are confused. The central column, which supports the minute anthers is described, as earlier, as an androgynophore, implying a fusion of androecium and gynoecium. Some descriptions claim a similar structure in the female flower, but this does not exist in the material I have examined. The function of the colleter-like organs on the disc of the male flower, which are conspicuous in fresh material, has not been elucidated. Kostermans comments that there could be some difficulty in establishing the number of anthers because of their small size, but concludes that there are four or five bithecate anthers per flower.

Nothing is known about the floral mechanism; flower visitors have not been reported. It is possible that the tuberculate disc, at the base of the male floral cup is a nectary, although it is absent from the female flowers. Pollination could be promoted by insect visitors which do not distinguish flowers with and without this presumed reward. Plants in cultivation, as at the Montgomery Botanical Center, Florida, set abundant viable seed.

Vegetative Features

Seedling. Germination hypogeal (Boerlage, 1898), usually with the ventral side of the fruit downward (Fig. B.67e, f), the fruit wall gaping ventrally (Fig. B.67g), the radicle extruded first and soon branched, the plumule extruded later and becoming erect with initial plumular nodes supporting stipule pairs (the leaf itself aborted). Seedling leaves somewhat narrower and longer (to much longer) than adult leaves. This method of germination seems unusual in the genus, where the rule is apparently epigeal germination. This correlates well with the distinction between wind- and water-dispersed species.

Growth. The architecture corresponds to Rauh' s model (Hallé et al., 1978) with somewhat differentiated branches. As is characteristic of Rauh's model, the growth of the axes is rhythmic and the shoots are distinctly articulate. Resting terminal buds develop a series of short bud scales, the scars of which remain conspicuous. Trees remain evergreen.

Root system. This is not elaborated, except for a few short buttresses at the base of the trunk (Fig. B.66), which may be extended horizontally in a sinuous way, described by Parkinson (1934) as like those of *Xylocarpus granatum* but generally larger close to the trunk.

Ecology and Geography

H. littoralis is a frequent member of the back mangal and may occupy the forest fringe or rocky shores. It seems intolerant of high salinities and does not occur in very exposed or poorly drained sites. This species has almost the full range of the genus, extending from east Africa and Madagascar to the Pacific. In Australia, it is common in Queensland and on the east coast to Cape York, but has been recorded in neither Western Australia nor the Northern Territory. Its northern limit is Hong Kong, and in its natural range it extends eastward into the Pacific as far as Fiji. In a study of genetic diversity, Jian et al. (2004) showed that genetic differentiation could be accounted for in terms both of geographic location and contrasted habitats, here between China and Australia, and marine versus terrestrial locations.

Systematics and Phylogeny

Kostermans (1959) suggests that the most closely related species is *H. dubia* Wallich. ex Kurz., which is an inland tree of the Khasia Hills at low elevations, probably distinguished by its dull rather than smooth and glossy fruit. Following Kostermans, it is possible to describe, within the family Sterculiaceae, a typological reduction series in fruit type, beginning with the *Sterculia* type and ending with *H. littoralis* (and similar species). In this series, there is a progressive change from animal dispersal of individual seeds from open carpels, to closed winged carpels (samaras) with wind-assisted dispersal, to closed, unwinged, and water-dispersed fruits (essentially large achenes). The sequence proceeds as follows:

1. *Sterculia*: Attractive seeds attached along the full margin of the open carpels.
2. *Firmiana*: Seeds along the full margin of the carpel, which is membranaceous and remains closed.
3. *Scaphium* and *Pterocymbium*: Only one seed at the base of the membranaceous wing-shaped carpel.

4.   *Hildegardia*: One seed at the base of the membranaceous, wing-shaped carpel.
5.   *Heritiera*: Some species with a one-seeded, winged carpel (samara); others with the wing reduced; in other species that are clearly water-dispersed, notable *H. littoralis*, the wing is represented by a ventral keel (but may be developed as a vestigial wing). The seeds of some of the wind-dispersed species may be fairly large so that dispersal cannot be over very great distances. On the other hand, water-dispersed species always have large seeds (e.g., *H. fomes* and *H. littoralis*). The status of inland and upland species with large, wingless seeds (e.g., *H. dubia*) is problematical.

The description of the fruit of *H. littoralis* as a "capsule" in Duke (2006) is inappropriate.

Heritiera fomes Buch.-Ham. 1800 [*Symes Kingdom Ava,* ed. 2, 3: 319-320]

This species is characteristic of the forest of the Sundarbans of the Ganges Brahmaputra Delta (Khuma District) and in the Irrawaddy Delta of Burma. The Sundarbans is said to get its name from this tree (vernacular "sundri"), which is characteristic of the region. *H. fomes* is considered to be the most commercially valuable of the two *Heritiera* species in India; it has properties superior to those of teak. The difficult access, which makes it uneconomical to exploit, may be an important factor in its conservation.

It is distinguished from *H. littoralis* in its fruit and is described by Parkinson (1934) as having the ability to produce "erect pointed pneumatophores like those of *Xylocarpus gangeticus* [= *X. mekongensis*], but more numerous and longer. Kostermans refers to these pneumatophores as "blind rootsuckers" and "perpendicular shoots from its roots," but it seems clear that they are morphologically root structures. Watson (1928) makes the same mistake in describing pneumatophores. Parkinson compares these two species of *Heritiera* with the two mangrove species of *Xylocarpus*; in both genera there are contrasted root structures in different species.

From the available descriptions of the habitat of this species, it would probably be classified as a mangrove if it were more widely distributed. It is said to grow "down to the edge of the tidal creeks" in Myanmar (Kostermans, 1959).

Heritiera globosa Kostermans 1959 [*Reinwardtia* 4(4): 484] (Fig. B.67j)

Lower leaf surface coppery. This species is known from Sarawak to Sabah but may have a wider distribution. It is distinguished from *H. littoralis* by its globose fruit, which has a shallow ventral crest extended at the distal end into a beak or vestigial wing, with the wing always slightly recurved in a characteristic manner. The seed apparently retains undigested endosperm at maturity, and the embryo has two distinct cotyledons but an obscure plumule and radicle. Details of germination are not known. It is described as occurring behind the tidal zone of the mangrove belt, but has been collected 70 km from the sea; even here, although the water is fresh, there is tidal fluctuation. The buttresses are said to be well developed, snake-like, and extending 2–4 m from the base of the trunk.

Family: Tiliaceae (Malvaceae: Brownlowioideae)

With the realignment of genera in the family Malvaceae sens. lat. (Alverson et al., 1999), the following genus is now included along with a number of other genera in a subfamily Brownlowioideae, chiefly tropical and now removed from a close association with some familiar temperate genera (e.g., *Tilia*).

Brownlowia Roxburgh 1819 [*Pl. Corom.* 3: 61, t. 265]

A genus of about 25 species widely distributed in Southeast Asia from Malaysia to the Solomon Islands. Two species are recorded in swamp forests and riverbanks and penetrate into back-mangal communities inundated by the highest tides. They are distinguished as follows:

1A. Tree to 10 m; leaves broad, cordate or ovate to 10 cm wide, petiole long (to 6 cm), slender with a distal fleshy or corky pulvinus; flowers in terminal, lax panicles. . . . . *B. argentata*
1B. Shrub to 2 m; leaves narrow, lanceolate to 5 cm wide at most, petiole short (1 cm) without a distinct pulvinus; flowers in congested, usually axillary clusters. . . . . *B. tersa*

Brownlowia argentata Kurz 1870 [*J. Asiat. Soc. Bengal* 39(2): 67]
See also Scheff, 1874 [*Nat.Tijd. Ned. Ind.* 34: 94]

This species has the distribution of the whole genus from the Malay Peninsula through Indonesia (but not Java and the Lesser Sunda Islands) to New Guinea and the Solomon Islands.

A shrub or small tree growing to 10 m, often branched below, with scaly or stringy bark. Leaves cordate to ovate with a conspicuous pulvinus immediately below the insertion of the blade, the lower leaf surface and twigs silvery and covered with a close indumentum of minute peltate scales. Petiole slender, 4–6 cm long, the blade up to 15 cm by 10 cm. Flowers in erect, terminal, but lax panicles. Flowers about 5 mm long with a five-lobed, scaly calyx; petals five, orange, stamens numerous. Fruit a woody capsule or nut, obliquely globose-ellipsoidal, bilobed to heart-shaped but asymmetrically inserted on a short (to 1 cm) thick stalk, fruit surface scaly like the leaves. Normally only one or two carpels per flower developing into fruits.

Brownlowia tersa (L.) Kosterm. 1959 [*Reinwardtia* 4(4): 73]

A shrub, scarcely 2 m tall, with narrow, lanceolate to elliptic-lanceolate leaves, 8–12 cm by 2–3 cm, the apex gradually narrowed, often bluntly rounded. Petiole 1–2 cm without a pulvinus. Flowers small, in compact axillary clusters.

This species has a distribution in the Malay Archipelago and the Philippines. It could most often be confused with *Camptostemon* (both have a scaly leaf), but it is distinguished by leaf shape, the bilobed asymmetrical fruit, and paniculate flower clusters.

References

Adams, D. C. and Tomlinson, P. B. 1979. *Acrostichum* in Florida. *Amer. Fern J.* **69**: 42–46.

Airy-Shaw, H. K. 1975. The Euphorbiaceae of Borneo. *Kew Bull.* Add. Series **4**: 1–245.

Allen, J. A., Krauss, K. W., and Hauff, R. D. 2003. Factors Limiting the Intertidal Distribution of the Mangrove Species *Xylocarpus granatum*. *Oecologia* **135**: 110–121.

Allen, P. H. 1956. *The Rain Forests of Golfo Dulce*. Gainesville, FL: University of Florida Press.

Aluri, R. J. S. 1990. Observations on the Floral Biology of Certain Mangroves. *Proc. Indian Nat. Sci. Acad.* B 56. **4**: 367–374.

Alverson, W. S., Whitlock, B. A., Nyffeler, R., et al. 1999. Phylogeny of the Core Malvales: Evidence from *ndh*F Sequence Data. *Amer. J. Bot.* **86**: 1476–1486.

Arathi, H. S., Ganeshaiah, K. N., Uma Skaaner, R., et al. 1999. Seed Abortion in *Pongama pinnata* (Fabaceae). *Amer. J. Bot.* **86**: 659–662.

Arebeláez-Cortez, E., Castillo-Cárdenas, M. F., Toro-Perea, N., et al. 2007. Genetic Structure of the Red Mangrove (*Rhizophora mangle* L.) on the Colombian Pacific Detected by Microsatellite Molecular Markers. *Hydrobiologia* **583**: 321–330.

Areschoug, F. W. C. 1902. Untersuchungen über den Blattbau der Mangrove-Pflanzen. *Bibl. Bot.* **56**: 1–90.

Arzt, T. 1936. Die Kutikula einiger afrikanischer Mangrove-Pflanzen. *Ber. d. bot. Gesell.* **54**: 247–260.

Ashton, P. S. 2014. *On the Forests of Tropical Asia - Lest the Memory Fade*. Kew, UK: Kew Publishing, Royal Botanic Gardens; Chicago, IL: University of Chicago Press.

Atkinson, M. R., Findlay, G. P., Hope, A. B., et al. 1967. Salt Regulation in the Mangrove *Rhizophora mucronata* Lam. and *Aegialitis annulata* R. Br. *Austr. J. Biol. Sci.* **20**: 589–599.

Attims, Y. and Cremer, G. 1967. Les radicelles capillaires des palétuviers dans une mangrove de Côte d'Ivoire. *Adansonia*, Series 2. **7**: 547–551.

Aubréville, A. 1964. Problèmes de Ia mangrove d'hier et d'aujourd'hui. *Adansonia, n.s.* **4**: 19–23.

Azuma, H., Toyota, M., Asakawa, Y., et al. 2002. Floral Scent Chemistry of Mangrove Plants. *J. Plant Res.* **115**: 47–53.

Backer, C. A. and Bakhuizen van den Brink, R. C. 1963–1968. *Flora of Java*. 3 vols. Groningen: N. V. P. Noordhoff.

Backer, C. A. and van Steenis, C. G. G. J. 1954. Sonneratiaceae. *Flora Malesiana* 1, **4**: 280–289.

Baillon, H. 1875. *Natural History of Plants* (English edn.). London, UK: L. Reeve.

Bakhuizen van den Brink, R. C. 1921. Revisio generis Avicenniae. *Bull. Jard. Bot. Buitenz.* Series 3, **3**(2): 199–226.

Bakhuizen van den Brink, R. C. 1924. Revisio Bombacacearum. *Bull. Jard. Bot. Buitenz.* Series 3, **6**(2): 161–240.

Ball, M. C. 1988. Salinity Tolerance in the Mangroves *Aegiceras corniculatum* and *Avicennia marina*. I. Water Use in Relation to Growth, Carbon Partitioning, and Salt Balance. *Austral. J. Plant Physiol.* **15**: 447–464.

Ball, M. C. 1996. Comparative Ecophysiology of Mangrove Forest and Tropical Lowland Moist Rainforest. In *Tropical Forest Plant Ecophysiology*. S. S. Mulkey, R. L. Chazdon, and A. P. Smith, eds. New York, NY: Chapman and Hall; Chap. 16, pp. 461–496.

Ball, M. C. and Farquhar, G. D. 1984a. Photosynthetic and Stomatal Responses of Two Mangrove Species, *Aegiceras corniculatum* and *Avicennia marina*, to Long Term Salinity and Humidity Conditions. *Plant Physiol.* **74**: 1–6.

Ball, M. C. and Farquhar, G. D. 1984b. Photosynthetic and Stomatal Responses of the Grey Mangrove, *Avicennia marina*, to Transient Salinity Conditions. *Plant Physiol.* **74**: 7–11.

Ballment, E. R., Smith III, T. J., and Stoddart, J. A. 1988. Sibling Species in the Mangrove Genus *Ceriops* (Rhizophoraceae), Detected using Biochemical Genetics. *Austral. Syst. Bot.* **1**: 391–397.

Barker, N. P., Harman, K. T., Ripley, B. S., et al. 2003. The Genetic Diversity of *Scaevola plumieri* (Goodeniaceae), an Indigenous Dune Coloniser, as Revealed by ISSR Fingerprinting. *S. Afr. J. Bot.* **68**: 532–541.

Barlow, B. 1966. A Revision of the Loranthaceae of Australia and New Zealand. *Austral. J. Bot.* **14**: 421–499.

Barth, H. 1982. The Biogeography of Mangroves. In *Tasks for Vegetation Science*, Vol. 2. D. N. Sen and K. S. Rajpurohit, eds.The Hague: Junk; Chap. 3.

Baum, D. A., Alverson, W. S, and Nyffeler, R. 1998. A Durian by any other Name: Taxonomy and Nomenclature of the Core Malvales. *Harvard Papers in Botany* **3**: 317–332.

Baylis, G. T. S. 1940. Leaf Anatomy of the New Zealand Mangrove. *Trans. R. Soc. N. Z.* **70**: 164–170.

Baylis, G. T. S. 1950. Root Systems of the New Zealand Mangrove. *Trans. R. Soc. N. Z.* **78**:509–514.

Beauvisage, L. 1918. *Étude anatomique de la famille des Ternstroemiacées*. Thesis. Toulouse: Université de Toulouse.

Beekman, E. M. 2011. *The Amboinese Herbal*. An English Translation of Georgius Everhardus Rumphius (Georg Eberhard Rumpf, 1627–1702). 6 vols. New Haven & London: Yale University Press and National Tropical Botanical Garden.

Behnke, H. D. 1988. Sieve Element Plastids and Systematic Relationships of Rhizophoraceae, Anisophylleaceae, and Allied Groups. *Ann. Mo. Bot. Gard.* **75**: 1387–1409.

Benecke, W. and Arnold, A. 1931. Kulturversuche mit Keimlingen von Mangrove-Pflanzen. *Planta* **14**: 471–481.

Bentham, G. and Hooker, J. D. 1862. *Genera Plantarum*, vol. I. London, UK: Reeve.

Benzing, D. H. 1990. *Vascular Epiphytes*. Cambridge, UK: Cambridge University Press.

Benzing, D. H. and Davidson, E. 1979. Oligotrophic *Tillandsia circinalis* Schld. (Bromeliacea): An Assessment of its Patterns of Mineral Association and Reproduction. *Amer. J. Bot.* **66**: 386–397.

Berjak, P. and Pammeter, N. W. 2006. From *Avicennia* to *Zizania*: Seed Recalcitrance in Perspective. *Ann. Bot.* **101**: 213–218.

Bhosale, L. J. and Shinde, L. S. 1983. Significance of Cryptovivipary in *Aegiceras corniculatum* (L.) Blanco. In *Biology and Ecology of Mangroves. Tasks for Vegetation Science* 8. H. J. Teas, ed. The Hague: Junk; Chap. 14.

Biebl, R. and Kinzel, H. 1965. Blattbau und Salzhausinhalt von *Laguncularia racemosa* (L.) Gaertn.f. und anderer Mangrovebaüme auf Puerto Rico. *Ost. Bot. Zeit.* **112**: 56–93.

Biswas, K. 1934. A Comparative Study of the Indian Species of *Avicennia*. *Notes R. Bot. Gard. Edinburgh* **89**: 159–166.

Black, J. M. 1913. The Flowering and Fruiting of *Pectinella antartica* (*Cymodocea antartica*). *Trans. Proc. R. Soc. S. Austral.* **37**: 1–5.

Blasco, F. 1977. Outlines of Ecology, Botany, and Forestry of the Mangals of the Indian Subcontinent. In *Ecosystems of the World. Vol. 1. Wet Coastal Ecosystems*. V. J. Chapman, ed. Amsterdam: Elsevier Scientific; Chap. 12.

Boerlage, J. G. 1898. Sur le maniche de flotter et Ia germination des fruits du *Heritiera littoralis* Dryand. *Ann. Jard. bot. Buitenz.* Suppl. **2**: 137–142.

Booberg, G. 1933. Die malayische Strandflora. Eine Revision der Schimperschen Artenliste. *Bot. Jahrb.* **66**: 1–38.

Borg, A. J., McDade, L. A., and Schönenberger, J. 2008. Molecular Phylogenetics and Morphological Evolution of Thunbergioideae (Acanthaceae). *Taxon* **57**: 811–822.

Borg, A. J. and Schönenberger, J. 2011. Comparative Floral Development and Structure of the Black Mangrove Genus *Avicennia* L. and Related Taxa in the Acanthaceae. *Int. J. Plant Sci.* **172**: 330–344.

Bowman, H. H. M. 1916. Physiological Studies in *Rhizophora*. *Proc. Nat. Acad. Sci. U.S.A.* **2**: 685–688.

Bowman, H. H. M. 1917. Ecology and Physiology of the Red Mangrove. *Proc. Amer. Philos. Soc.* **56**: 589–672.

Brenner, W. 1902. Ueber die Luftwurzeln von *Avicennia tomentosa*. *Ber d. bot. Gesell.* **20**: 175–189.

Breteler, F. J. 1969. The Atlantic Species of *Rhizophora*. *Acta Bot. Neerl.* **18**: 434–444.

Breteler, F. J. 1977. America's Pacific Species of *Rhizophora*. *Acta Bot. Neerl.* **26**: 225–230.

Briggs, B. G. and Johnson, L. A. S. 1979. Evolution of the Myrtaceae – Evidence from Inflorescence Structure. *Proc. Linn. Soc. N. S. W.* **102**(4): 157–256.

Brown, F. H. B. 1935. Flora of Southeastern Polynesia. III. Dicotyledons. *Bull. Bishop Mus.* **130**: 1–386.

Brown, J. M. A., Outred, H., and Hill, F. C. 1969. Respiratory Metabolism in Mangrove Seedlings. *Plant Physiol.* **44**(2): 287–294.

Brown, W. H. and Fischer, A. F. 1920. Philippine Mangrove Swamps. In *Minor Products of Philippine forests*. W. H. Brown, ed. *Vol. 1. Phil. Bur. Forest. Bull.* **22**: 1–125.

Brown, W. H. and Merrill, E. D. 1920. Philippine Palms and Palm Products. In: *Minor Products of Philippine forests*. W. H. Brown, ed. *Vol. 1. Phil. Bur. Forest.* Bull. **22**: 127–248.

Browne, P. 1756. *The Civil and Natural History of Jamaica*. London: published privately.

Budowski, G. 1965. Distribution of Tropical American Rain Forest Species in the Light of Successional Processes. *Turrialba (Costa Rica)* **15**: 40–42.

Bunt, J. S. and Williams, W. T. 1981. Vegetational Relationships in the Mangroves of Tropical Australia. *Mar. Ecol. Prog. Ser.* **4**: 349–359.

Bunt, J. S., Williams, W. T., and Duke, N. C. 1982. Mangrove Distributions in Northeast Australia. *J. Biogeogr.* **9**: 111–120.

Burkill, I. H. 1935. *A Dictionary of the Economic Products of the Malay Peninsula*. 2 vols. Oxford: Oxford University Press.

Byrnes, N. B. 1977. A Revision of Combretaceae in Australia. *Contrib. Queensl. Herb.* **20**: 1–88.

Camilleri, J. C. and Ribi, G. 1983. Leaf Thickness of Mangroves (*Rhizophora mangle*) Growing in Different Salinities. *Biotropica* **15**: 139–141.

Camilleri, J. C. and Ribi, G. 1844. *Prodromus systematis naturalis*. Part **8**. Paris: Fortin Masson et Soc.

Carey, G. 1934. Further Investigations on the Embryology of Viviparous Seeds. *Proc. Linn. Soc. N.S.W.* **59**: 392–410.

Carey, G. and Fraser, L. 1932. The Embryology and Seedling Development of *Aegiceras majus* Gaert. *Proc. Linn. Soc. N.S.W.* **57**: 341–360.

Carolin, R. C. 1960. The Structures Involved in the Presentation of Pollen to Visiting Insects in the Order Campanales. *Proc. Linn. Soc. N.S.W.* **85**: 197–207.

Castillo-Cardenas, M.F., Diaz-Gonzales, F., Ceron-Souza, I., et al. 2015. Jumping a Geographic Barrier: Diversification of the Mangrove Species *Pelliciera rhizophorae* (Tetrameristicaceae) across the Central American Isthmus. *Tree Genet. Genom.* **11**: art. 822.

Cerón-Souza, I., Bermingham, E., Rivera-Ocasio, E., et al. 2012. Comparative Genetic Structure of Two Mangrove Species in Caribbean and Pacific Estuaries of Panama. *BMC Evol. Biol.* **12**: 205.

Cerón-Souza, I., Rivera-Ocasio, E., Medina, E., et al. 2010. Hybridization and Introgression in New World Red Mangroves, *Rhizophora* (Rhizophoraceae). *Amer. J. Bot.* **97**: 945–957.

Cerón-Souza, I., Toro-Perea, N., and Cárdenas-Henao, H. 2005. Population Genetic Structure of Two Mangrove Species on the Colombian Pacific Coast: *Avicennia germinans* (Avicenniaceae). *Biotropica* **37**: 258–265.

Chai, P. K. 1982. *Ecological Studies of Mangrove Forest in Sarawak*. Ph.D. Thesis. Kuala Lumpur: University of Malaysia.

Chandler, M. E. J. 1957. Note on the Occurrence of Mangrove in the London Clay. *Proc. Geol. Assoc.* **62**: 271–272.

Chapman, V. J. 1940. The Functions of the Pneumatophores of *Avicennia nitida* Jacq. *Proc. Linn. Soc. Lond.* **152**(3): 228–233.

Chapman, V. J. 1947. Secondary Thickening and Lenticels in *Avicennia nitida* Jacq. *Proc. Linn. Soc. Lond.* **158**: 2–6.

Chapman, V. J. 1962a. Respiration Studies of Mangrove Seedlings, I. *Bull. Mar. Sci. Gulf. Caribb.* **12**(1): 137–167.

Chapman, V. J. 1962b. Respiration Studies of Mangrove seedlings, II. *Bull. Mar. Sci. Gulf. Caribb.* **12**(2):245–263.

Chapman, V. J. 1976. *Mangrove Vegetation*. Valduz: Cramer.

Chapman, V. J. 1977a. Ecosystems of the World. Vol. 1. Wet Coastal Ecosystems. Amsterdam: Elsevier Scientific.

Chapman, V. J. 1977b. Africa B, The Remainder of Africa. In Ecosystems of the World. Vol. 1. Wet Coastal Ecosystems. V. J. Chapman. ed. Amsterdam: Elsevier Scientific; Chap. 11.

Chapman, V. J. 1977c. Wet Coastal Formations of Indo-Malesia and Papua New Guinea. In Ecosystems of the World. Vol. 1. Wet Coastal Ecosystems. V. J. Chapman, ed. Amsterdam: Elsevier Scientific, Chap. 13.

Chen, X. Y. 2000. Effects of Plant Density and Age on the Mating System of *Kandelia candel* Druce (Rhizophoraceae), a Viviparous Mangrove Species. *Hydrobiologia* **432**: 189–193.

Chomicki, G., Bidel, L. P. R., Baker, W. J., et al. 2014. Palm Snorkelling: Leaf Bases as Aeration Structures in the Mangrove Palm (*Nypa fruticans*). *Bot. J. Linn. Soc.* **174**: 257–270.

Christensen, B. 1978. Biomass and Primary Production of *Rhizophora apiculata* Bl. In *A Mangrove in Southern Thailand. Aquat. Bot.* **4**: 43–52.

Christensen, B. 1983. Mangroves – What are they Worth? *Unasylva* **35**: 2–15.

Christensen, B. and Wium-Anderson, S. 1977. Seasonal Growth of Mangrove Trees in Southern Thailand, I. The Phenology of *Rhizophora apiculata* Bl. *Aquat. Bot.* **3**: 281–286.

Cintron, G., Lugo, A. E., Pool, D. J., et al. 1978. Mangroves of Arid Environments in Puerto Rico and Adjacent Islands. *Biotropica.* **10**: 110–121.

Clarke, P. J. and Myerscough, P. J. 1991. Bouyancy of *Avicennia marina* Propagules in Southeastern Australia. *Austral. J. Bot.* **39**: 77–83.

Clough, B. F., Boto, K. G., and Attiwill, P. M. 1983. Mangroves and Sewage: A Re-evaluation. In *Biology and Ecology of Mangroves. Tasks for Vegetation Science* 8. H. J. Teas, ed. The Hague: Junk; Chap. 17.

Clough, B. F. 1992. Primary Productivity and Growth of Mangrove Forests. In *Coastal and Estuarine Studies.* A. I. Robertson and D. M. Alongi, eds. Washington, DC: American Geophysical Union, Chap. 8.

Collins, J. P., Berkelhamer, R. C., and Mesler, M. 1977. Notes on the Natural History of the Mangrove *Pelliciera rhizophorae* Tr. & Pl. (Theaceae). *Brenesia* **10/11**: 17–29.

Compère, P. 1963. The Correct Name of the Afro-American Black Mangrove. *Taxon.* **12**: 150–152.

Connor, D. I. 1969. Growth of Grey Mangrove *(Avicennia marina)* in Nutrient Culture. *Biotropica* **1**: 36–40.

Cook, M. T. 1907. The Embryology of *Rhizophora mangle. Bull. Torrey Bot. Club* **34**: 271–277.

Corner, E. J. H. 1939. Notes on the Systematy and Distribution of Malayan Phanerogams, III. *Gard. Bull. Straits Settl.* **10**:239–328.

Corner, E. J. H. 1966. *The Natural History of Palms.* Berkeley and Los Angeles. CA: University of California Press.

Corner, E. J. H. 1976. *The Seeds of Dicotyledons.* 2 vols. Cambridge, UK: Cambridge University Press.

Corner, E. J. H. 1978. The Freshwater Swamp-forest of South Johore and Singapore. *Gard. Bull. Singapore*, Suppl. **1**: 1–266.

Correll, D. S. and Correll, H. B. 1982. *Flora of the Bahama Archipelago.* Valduz: Cramer.

Cowan, R. S. and Polhill, R. M. 1981. Tribe 4, Detarieae. In *Advances in Legume Systematics*. R. M. Polhill and P. H. Raven, eds. Royal Botanic Gardens, Kew: Ministry of Agriculture and Fisheries; pp. 117–134.

Craighead, F. C. 1971. *The Trees of South Florida .I. The Natural Environments and their Succession.* Coral Gables, FL: University of Miami Press.

Cridland, A. A. 1964. *Amyelon* in American Coal-balls. *Palaeontology* **7**: 186–209.

Cruden, R. W. 1977. Pollen-ovule Ratios: A Conservative Indicator of Breeding Systems in Flowering Plants. *Evolution* **31**: 32–46.

Dacey, J. W. H. 1980. Internal Winds in Waterlilies: An Adaptation for Life in Anaerobic Sediments. *Science* **203**: 1253–1255.

Dacey, J. W. H. 1981. Pressurized Ventilation in the Yellow Waterlily. *Ecology* **62**: 1137–1147.

Dahlgren, R. M. T. 1988. Rhizophoraceae and Anisophylleaceae: Summary Statement, Relationships. *Ann. Mo. Bot. Gard.* **75**: 1259–1277.

Dandy, J. E. and Exell, A. W. 1938. On the Nomenclature of Three Species of *Caesalpinia. J. Bot. Lond.* **76**: 175–183.

Dangremond, E. M. and Feller, I. C. 2014. Functional Traits and Nutrient Limitations in the Rare Mangrove *Pelliciera rhizophorae. Aquatic Botany* **116**: 1–7.

Das, S. and Vincent, J. R. 2009. Mangroves Protected Villages and Reduced Death Toll during Indian Supercyclone. *Proc. Nat. Acad. Sci.* **106**(18): 7357–7360.

Davey, J. E. 1975. Notes on the Mechanism of Pollen Release in *Bruguiera gymnorrhiza. J. S. Afr. Bot.* **41**: 269–272.

Davis, J. H. 1940. The Ecology and Geologic Role of Mangroves in Florida. *Pap. Tortugas Lab.* **32** *(Publ. Carn. Inst.* No. 517): 303–341.

De Candolle, A. C. 1841. Second memoire sur Ia famille des Myrsinaceae. *Ann. Sci. Nat.* Series 2, **16**: 1–46.

De Menezes, N. L. 2006. Rhizophores in *Rhizophora mangle* L.: An Alternative Explanation Interpretation of so-called "Aerial Roots." *An. Acad. Bras. Cienc.* **78**(2): 213–226.

De Vogel, E. F. 1980. *Seedlings of Dicotyledons*. Wageningen: Centre for Agricultural Publication and Documentation.

Den Hartog, C. 1970. The Sea-Grasses of the World. *Verhandl. Kon. Ned. Akad. Wetensch. Nat.* **59**(1): 1–275.

Deng, S. L., Huang, Y. L., He, H. H., et al. 2009. Genetic Diversity of *Aegiceras corniculatum* (Myrsinaceae) Revealed by AFLP. *Aquat. Bot.* **90**: 275–281.

Ding Hou. 1957. A Conspectus of the Genus *Bruguiera* (Rhizophoraceae). *Nova Guinea n.s.* **8**(1): 163–171.

Ding Hou. 1958. Rhizophoraceae. *Flora Malesiana*, Series 1, **5**: 429–493.

Ding Hou. 1960. A Review of the Genus *Rhizophora*. *Blumea* **10**(2): 625–634.

Ding Hou. 1963. Celastraceae. *Flora Malesiana*, Series 1, **6**(2): 227–291.

Ding Hou. 1977. Anacardiaceae. *Flora Malesiana*, Series 1, **8**: 395–548.

Docters van Leeuwen, W. 1911. Ueber die Ursache der weiderholter Versweigung der Stiiltzwurzeln von *Rhizophora*. *Ber. d. bot. Gesell.* **29**: 476–478.

Dodd, R. S., Afzal-Rafi, Z., Kashani, N., et al. 2002. Land Barriers and Open Oceans: Effects on Gene Diversity and Population Structure in *Avicennia germinans* L. (Avicenniaceae). *Mol. Ecol.* **11**: 1327–1338.

Dransfield, J., Uhl, N. W., Asmussen, C. B., et al. 2008. *Genera Palmarum: The Evolution and Classification of Palms*. Kew, UK: Royal Botanic Gardens, Kew Publishing.

Duke, N. C. 1984. A Mangrove Hybrid, *Sonneratia* x *gulngai* (Sonneratiaceae) from Northeastern Australia. *Austrobaileya*. **2**: 103–105.

Duke, N. C. 1988. An Endemic Mangrove Species *Avicennia integra* sp. nov. (Avicenniaceae) in Northern Australia. *Austral. Syst. Bot.* **1**: 177–180.

Duke, N. C. 1990. Morphological Variation in the Mangrove Genus *Avicennia* in Australasia: Systematic and Ecological Considerations. *Austral. Syst. Bot.* **3**: 221–239.

Duke, N. C. 1991a. A Systematic Revision of the Mangrove Genus *Avicennia* (Avicenniaceae) in Australasia. *Austral. Syst. Bot.* **4**: 299–324.

Duke, N. C. 1991b. *Nypa* in the Mangroves of Panama in Central America: Introduced or Relic. *Principes* **35**: 127–132.

Duke, N. C. 1992. Mangrove Floristics and Biogeography. In *Coastal and Estuarine Studies*. A. I. Robertson and D. M. Alongi, eds. Washington, DC: American Geophysical Union; Chap. 4, pp. 63–99.

Duke, N. C. 1994. A Mangrove Hybrid, *Sonneratia x urama* (Sonneratiaceae) from Northern Australia and Southern New Guinea. *Austral. System. Botany*. **7**: 521–526.

Duke, N. C. 2006. *Australia's Mangroves: The Authoritative Guide to Australia's Mangrove Plants*. Brisbane: University of Queensland.

Duke, N. C. 2010. Overlap of Eastern and Western Mangroves in the South-western Pacific: Hybridization of all Three *Rhizophora* Combinations in New Caledonia. *Blumea* **55**: 171–188.

Duke, N. C., Ball, M. C., and Ellison, J. C. 1998. Factors Influencing Biodiversity and Distributional Gradients in Mangroves. *Global Ecol. Biogeogr. Lett.* **7**: 27–47.

Duke, N. C., Benzie, J. H., Goodall, J. A., et al. 1998. Genetic Structure and Evolution of Species of the Mangrove Genus *Avicennia* (Avicenniaceae) in the Indo-west Pacific. *Evolution* **52**: 1612–1626.

Duke, N. C., Birch, W. R., and Williams, W. T. 1981. Growth Rings and Rainfall Correlations in a Mangrove Tree of the Genus *Diospyros* (Ebenaceae). *Austral. J. Bot.* **29**: 135–142.

Duke, N. C. and Bunt, J. S. 1979. The Genus *Rhizophora* (Rhizophoraceae) in Northeastern Australia. *Austral. J. Bot.* **27**: 657–678.

Duke, N. C., Bunt, J. S., and Williams, W. T. 1984. Observations on the Floral and Vegetative Phenologies of North-eastern Australian Mangroves *Austral. J. Bot.* **32**: 87–99.

Duke, N. C. and Ge, X-J. 2011. *Bruguiera* (Rhizophoraceae) in the Indo-West Pacific: A Morphometric Assessment of Hybridisation within Single-flowered Taxa. *Blumea* **56**: 36–48.

Duke, N. C. and Jacques, B. R. 1987. A Systematic Revision of the Mangrove Genus *Sonneratia* (Sonneratiaceae) in Australasia. *Blumea* **32**: 277–302.

Duke, N. C., Lo, E. Y. Y., and Sun, M. 2002. Global Discontinuities of Mangroves – Emerging Patterns in the Evolution of *Rhizophora* L. (Rhizophoraceae). *Trees Struct. Funct.* **16**: 65–79.

Duke, N. C. and Pinzon, M. Z. S. 1992. Aging *Rhizophora* Seedlings from Leaf Scar Nodes: A Technique for Studying Recruitment and Growth in Mangrove Forests. *Biotropica* **24**: 173–186.

Egler, F. E. 1948. The Dispersal and Establishment of Red Mangroves, *Rhizophora*, in Florida. *Caribb. For.* **9**(4): 299–319.

Egler, F. E. 1952. Southeast Saline Everglades Vegetation, Florida, and its Management. *Vegetatio* **3**: 213–265.

Ellison, A. M. and Farnsworth, E. J. 1997. Simulated Sea Level Change Alters Anatomy, Physiology, Growth, and Reproduction of Red Mangrove (*Rhizophora mangle* L.). *Oecologia* **112**: 435–446.

Elmquist, T. and Cox, P. A. 1996. The Evolution of Vivipary in Flowering Plants. *Oikos* **77**: 3–9.

Emura, N., Denda, T., Sakai, M., et al. 2014. Dimorphism of the Seed-dispersing Organ in a Pantropical Coastal Plant, *Scaevola taccada*; Heterogenous Population Structures across Islands. *Ecol. Res.* **29**: 733–740.

Endress, P. K. 1999. Symmetry in Flowers: Diversity and Evolution. *Int. J. Plant Sci.* **160**(Suppl) S3–23: 489–503.

Eshbaugh, W. H. 2014. The Flora of the Bahamas, Donovan Correll and the Miami University Connection. *Bot. Rev.* **80**: 184–203.

Essig, F. B. 1973. Pollination in some New Guinea Palms. *Principes* **17**: 75–83.

Estrada, G. C. O., Cellado, C. H., Soaves, M. L. G., et al. 2008. Annual Growth Rings in the Mangrove *Laguncularia racemosa* (Combretaceae). *Trees – Struct. Function* **22**: 663–670.

Ewel, J. 1980. Tropical Succession: Manifold Routes to Maturity. *Biotropica* **12**(suppl.): 2–7.

Exell, A. W. 1954. Combretaceae. *Flora Malesiana*, Series 1, **4**: 533–589.

Exell, A. W. and Stace, C. 1966. Revision of Combretaceae. *Bol. Soc. Broteriana* **40**(2): 5–25.

Fahn, A. 1979. *Secretory Tissues in Plants*. London, UK: Academic Press. [See also Fahn, 2000. *Adv. Bot. Res.* **31**: 37–75.]

Fahn, A. and Shimony, C. 1977. Development of Glandular and Nonglandular Leaf Hairs of *Avicennia marina* (Forsskal) Vierh. *Bot. J. Linn. Soc.* **74**: 37–46.

Farnsworth, E. J. and Ellison, A. M. 1991. Patterns of herbivory in Belizean mangrove swamps. *Biotropica* **23**: 555–567.

Fisher, J. B. 1978. A Quantitative Study of *Terminalia* Branching. In *Tropical Trees as Living Systems*. P. B. Tomlinson and M. H. Zimmermann, eds. Cambridge, UK: Cambridge University Press, Chap. 13.

Fisher, J. B. and Honda, H. 1977. Computer Simulation of Branching Pattern and Geometry in *Terminalia* (Combretaceae), a Tropical Tree. *Bot. Gaz.* **138**: 337–384.

Fisher, J. B. and Honda, H. 1979a. Branch Geometry and Effective Leaf Area: A Study of *Terminalia-branching* Pattern, 1. Theoretical Trees. *Amer. J. Bot.* **66**: 633–644.

Fisher, J. B. and Honda, H. 1979b. Branch Geometry and Effective Leaf Area: A Study of *Terminalia-branching*, 2. Survey of Real Trees. *Amer. J. Bot.* **66**: 645–655.

Fisher, J. B. and Stevenson, J. W. 1981. Occurrence of Reaction Wood in Branches of Dicotyledons and its Role in Tree Architecture. *Bot. Gaz.* **142**: 82–95.

Fisher, J. B. and Tomlinson, P. B. 2002. Tension Wood Fibers are Related to Gravitropic Movement of Red Mangrove (*Rhizophora mangle*) seedlings. *J. Plant Res.* **115**: 39–45.

Fisher, J. B. and Tomlinson, P. B. 2012. How Red Mangrove Seedlings stand up – An Answer to Cheeseman (2012). *Plant & Soil* **355**: 1400–1407.

Flowers, T. J., Troke, P. F., and Yeo, A. R. 1977. The Mechanism of Salt Tolerance in Halophytes. *Ann. Rev. Plant Physiol.* **28**: 89–121.

Fong, F. W. 1987a. *Studies on the Population Structure, Growth Dynamics, and Resource Importance of Nipa Palm (Nypa fruticans* Wurmb.) Ph.D. Thesis. Kuala Lumpur: University of Malaya.

Fong, F. W. 1987b. An Unconventional Fuel Crop. *Principes* **31**: 64–67.

Fong, F. W. 1989. The Apung Palm: Traditional Technique and Alcohol Extraction in Sarawak. *Principes* **33**: 21–26.

Fosberg, F. R. 1939. *Diospyros ferrea* (Ebenaceae) in Hawaii. *Occas. Pap. Bernice Pauahi Bishop Mus.* **15**:119–131.

Fosberg, F. R. 1975. Phytogeography of Micronesian Mangroves. In *Proceedings of the International Symposium on Biology and Management of Mangroves.* G. E. Walsh, S. C. Snedaker, and H. J. Teas., eds. Gainesville, FL: Institute of Food and Agricultural Sciences, University of Florida.

Fosberg, F. R. and Sachet, M.-H. 1972. *Thespesia populnea* (L.) Solander ex Correa and *Thespesia populneoides* (Roxburgh) Kosteletsky (Malvaceae). *Smithson. Contrib. Bot.* **7**: 1–13.

Fourqurean, J. W., Smith, T. J., Possley, J., et al. 2009. Are Mangroves in the Tropical Atlantic Ripe for Invasion? Exotic Mangroves in the Forests of South Florida. *Biol. Invasions* **12**: 2509–2522.

Foxworthy, F. W. 1910. Distribution and Utilization of Mangrove Swamps in Malay. *Ann. Jard. Bot. Buitenz.* Series **2**, suppl. 3, part 1: 319–344.

Fryxell, P. A. 2001. Malvaceae. *Contributions from the University of Michigan Herbarium* **23**: 225–270.

Garcia de Lopez, I. 1978. Revision del genero *Acrostichum* en Ia Republica Dominicana. *Moscosoa* **1**: 64–70.

Ge, X. and Sun, M. 1999. Reproductive and Genetic Diversity of a Cryptoviviparous Mangrove *Aegiceras corniculatum* (Myrsynaceae) using Allozyme and Intersimple Sequence Repeat (ISSR) Analysis. *Mol. Ecol.* **8**: 2061–2069.

Gehrmann, K. 1911. Zur Blutenbiologie der Rhizophoraceae. *Ber. d. bot. Gesell.* **29**: 308–318.

Geissler, N., Schnetter, R., and Schnetter, M. 2002. The Pnemathodes of *Laguncularia racemosa*: Little Known Rootlets of Surprising Structure, and Notes on a New Fluorescent Dye for Lipophilic Substances. *Plant Biol.* **4**: 729–739.

Gentry, A. H. 1982. Phytogeographic Patterns as Evidence for a Choco Refuge. in *Biological diversification in the tropics.* Gh. T. Prance, ed. New York, NY: Columbia University Press; pp. 112–136.

Giang, Q., Hong, P. N., Tuan, M. S., et al. 2003. Genetic Variation of *Avicennia marina* (Forsk.) Vierh. (Avicenniaceae) in Vietnam Revealed by Microsatellites and AFLP Markers. *Genes Genet. Syst.* **78**: 399–407.

Giang, L. H., Geada, G. L., Hong, P. N., et al. 2006. Genetic Variation of Two Mangrove Species in *Kandelia* (Rhizophoraceae) in Vietnam and Surrounding Area Revealed by Mocrosatellite Markers. *Int. J. Plant Sci.* **167**: 291–298.

Gill, A. M. 1971. Endogenous Control of Growth-ring Development in *Avicennia*. *For. Sci.* **17**: 462–465.

Gill, A M. and Tomlinson, P. B. 1969. Studies on the Growth of Red Mangrove *(Rhizophora mangle* L.). I. Habit and General Morphology. *Biotropica* **1**: 1–9.

Gill, A M. and Tomlinson, P. B. 1971a. Studies on the Growth of Red Mangrove *(Rhizophora mangle* L.). 2. Growth and Differentiation of Aerial Roots. *Biotropica* **3**: 63–77.

Gill, A M. and Tomlinson, P. B. 1971b. Studies on the Growth of Red Mangrove *(Rhizophora mangle* L.). 3. Phenology of the Shoot. *Biotropica* **3**: 109–124.

Gill, A M. and Tomlinson, P. B. 1975. Aerial Roots: an Array of Forms and Functions. in *The Development and Function of Roots*. J. G. Torrey and D. T. Clarkson, eds. London, UK: Academic Press; Chap. 12.

Gill, A M. and Tomlinson, P. B. 1977. Studies on the Growth of Red Mangrove *(Rhizophora mangle* L.). 4. The Adult Root System. *Biotropica* **9**: 145–155.

Gill, L. S. and Kyauka, P. S. 1977. Heterostyly in *Pemphis acidula* Forst. (Lythraceae) in Tanzania. *Adansonia*. **17**(2): 139–146.

Goebel, K. 1886. Ueber die Luftwurzeln von *Sonneratia*. *Ber. d. bot. Gesell.* **4**: 249–255.

Gomez-Pompa, A. and Vazquez-Yanes, C. 1974. Studies on the Secondary Succession of Tropical Lowlands: The Life Cycle of Secondary Species. *Proc. First Int. Cong. Ecology*; pp. 336–342.

Gonzalez, V. V., Solis, S. M., and Ferrucci, M. 2014. Anatomia reproductiva en flores estaminades y pistilades de *Allophylus edulis* (Sapindaceae). *Bol. Soc. Argentina Bot.* **49**: 207–216.

Graham, A. 1977. New Records of *Pellicieria* (Theaceae/Pelliceriaceae) in the Tertiary of the Caribbean. *Biotropica* **9**: 48–52.

Graham, A. 2006. Paleobotanical Evidence and Molecular Data in Reconstructing the Historical Phytogeography of Rhizophoraceae. *Ann. Mo. Bot. Gard.* **93**: 327–334.

Graham, S. 1964. The Genera of Rhizophoraceae and Combretaceae in the Southeastern United States. *J. Arnold. Arbor.* **45**: 285–301.

Graham, S. and Graham, A. 2014. Fruit and Seed Morphology of the Lythraceae. *Int. J. Plant Sci.* **175**: 212–240.

Graham, S. A., Hall, J., Sytsma, K., et al. 2005. Phylogenetic Analysis of the Lythraceae Based on Four Gene Regions and Morphology. *Int. J. Plant. Sci.* **166**: 995–1017.

Greenway, H. and Munns, R. 1980. Mechanisms of Salt Tolerance in Nonhalophytes. *Ann. Rev. Plant Physiol.* **31**: 149–190.

Guo, M., Zhou, R., Huang, Y., et al. 2011. Molecular Confirmation of Natural Hybridization between *Lumnitzera racemosa* and *Lumnitzera littorea*. *Aquat. Bot.* **95**: 59–64.

Guppy, H. B. 1906. *Observations of a Naturalist in the Pacific between 1896 & 1899, Plant Dispersal*, Vol. 2. London, UK: Macmillan; pp. 627.

Guppy, H. B. 1917. *Plant Seeds and Currents in the West Indies and Azores*. London, UK: Williams & Norgate; pp. 531.

Haberlandt, G. 1910. *Eine botanische Tropenreise*. Leipzig: Wilhelm Engelmann.

Haberlandt, G. 1928. *Physiological Plant Anatomy*. London, UK: Macmillan.

Hallé, F., Oldeman, R. A. A., and Tomlinson, P. B. 1978. *Tropical Trees and Forest: an Architectural Analysis*. Berlin: Springer-Verlag.

Hanagata, N., Takemura, T., Karube, I., et al. 1999. Salt/Water Relationships in Mangroves. *Israel J. Plant Sci.* **47**: 63–76.

Hattink, T. A. 1974. A Revision of Malesian *Caesalpinia,* including *Mezoneuron* (Leguminosae-Caesalpiniaceae). *Reinwardtia* **9**(1): 1–69.

Hemsley, W. B. 1879–1888. Botany. In *Biologia Central-Americana*. 5 vols. London, UK: Porter and Dulau.

Hileman, L. C. 2014. Trends in Flower Symmetry Revealed through Phylogenetic and Developmental Genetic Advances. *Phil. Trans. R. Soc.* B **369**. Doi: 10.1098/rstb.2013.0348.

Holbrook, N. M. and Putz, F. E. 1982. Vegetative Seaward Expansion of *Sonneratia alba* Trees in a Malaysian Mangrove Forest. *Malay. For.* **45**: 278–281.

Holttum, R. E. 1953. *Ferns of Malaya*. Singapore: Government Publishing Singapore.

Hoppe, L. E. 2005. *The Pollination Biology and Biogeography of the Mangrove Palm* Nypa fruticans *Wurmb. (Arecaceae)*. Master's Thesis. Aarhus, Denmark: University of Aarhus.

Hosakawa, T., Tagawa, H., and Chapman, V. J. 1977. Mangals of Micronesia, Taiwan, Japan, the Philippines and Oceania. In Ecosystems of the World. Vol. 1. Wet Coastal Ecosystems. Amsterdam: Elsevier Scientific, Chap. 14.

Howard, R. A. 1981. Three Experiences with Manchineel *(Hippomane* spp., Euphorbiaceae). *Biotropica* **13**: 224–227.

Howe, M. A. 1911. A Little Known Mangrove of Panama. *J. N.Y. Bot. Gard.* **12**: 61–72.

Huang, S. and Chen, Y.-C. 2000. Patterns of Genetic Variation of *Kandelia candel* among Populations around South China Sea. In T.-Y. Chiang and T.-W. Hsu, eds. *Wetland Biodiversity; Proceedings. Taiwan Endemic Species Research Institute*, Nantou; pp. 59–64.

Huggett, B. and Tomlinson, P. B. 2010. Aspects of Vessel Dimensions in the Aerial Roots of Epiphytic Araceae. *Inter. J. Pl. Sc.* **171**: 362–369.

Infante-Mata, D., Moreno-Casasola, P., and Madero-Vega, C. 2014. *Pachira aquatica*, as Indicator of Mangrove Limit. *Rev. Mexic. Biodiversidad* **85**: 143–160.

Islam, M. S., Lian, C. L., Kameyama, N., et al. 2004. Development of Microsatellite Markers in *Rhizophora stylosa* using a Dual-Suppression-Polymerase Chain Reaction Technique. *Mol. Ecol. Notes* **4**: 110–112.

Islam, M. S., Lian, C. L., Kameyama, N., et al. 2006a. Development and Characterisation of Ten New Microsatellite Markers in Mangrove Tree Species *Bruguiera gymnorrhiza*(L.). *Mol. Ecol. Notes* **6**: 30–32.

Islam, M. S., Tao, J. M., Geng, Q. F., et al. 2006b. Isolation and Characterisation of Eight Compound Microsatellite Markers in a Mangrove *Kandelia candel* (L.) Druce. *Mol. Ecol. Notes* **6**:1111–1113.

Janssonius, H. H. 1950. The Vessels in the Wood of Javan Mangrove Trees. *Blumea* **6**: 465–469.

Jena, S. N., Verma, S., Nair, K. N., et al. 2015. Genetic Diversity and Population Structure of the Mangrove Lime *(Merope angulata)* in India Revealed by AFLP and ISSR Markers. *Aquatic Bot.* **120**: 260–267.

Jenik, J. 1970. Root System of Tropical Trees. 5. The Peg Roots and the Pneumathodes of *Laguncularia racemosa* Gaertn. *Preslia* **42**: 105–113.

Jenik, J. 1978. Roots and Root Systems in Tropical Trees: Morphologic and Ecologic Aspects. In Tropical Trees as Living Systems. P. B. Tomlinson and M. H. Zimmermann, eds. Cambridge, UK: Cambridge University Press; Chap. 14.

Jian, S. G., Tang, T., Zhong,Y., and Shi, S. B. 2004. Variation in Inter-simple Sequence Repeat (ISSR) in Mangrove and Non-mangrove Populations of *Heritiera littoralis* (Sterculiaceae) from China and Australia. *Aquatic Bot.* **79**: 75–86.

Jiménez, J. A. 1984. A Hypothesis to Explain the Reduced Range of the Mangrove *Pelliciera rhizophorae* Tr. & Pl. *Biotropica* **16**: 304–308.

Jiménez, J. A. 1987. A Clarification on the Existence of *Rhizophora* species along the Pacific Coast of Central America. *Brenesia* **28**: 25–32.

Jiménez, J. A., Lugo, A. E., and Cintron, G. 1985. Tree Mortality in Mangrove Forests. *Biotropica* **17**: 177–185.

Jiménez, J. A. and Soto, R. 1985. Patrones regionales en Ia estructura y composición floristica de los manglares de Ia costa Pacifica de Costa Rica. *Rev. Biol. Trop.* **33**: 25–37.

Johnston, I. M. 1949. The Botany of San Jose Island (Gulf of Panama). *Sargentia* **8**: 1–306.

Johnstone, I. M. 1981. Consumption of Leaves by Herbivores in Mixed Mangrove Stands. *Biotropica* **13**: 252–259.

Johnstone, I. M. 1983. Mangrove Succession and Climax. In *Biology and Ecology of Mangroves. Tasks for Vegetation Science* 8. H. J. Teas, ed. The Hague: Junk; Chap. 12.

Jones, W. T. 1971. The Field Identification and Distribution of Mangroves in Eastern Australia. *Queensl. Nat.* **20**: 35–51.

Jonker, F. P. 1959. The Genus *Rhizophora* in Suriname. *Acta. Bot. Neerl.* **8**: 58–60.

Joshi, G. V., Pimplaskar, M., and Bhosale, L. J. 1972. Physiological Studies in Germination of Mangroves. *Bot. Mar.* **15**: 91–95.

Judd, W. S. and Manchester, S. R. 1997. Circumscription of Malvaceae (Malvales) as Determined by a Preliminary Analysis of Morphological, Anatomical, Palynological and Chemical Characters. *Brittonia* **49**: 384–405.

Juncosa, A. M. 1982a. Embryo and Seedling Development in the Rhizophoraceae. Ph.D. Thesis. Durham, NC: Duke University.

Juncosa, A. M. 1982b. Developmental Morphology of the Embryo and Seedling of *Rhizophora mangle* L. (Rhizophoraceae). *Amer. J. Bot.* **69**: 1599–1611.

Juncosa, A. M. 1984a. Embryogenesis and Seedling Development in *Cassipourea elliptica* (Sw.) Poir. (Rhizophoraceae). *Amer. J. Bot.* **71**: 170–179.

Juncosa, A. M. 1984b. Embryogenesis and Seedling Developmental Morphology of the Seedling in *Bruguiera exaristata* Ding Hou (Rhizophoraceae). *Amer. J. Bot.* **71**: 180–191.

Juncosa, A. M. and Tomlinson, P. B. 1987. Floral Development in Mangrove Rhizophoraceae *Amer. J. Bot.* **74**: 1263–1277.

Juncosa, A. M. 1988a. A Historical and Taxonomic Synopsis of Rhizophoraceae and Anisophylleaceae. *Ann. Mo. Bot. Gard.* **75**: 1278–1295.

Juncosa, A. M. 1988b. Systematic Comparison and Some Biological Characteristics of Rhizophoraceae and Anisophylleaceae. *Ann. Mo. Bot. Gard.* **75**: 1296–1318.

Karsten, G. 1890. Ueber die Mangrove-Vegetation im malayischen Archipel. *Ber. d. bot. Gesell.* **8**: 49–55.

Karsten, G. 1891. Ueber die Mangrove-Vegetation im malayischen Archipel. Eine morphologisch- biologische Studie. *Bibl. Bot.* **22**:71.

Kearney, T. H. 1954. Notes on Malvaceae V. *Leafl. West. Bot.* **7**: 118–119.

Keating, R. C. and Randrianasolo, V. 1988. The Contribution of Leaf Architecture and Wood Anatomy to the Classification of Rhizophoraceae and Anisophylleaceae. *Ann. Mo. Bot. Gard.* **75**: 1343–1368.

Keay, R. W. J. 1953. *Rhizophora* in West Africa. *Kew Bull.* **1953**: 121–127.

Kenneally, K. F., Wilson, P. G., and Semeniuk, J. 1978. A New Character for Distinguishing Vegetative Material of the Mangrove Genera *Bruguiera* and *Rhizophora* (Rhizophoraceae). *Nuytsia* **4**: 178–180.

Khosla, P. K. and Styles, B. T. 1975. Karyological Studies and Chromosomal Evolution in the Meliaceae. *Silvae Genetica* **24**: 73–83.

Kipp-Goller, A. 1940. Ueber Bau und Entwicklung der viviparen Mangrovekeimlinge. *Z. Bot.* **35**: 1–40.

Klekowski, E. J., Lowenfeld, R., and Hepler, P. K. 1994a. Mangrove Genetics II. Outcrossing and Lower Spontaneous Mutation Rates in Puerto Rican *Rhizophora*. *Int. J. Plant Sci.* **155**: 373–381.

Klekowski, E. J., Corredor, J. E., Morell, J. M., et al. 1994b. Petroleum Pollution and Mutation in Mangroves. *Mar. Pollution Bull.* **28**: 166–169.

Klekowski, E. J., Corredor, J. E., Lowenfeld, R., et al. 1994c. Using Mangroves to Screen for Mutagens in Tropical Marine Environments. *Mar. Pollution Bull.* **28**: 346–350.

Knapp-van Meeuwen, M. S. 1970. A Revision of Four Genera of the Tribe Leguminosae Caesalpinioideae- Cynometreae in Indomalesia and the Pacific. *Blumea* **18**(1): 1–52.

Knevel, I. C. and Lubka R. A. 2004. Reproductive Phenology of *Scaevola plumieri*: A Key Coloniser of the Coastal Foredunes of South Africa. *Plant Ecol.* **175**: 132–145.

Kobuski, C. E. 1951. Studies in the Theaceae XXII, The genus *Pelliciera*. *J. Arnold Arbor.* **32**: 256–262.

Kojit T., Tamura, M., Tateischi Y., et al. 2013. Strong Genetic Structure over the American Continents and Transoceanic Dispersal in the Mangrove Genus *Rhizophora* (Rhizophoraceae) Revealed by Broad Scale Nuclear and Chloroplast DNA Anatomy. *Amer. J. Bot.* **100**: 1191–1201.

Kondo, K., Nakamura, T., Tsuruda, K., et al. 1987. Pollination in *Bruguiera gymnorrhiza* and *Rhizophora mucronata* (Rhizophoraceae) in Ishigaki Island, the Ryukyu Islands, Japan. *Biotropica* **19**: 377–380.

Kostermans, A. J. G. J. 1959. Monograph of the Genus *Heritiera* Aiton (Stercul.). *Reinwardtia* **4**: 465–483.

Krauss, K. W. and Allen. J. A. 2003. Factors Influencing the Regeneration of the Mangrove *Bruguiera gymnorrhiza* (L.) Lamk. on a Tropical Pacific Island. *For. Ecol. Manag.* **176**: 49–60.

Krauss, K. W., McKee, K. L., Catherine, E., et al. 2014. How Mangrove Forests Adjust to Rising Sea Levels. *New Phytol.* **202**: 19–34.

Kryger, L. and Lee, S. K. 1995. Effects of Soil Ageing on the Accumulation of Hydrogen Sulphide and Metallic Sulphides in Mangrove Areas in Singapore. *Environ. Intern.* **21**: 85–92.

Kryger, L. and Lee, S. K. 1996. Effects of Mangrove Soil Ageing on the Accumulation of Hydrogen Sulphide in Roots of *Avicennia*. *Biogeochemistry* **35**: 367–375.

Kubitzki, K. 2004. Pellicieraceae. In *1990-. Families and Genera of Vascular Plants*. K. Kubitzki, ed. Heidelberg: Springer; VII: 297–299.

Kubitzki, K. 2011. Rutaceae. In *1990-. Families and Genera of Vascular Plants*. K. Kubitzki, ed. Heidelberg: Springer; VII: 276–356.

Lacerda, L. D. 1981. Mangrove Wood Pulp, an Alternative Food Source for the Tree Crab *Aratus pisonii*. *Biotropica* **13**: 317.

Lacerda, L. D. and May, D. V. 1982. Evolution of a New Community Type during the Degradation of a Mangrove Ecosystem. *Biotropica* **14**: 238–239.

Lacerda, L. D., Carvalho, C. E. V., Tanizaki, K. F., et al. 1993. The Biogeochemistry and Trace Metals Distribution of Mangrove Rhizospheres. *Biotropica* **25**: 252–257.

Landry, C. and Rathke, B. J. 2007. Do Inbreeding Depression and Relative Male Fitness explain the Maintenance of Androdioecy in White Mangrove, *Laguncularia racemosa* (Combretaceae). *New Phytol.* **176**: 891–901.

Landry, C. L., Rathke, B. J., and Kass, L. B. 2009. Distribution of Androdioecious and Hermaphroditic Populations of the Mangrove *Laguncularia racemosa* (Combretaceae) in Florida. *J. Trop. Ecol.* **25**: 75–83.

Larue, C. D. and Muzik, T. J. 1951. Does the Mangrove really Plant its Seedlings? *Science* **114**: 661–662.

Larue, C. D. and Muzik, T. J. 1954. Growth Regeneration and Precocious Rooting in *Rhizophora mangle*. *Pap. Mich. Acad. Sci. Arts. Lett. Part 1, Botany and Forestry*. **39**: 9–29.

Lawrence, D. B. 1949. Self-erecting Habit of Seedling Red Mangroves *(Rhizophora mangle* L.). *Amer. J. Bot*. **36**(5): 426–427.

Lee, D. 1980. *The Sinking Ark: Environmental Problems in Malaysia and Southeast Asia*. Kuala Lumpur: Heinemann.

Lee, S. K., Tan, W. H., and Havanond, S. 1996. Regeneration and Colonisation of Mangrove on Clay-filled Reclaimed Land in Singapore. *Hydrobiologia* **319**: 23–35.

Leenhouts, P. W. 1967. A Conspectus of the Genus *Allophylus* (Sapindaceae). *Blumea* **15**: 301–358.

Lersten, N. R. and Curtis, J. D. 1974. Colleter Anatomy in Red Mangrove, *Rhizophora mangle* (Rhizophoraceae). *Can. J. Bot*. **52**: 2277–2280.

Lewis, D. and Rao, A. N. 1971. Evolution of Dimorphism and Population Polymorphism in *Pemphis acidula* Forst. *Proc. R. Soc. Lond. B* **178**: 79–94.

Lewis, R. R. 1983. Impact of Oil Spills on Mangrove Forests. In *Biology and Ecology of Mangroves. Tasks for Vegetation Science* 8. H. J. Teas, ed. The Hague: Junk; Chap. 19.

Li, H. and Chen, G. 2008. Genetic Relationships among Species in the Genus *Sonneratia* in China as Revealed by Inter-simple Sequence Repeat (ISSR) markers. *Biochem. Syst. Ecol*. **36**: 392–398.

Liebau, O. 1914. Beiträge zur Anatomie und Morphologie der Mangrove-Pflanzen, insbesondere ihres Wurzelsystems. *Beit. Biol. Pflanz*. **12**: 181–213.

Linnaeus, C. 1753. *Species plantarum*. Stockholm: Holmiae.

Lledo, M. D., Karis, P. O., Crespo, M. B., et al. 2001. Phylogenetic Position and Taxonomic Status of the Genus *Aegialitis* and Subfamilies Staticoideae and Plumbaginoideae: Evidence from Plastid DNA Sequences and Morphology. *Plant Syst. Evol*. **229**: 107–124.

Lloyd, R. M. 1980. Reproductive Biology and Gametophyte Morphology of New World Populations of *Acrostichum aureum*. *Amer. Fern. J*. **70**: 99–110.

Lloyd, R. M. and Gregg, T. L. 1975. Reproductive Biology and Gametophyte Morphology of *Acrostichum danaeifolium* from Mexico. *Amer. Fern. J*. **65**: 105–120.

Longman, K. A. and Jenik, J. 1974. *Tropical Forest and its Environment*. London, UK: Longman Group.

López-Portillo, J., Ewers, F. W., Méndez-Alonzo, R., et al. 2014. Dynamic Control of Osmolality and Ionic Composition of the Xylem Sap in Two Mangrove Species. *Am. J. Bot*. **101**: 1013–1022.

Lot-Helgueras, A., Vazquez-Yanes, C., and Menendez, F. 1975. Physiognomy and Floristic Changes near the Northern Limit of Mangroves in the Gulf Coast of Mexico. In *Proceedings of the International Symposium on Biology and Management of Mangroves* G. E. Walsh, S. C. Snedaker, and H. J. Teas, eds. Gainesville, FL: Institute of Food and Agricultural Sciences, University of Florida; pp. 52–61.

Lotschert, W. and Liemann, F. 1967. Die Salzspeicherung im Keimling von *Rhizophora mangle* L. während der Entwicklung auf der Mutterpflanzen. *Planta* **77**: 142–156.

Lowenfeld, R. and Klekowkski, E. J. 1992. Mangrove Genetics I. Mating System and Mutation Rates of *Rhizophora mangle* in Florida and San Salvadore Island, Bahamas. *Int. J. Plant Sci*. **153**: 394–399.

Lugo, A. E. 1980. Mangrove Ecosystems: Successional or Steady State. *Biotropica* **12**(suppl.): 65–72.

Lugo, A. E. and Snedaker, S. C. 1974. The Ecology of Mangroves. *Ann. Rev. Ecol. Syst*. **5**: 39–64.

Luther, D. and Greenberg, R. 2009. Mangroves: A Global Perspective on the Evolution and Conservation of their Terrestrial Vertebrates. *BioScience* **59**: 602–612.

Mabberley, D. G., Pannell, C. H., and Sing, A. M. 1995. Meliaceae. In *Flora Malesianal: Iliaceae*. Series 1, Vol. 12, pp. 1–407.

MacArthur, R. H. and Wilson, E. O. 1967. *The Theory of Island Biogeography*. Princeton, NJ: Princeton University Press.

MacNae, W. 1968. A General Account of the Fauna and Flora of Mangrove Swamps and Forests in the Indo-West Pacific Region. *Adv. Mar. Biol.* **6**: 73–270.

MacNae, W. and Fosberg, F. R. 1981. Sonneratiaceae. In *A Revised Handbook to the Flora of Ceylon*, Vol. 3. M. D. Dassanayake, ed. New Delhi: American Publishing.

Maguire, T. L., Edwards, K. J., Saenger, P., and Henry, R. 2000a. Characterization and Analysis of Microsatellite Loci in a Mangrove Species *Avicennia marina* (Forsk.) Vierh. (Avicenniaceae). *Theor. Appl. Genet.* **101**: 279–285.

Maguire, T. L., Peakall, R., Saenger, P., and Henry, R. 2002. Comparative Analysis of Genetic Diversity in the Mangrove Species *Avicennia marina* (Forsk.) Vierh. (Avicenniaceae) Detected by AFLPs and SSRs. *Theor. Appl. Genet.* **104**: 388–398.

Maguire, T. L., Saenger, P., Baverstock, P., et al. 2000b. Microsatellite Analysis of genetic structure in the mangrove species *Avicennia marina* (Forsk.) Vierh. (Avicenniaceae). *Mol. Ecol.* **9**: 1853–1862.

Mancina, C. A., Balseiro, F., and Herrera, L. G. 2005. Pollen Digestion by Nectarivorous and Frugivorous Bats. *Mammal. Biol.* **70**: 282–290.

Markgraf, F. 1927. Die Apocynaceen von Neue-Guinea, 117. In *Beiträge zur Flora von Papuasien*, 14. C. Lauterbach, ed. Leipzig: Max Weg. (Sonderabdruck an Engler, *Bot. Jahrb.* **61**: 1927.)

Markley, J. L., McMillan, C., and Thompson, G. A. 1982. Latitudinal Differentiation in Response to Chilling Temperatures among Populations of Three Mangroves, *Avicennia germinans*, *Laguncularia racemosa*, and *Rhizophora mangle*, from the Western Tropical Atlantic and Pacific Panama. *Can. J. Bot.* **60**: 2704–2715.

Marshall, A. G. 1983. Bats, Flowers and Fruit: Evolutionary Relationships in the Old World. *Biol. J. Linn. Soc.* **20**: 125–135.

Masters, H. M. 1872. *Camptostemon schultzii. Hooker's lc. Pl.* **12**: 18 (Table 111a).

Maurin, O., Chase, M. W., Jordan, M., and van der Bank, M. 2010. Phylogenetic Relationships of Combretaceae Inferred from Nuclear and Plastid DNA Sequence Data: Implications for Generic Classification. *Bot. J. Linn. Soc.* **162**: 453–476.

Maury, P. 1886. Études sur l'organisation et la distribution géographiques des Plumbaginacées. Thesis, Facultés des Sciences, Paris. G. Masson, Paris.

Maxwell, G. S. 1995. Ecogeographic Variation in *Kandelia candel* from Brunei, Hong Kong, and Thailand. *Hydrobiologia* **295**: 59–65.

McCoy, E. D. and Heck, K. L. 1976. Biogeography of Corals, Seagrasses and Mangroves: An Alternative to the Centre of Origin Concept. *Syst. Zool.* **25**: 201–210.

McDade, L. A., Daniel, T. F., and Kiel, C. A. 2008. Toward a Comprehensive Understanding of Phylogenetic Relationships among Lineages of Acanthaceae s.l. (Lamiales). *Amer. J. Bot.* **95**: 1136–1152.

McDade, L. A., Daniel, T. F., and Kiel, C. A. 1975. Adaptive Differentiation to Chilling in Mangrove Populations. In *Proceedings of the International Symposium on Biology and Management of Mangroves*. G. E. Walsh, S. C. Snedaker, and H. J. Teas, eds. Gainesville, FL: Institute of Food and Agricultural Sciences, University of Florida; pp. 62–68.

Mehltreter, K. and Palacios-Ricos, M. 2003. Phenological Studies of *Acrosticum danaeifolium* (Pteridaceae, Pteridophyta) at a Mangrove Site on the Gulf of Mexico. *J. Trop. Ecol.* **19**: 155–162.

Méndez-Alonzo, R., Moctezuma, C., Ordenez, V. R., et al. 2015. The Root Biomechanics in *Rhizophora mangle*; Anatomy, Morphology and Ecology of Mangrove Flying Buttresses. *Annals of Botany* **115**: 833–840.

Mepham, R. H. 1983. Mangrove Floras of the Southern Continents. Part 1: The Geographical Origin of Indo-Pacific Mangrove Genera and the Development and Present Status of the Australian Mangroves. *S. Afr. J. Bot.* **2**: 1–8.

Merrill, E. D. 1917. *An Interpretation of Rumphius' Herbarium Amboinense.* Pub. 9, Dept. Agric. Nat. Resources Bureau of Science, Manila.

Merrill, E. D. 1918. Species Blancoanae. A Critical Revision of the Philippine Species of Plants described by Blanco and by Llanos. *Bur. Sci. Publ. Manila* **12**: 1–423.

Metcalfe, C. R. and Chalk, L. 1950. *Anatomy of the Dicotyledons* (2 Vols.) Oxford, UK: Clarendon Press.

Mez, C. 1902. Myrsinaceae. In *Engler's Das Pflanzenreich*, **4**, p. 236. Leipzig: Wilhelm.

Miers, J. 1880. On the Barringtoniaceae. *Trans. Linn. Soc. Bot.*. Series 2, **1**: 47–118.

Mohd-Azian, J., Noske, R. A., and Lawes, M. J. 2014. Resource Partitioning by Mangrove Bird Communities in North Australia. *Biotropica* **46**: 331–340.

Moldenke, H. N. 1960. Materials Towards a Monograph of the Genus *Avicennia* L. I., II. *Phytologia* **7**: 123–168, 179–252, 259–263.

Moldenke, H. N. 1967. Additional Notes on the Genus *Avicennia*, I., II. *Phytologia* **14**: 301–320, 326–336.

Moon, G. J., Clough, B. F., Peterson, C. P., et al. 1986. Apoplastic and Symplastic Pathways in *Avicennia marina* (Forsk.) Vierh. Roots Revealed by Fluorescent Tracer Dyes. *Austral. J. Plant Physiol.* **13**: 637–648.

Mori, M. M., Zucchi, M. I., and Sampolo, I. 2015. Species Distribution and Introgressive Hybridisation of Two *Avicennia* Species from the Western Hemisphere Unveiled by Phylogeographic Patterns. *BMC Evol. Biol.* **15**: 61.

Mukherjee, J. and Chanda, S. 1973. Biosynthesis of *Avicennia* L. in Relation to Taxonomy. *Geophytology* **3**: 85–88.

Mullan, D. P. 1931a. Observations on the Water-storing Devices in the Leaves of some Indian Halophytes. *J. Ind. Bot. Soc.* **10**: 126–132.

Mullan, D. P. 1931b. On the Occurrence of Glandular Hairs (Salt Glands) on the Leaves of some Indian Halophytes. *J. Ind. Bot. Soc.* **10**: 184–189.

Muller, J. 1969. A Palynological Study of the Genus *Sonneratia* (Sonneratiaceae) *Pollen et Spores* **11**: 223–298.

Muller, J. 1981. Fossil Pollen Records of Extant Angiosperms. *Bot. Rev.* **47**: 1–142.

Muller, J. and Hou Liu, S. Y. 1966. Hybrids and Chromosomes in the Genus *Sonneratia*. *Blumea* **14**(2): 337–343.

Muller, J. and van Steenis, C. G. G. J. 1968. The Genus *Sonneratia* in Australia. *N. Queensl. Nat.* **34**: 6–8.

Murphy, D. H. 1990. The Natural History of Insect Herbivory on Mangrove Trees in and near Singapore. *Raffles Bull. Zool.* **38**: 119–203.

Nadia, T. de L. and Machado, I. C. 2014. Interpopulational Variation in the Sexual and Pollination Systems of Two Combretaceae Species in Brazilian Mangroves. *Aquatic Bot.* **114**: 35–41.

Nakanischi, S. 1964. An Epiphytic Community on the Mangrove Tree, *Kandelia candel*. *Hikobia* **4**: 124.

Nettel, A. and Dodd, R. S. 2007. Drifting Propagules and Receding Swamps: Genetic Footprints of Mangrove Recolonization and Dispersal along Tropical Coasts. *Evolution* **61**: 958–971.

Nettel, A., Rafi, F., and Dodd, R. S. 2005. Characterization of Microsatellite Markers for the Mangrove Tree *Avicennia germinans* L. (Avicenniaceae). *Mol. Ecol. Notes* **5**: 103–105.

Ng, F. S. P. 1978. Strategies of Establishment in Malayan Forest Trees. In *Tropical Trees as Living Systems*. P. B. Tomlinson and M. H. Zimmermann, eds. Cambridge, UK: Cambridge University Press; Chap. 5.

Ng, F. S. P. 2014. *Tropical Forest Fruits, Seeds and Seedlings*. Malay Forest Records No.52. Forest Research Institute of Malaysia.

Ng, W. L., Chan, H. T., and Szmidt, A. E. 2014. Molecular Identification of Natural Mangrove Hybrids of *Rhizophora* in Peninsular Malaysia. *Tree Genet. Genom.* **9**: 1151–1160.

Ng, W. L., Onishi, Y., Inomata, N., et al. 2015. Closely Related and Sympatric but not all the Same: Genetic Variation of Indo-West Pacific *Rhizophora* Mangroves across the Malay Peninsula. *Conserv. Genet.* **16**: 137–150.

Ngyen, H. T., Stanton, D. E., Schmitz, N. et al. 2015. Growth Responses of the Mangrove *Avicennia marina* to Salinity: Development and Function of Shoot Hydraulic Systems require Saline Conditions. *Ann. Bot.* **115**: 397–407.

Nickerson, N. H. and Thibodeau, F. R. 1985. Association between Porewater Sulfide Concentrations and the Distributions of Mangroves. *Biogeochemistry* **1**: 183–192.

Noamesi, G. K. 1958. *A Revision of the Xylocarpeae (Meliaceae)*. Thesis. Madison Madison, WI: University of Wisconsin.

Noske, R. A. 1993. *Bruguiera hainesii*: Another Bird-pollinated Mangrove? *Biotropica* **25**: 481–483.

Nyffeler, R. and Baum, D A. 2000. Phylogenetic Relationships of the Durians (Bombacaceae–Durioneae or /Malvaceae/Helicterioideae/Durioneae) based on Chloroplast and Nuclear Ribosomal DNA Sequences. *Plant Syst. Evol.* **224**: 55–82.

Ogita, S., Yeung, E. C., and Sasamoto, H. 2004. Histological Analysis in Shoot Organization from Explants of *Kandelia candel*. *J. Plant Res.* **117**: 457–464.

Ogura, Y. 1940. On the Types of Abnormal Roots in Mangrove and Swamp Plants. *Bot. Mag. Tokyo* **54**: 389–404.

Ohira, W., Honda, K., Nagai, M., et al. 2013. Mangrove Stilt-root Morphology Modelling for Estimating Hydraulic Drag in Tsunami Inundation Simulation. *Trees Struct. Func.* **27**: 141–148.

Olexa, M. T. and Freeman, T. E. 1978. A Gall Disease of Red Mangrove Caused by *Cylindrocarpon didymum*. *Plant Dis. Rep.* **62**: 283–285.

Ong, J. F., Gong W. K., and Wong, C.H. 2004. Allometry and Partitioning of the Mangrove *Rhizophora apiculata*. *Forest Ecol. Management* **188**: 395–408.

Orihuela, B., Diaz, H., and Conde, J. E. 1991. Mass Mortality in a Mangrove Roots Fouling Community in a Hypersaline Tropical Lagoon. *Biotropica* **23**: 592–601.

Päivökee, A. E. A. 1984. Tapping Patterns in the Nipa Palm (*Nypa fruticans* Wurmb.). *Principes* **28**: 132–137.

Pandey, C. N., Pandey, R., and Jain, B. K. 2010. Reproductive Phenology of *Rhizophora apiculata* (L.) Lamk. (Rhizophoraceae) in the Gulf of Kachchh, Gujarat, India. *Phytomorphology* **60**: 91–100.

Pannier, F. 1959. El efecto de distintas concentraciones salinas sobre el desarrollo de *Rhizophora mangle* L. *Acta Cient. Venezolana* **10**(3): 68–78.

Pannier, F. 1962. Estudio fisiológico sobre Ia viviparía de *Rhizophora mangle* L. *Acta Cient. Venezolana* **13**(6): 184–197.

Pannier, F. and Pannier, R. F. 1975. Physiology of Vivipary in *Rhizophora mangle* L. *Proceedings of the International Symposium on Biology and Management of Mangroves*.

Vol. 2. G. E. Walsh, S. C. Snedaker, and H. J. Teas., eds. Gainesville, FL: Institute of Food and Agricultural Sciences, University of Florida; pp. 632–639.

Pannier, F. and Rodriquez, M. del P. 1967. The ß-inhibitor Complex and its Relation to Vivipary in *Rhizophora mangle* L. *Int. Rev. Ges. Hydrobiol.* **52**: 783–792.

Panshin, A. J. 1932. An Anatomical Study of the Woods of the Philippine Mangrove Swamps. *Philipp. J. Sci.* **48**(2): 143–205.

Parani, M., Rao, C. S., Mathan, N., et al. 1997. Molecular Phylogeny of Mangroves. III. Parentage Analysis of a *Rhizophora* Hybrid using Random Amplified Polymorphic DNA (RAPD) and Restriction Length Polymorphism (RFLP) Markers. *Aquat. Bot.* **58**: 165–172.

Parida, A. K. and Das, A. B. 2005. Salt Tolerance and Salinity Effects on Plants: A Review. *Ecotoxicology Env. Safety* **60**: 324–349.

Parkinson, C. E. 1934. The Indian Species of *Xylocarpus. Ind. For.* **60**: 136–143.

Pascarella, J. B. 1997a. Pollination Ecology of *Ardisia escallonioides* (Myrsinaceae). *Castanea* **62**: 1–7.

Pascarella, J. B. 1997b. Breeding Systems of *Ardisia* Sw. (Myrsinaceae). *Brittonia* **49**: 45–53.

Patra, J. K. and Thatoi, H. N. 2011. Metabolic diversity and bioactivity of mangrove plants: a review. *Acta Physiol. Plant.* **33**: 1051–1061.

Pax, F. and Hoffmann, K. 1912. Euphorbiaceae. In *Engler's Das Pflanzenreich* **52** (4.147 5 Hippomaneae). Leipzig: Wilhelm Engelmann.

Pax, F. and Hoffmann, K. 1931. Euphorbiaceae. In Engler and Prantl, *Die natürlichen Pflanzen-familien*, edn. 2, Bd. **19c**: 11–233. Leipzig. Wilhelm Engelmann.

Payens, J. P. D. W. 1967. A Monograph of the Genus *Barringtonia* (Lecythidaceae). *Blumea* **15**: 157–263.

Pennington, T. D. 2004. Sapotaceae. In *1990-. Families and Genera of Vascular Plants.* K. Kubitzki, ed. **VI**: 390–421.

Pennington, T. D. and Styles, B. T. 1975. A Generic Monograph of the Meliaceae. *Blumea* **22**: 419–540.

Percival, M. and Womersley, J. S. 1975. *Floristics and Ecology of the Mangrove Vegetation of Papua New Guinea. Bot. Bull.* No. 8. Dept. of Forests, Division of Botany, Lae, Papua New Guinea.

Pil, M. P., Boeger, M. R. T., Muschner, V. C., et al. 2011. Post-glacial North-south Expansion of Populations of *Rhizophora mangle* (Rhizophoraceae) along the Brazilian Coast Revealed by Microsatellite Anaysis. *Amer. J. Bot.* **98**: 1031–1039.

Pitot, A. 1958. Rhizophores et racines chez *Rhizophora* sp. *Bull. lnst. Fr. Afr. Noire* **20**: 1103–1138.

Polhill, R. M. 1971. Some Observations on Generic Limits in Dalbergieae–Lonchocarpineae Benth. (Leguminosae). *Kew Bull.* **25**: 259–273.

Prado, J., Rodrigues, C. del N., Salatino, A., and Salatino, M. 2007. Phylogenetic Relationships among Pteridaceae, including Brazilian Species, Inferred from *rbcL* Sequences. *Taxon* **56**: 355–368.

Prance, G. T. 1975. Revisao taxonomica das especies amazonicas de Rhizophoraceae. *Acta Amazonica* **5**(1): 5–22.

Pridgeon, A. M. 2014. *Anatomy of the Monocotyledons. Introduction* in Stern, W.L. 2014. Orchidaceae vol. X. M. Gregory and D.F. Cutler, eds. Oxford, UK: Oxford University Press.

Primack, R. B., Duke, N. C., and Tomlinson, P. B. 1981. Floral Morphology in Relation to Pollination Ecology in Five Queensland Coastal Plants. *Austrobaileya* **4**: 346–355.

Primack, R. B. and Tomlinson, P. B. 1978. Sugar Secretions from the Buds of *Rhizophora*. *Biotropica* **10**(1): 74–45.

Primack, R. B. and Tomlinson, P. B. 1980. Variation in Tropical Forest Breeding Systems. *Biotropica* **12**: 229–231.

Purnobasuki, H. and Suzuki, M. 2004. Aerenchyma Formation and Porosity in Root of a Mangrove Plant, *Sonneratia alba* (Lythraceae). *J. Plant Res.* **117**: 465–472.

Purnobasuki, H. and Suzuki, M. 2005. Aerenchyma Tissue Development and Gas-pathway Structure in Root of *Avicennia marina* (Forsk.) Vierh. *J. Plant Res.* **118**: 285–294.

Rabinowitz, D. 1978a. Dispersal Properties of Mangrove Propagules. *Biotropica* **10**: 47–57.

Rabinowitz, D. 1978b. Early Growth of Mangrove Seedlings in Panama, and an Hypothesis Concerning the Relationship of Dispersal and Zonation. *J. Biogeogr.* **5**: 113–133.

Rabinowitz, D. 1978c. Mortality and Initial Propagule Size in Mangrove Seedlings in Panama. *J. Ecol.* **66**: 45–51.

Radlkofer, L. 1932. Sapindaceae. In *Das Pflanzenreich.* A. Engler, ed. Heft **98b**, 4. Leipzig: Wilhelm Engelmann; p. 165.

Rains, D. W. and Epstein, E. 1967. Preferential Absorption of Potassium by Leaf Tissue of the Mangrove *Avicennia marina:* An Aspect of Halophytic Competence in Coping with Salt. *Austral. J. Bioi. Sci.* **20**: 847–857.

Raju, A. J. S. and Rajesh, B. 2014. Pollination Ecology of Chengam *Scyphiphora hydrophyllacea* C. F. Gaertn. (Magnoliales; Rubiales; Rubiaceae), a Non-viviparous Evergreen Tree Species. *J. Threatened Taxa* **6**: 6668–6676.

Rao, A. N. 1971. Morphology and Morphogenesis of Foliar Sclereids in *Aegiceras corniculatum. Isr. J. Bot.* **20**: 124–132.

Raven, P. H. and Tomlinson, P. B. 1988. Rhizophoraceae and Anisophylleaceae: A Symposium. *Ann. Mo. Bot. Gard.* **75**: 1257.

Raymond, A. and Phillips, T. L. 1983. Evidence for an Upper Carboniferous Mangrove Community. In *The Biology and Ecology of Mangroves. Tasks for Vegetation Science* 8. H. J. Teas, ed. The Hague: Junk; Chap. 2.

Razafimandimbison, S. G., Kainulainen, K., Wong, K., et al. 2011. Molecular Support for a Basal Grade of Morphologically Distinct, Monotypic Genera in the Species-rich Vanguverieae Alliance (Rubiaceae, Ixorioidea): its Systematic and Conservation Implications. *Taxon* **60**: 941–952.

Reef, R. and Lovelock, C. E. 2015. Regulation of Water Balance in Mangroves. *Ann. Bot.* **115**: 385–395.

Rehm, A. and Humm, H. J. 1973. *Sphaeroma terebrans:* A Threat to the Mangroves of Southwestern Florida. *Science* **182**: 173–174.

Reimold, R. J. 1977. Mangals and Salt Marshes of Eastern United States. In Ecosystems of the World. Vol. 1. Wet Coastal Ecosystems. V. J. Chapman, ed. Amsterdam: Elsevier Scientific; Chap. 7.

Retallack, G. and Dilcher, D. L. 1981. A Coastal Hypothesis for the Dispersal and Rise to Dominance of Flowering Plants. In *Paleobotany, Paleoecology, and Evolution.* K. J. Niklas, ed. New York, NY: Praeger; pp. 27–77.

Ribi, G. 1982. Does the Wood Boring Isopod *Sphaeroma terebrans* Benefit Red Mangroves *(Rhizophora mangle)? Bull. Mar. Sci. Miami* **31**: 925–928.

Ridley, H. N. 1930. *The Dispersal of Plants throughout the World.* Ashford, UK: Reeve & Co.

Ridley, H. N. 1938. Notes on *Xylocarpus. Kew Bull.* **1938**: 288–292.

Robert, E. M. R., Jambia, A. H., Schmitz, N., et al. 2014. How to Catch the Patch? A Dendrometer Study of the Radial Increment through Successive Cambia in the Mangrove *Avicennia. Ann. Bot.* **113**: 741–752.

Robert, E. M. R., Schmitz, N., Okello, J. A., et al. 2011. Mangrove Growth Rings - Fact or Fiction? *Trees-Struct. Funct.* **25**: 49–58.

Rollet, B. 1981. *Bibliography on Mangrove Research, 1600–1975.* Paris: UNESCO.

Rosero-Galindo, C., Gaitan-Solis E., Cardenas-Henao H., et al. 2002. Polymorphic Microsatellites in a Mangrove Species, *Rhizophora mangle* L. *Mol. Ecol. Notes* **2**: 281–283.

Roth, I. 1965. Histogenese der Lentizellen am Hypokotyl von *Rhizophora mangle* L. *Ost. Bot. Z.* **112**: 640–653.

Roth, I. 1992. Hurricanes and Mangrove Regeneration: Effects of Hurricane Joan, October 1988, on the Vegetation of Isla del Venado, Bluefields, Nicaragua. *Biotropica* **24**: 375–384.

Roth, L. C. and Grijalva, A. 1991. New record of the mangrove, *Pelliciera rhizophorae*, on the Caribbean coast of Nicaragua. *Rhodora* **93**: 183–186.

Rumphius, G. E. 1741–1755. *Herbarium Amboinense.* Vol. 2, 4. J. Burmann, ed. Amsterdam: Meinard Uytwerf.

Saenger, P., Specht, M. M., Specht, R. L., and Chapman, V. J. 1977. Mangal and Coastal Salt-Marsh Communities in Australasia. In *Ecosystems of the World. Vol. 1. Wet Coastal Ecosystems.* Amsterdam: Elsevier Scientific; Chap. 15.

Salvoza, F. M. 1936. *Rhizophora. Nat. Appl. Sci. Bull. Un. Philipp.* **5**: 179–237.

Sandovan-Castro, E., Dodd, R. S., Riosmena-Rodriguez, R., et al. 2014. Post-glacial Expansion and Population Divergence of Mangrove Species *Avicennia germinans* (L.) Stearn and *Rhizophora mangle* L. along the Mexican Coast. *PLoS ONE* **9**(4); doi: 10.1371/journal.pone.0093358.

Sandwith, N. Y. 1940. Contributions to the Flora of Tropical America: 44. Further Notes on Tropical American Bignoniaceae. *Kew Bull.* **1940**: 302–304.

Santos, J. K. 1932. The Laticiferous Vessels and other Anatomical Structures of *Excoecaria agallocha. Philipp. J. Sci.* **47**: 295–304.

Savory, H. J. 1953. A Note on the Ecology of *Rhizophora* in Nigeria. *Kew Bull.* **1953**: 127–128.

Schenk, H. 1889. Ueber die Luftwurzeln von *Avicennia tomentosa* und *Laguncularia racemosa. Flora* **72**: 83–88.

Schimper, A. F. W. 1891. *Botanische Mittheilungen aus den Tropen. Heft 3: Die Indomalayische Strandflora.* Jena: Gustav Fischer.

Schmid, R. 1980. Comparative Anatomy and Morphology of *Psiloxylon* and *Heteropyxis*, and the Subfamilial and Tribal Classification of Myrtaceae. *Taxon* **29**: 559–595.

Schmidt, J. 1903. *Bidrag till kundskab om skuddene has den gamle verdens mangrovetraeer.* Copenhagen.

Schmitz, N., Verheyden, A., Beeckman, H., et al. 2006. Influence of Salinity Gradient on the Vessel Characters of the Mangrove Species *Rhizophora mucronata. Ann. Bot.* **98**: 1321–1330.

Schnetter, M.-L. 1978. Der Einfluss von Ausserfaktoren auf die Struktur des Blattes von *Avicennia germinans* (L.) L. unter natürlichen Bedingungen. *Beitr. Biol. Pflanz.* **54**: 13–28.

Scholander, P. F. 1968. How Mangroves Desalinate Seawater. *Physiol. Plant* **21**: 251–261.

Scholander, P. F., Hammel, H. T., Bradstreet, E. D., and Hemmingsen, E. A. 1965. Sap Pressure in Vascular Plants. *Science* **148**: 339–340.

Scholander, P. F., Hammel, H. T., Hemmingsen, E. A., et al. 1964. Hydrostatic Pressure and Osmotic Potential in Leaves of Mangroves and Some Other Plants. *Proc. Nat. Acad. Sci. U.S.A.* **52**: 119–125.

Scholander, P. F., Hammel, H. T., Hemmingsen, E. A., and Garey, W. 1962. Salt Balance in Mangroves. *Plant Physiol.* **37**: 722–729.

Scholander, P. F., van Dam, L., and Scholander, S. I. 1955. Gas Exchange in the Roots of Mangroves. *Amer. J. Bot.* **42**: 92–98.

Schumann, K. 1891. Rubiaceae. In *Die natürlichen Pflanzenfamilien*. 1, t. **IV**. ab. 4. A. Engler and K. Prantl, eds. Leipzig: Wilhelm Engelmann; pp. 1–156.

Schwarzbach, A. E. 2014. Rhizophoraceae. In *2009–. Families and Genera of Vascular Plants*. K. Kubitzki, ed. Heidelberg: Springer; Vol. XI: 283–295.

Schwarzbach, A. E. and McDade, L. 2002. Phylogenetic Relationships of the Mangrove Family Avicenniaceae based on Chloroplast and Nuclear Ribosomal DNA Sequence. *Syst. Bot.* **27**: 84–98.

Schwarzbach, A. E. and Ricklefs, R. E. 2000. Systematic Affinities with Rhizophoraceae and Anisophylleaceae, and Intergeneric Relationships within Rhizophoraceae, based on Chloroplast DNA, Nuclear Ribosomal DNA, and Morphology. *Amer. J. Bot.* **87**: 547–564.

Seago, J. L., Marsh, L. C., Stevens, K. J., et al. 2005. A Re-examination of the Root Cortex in Wetland Flowering Plants with Respect to Aerenchyma. *Ann. Bot.* **96**: 565–579.

Seetharaman, K. and Kandasamy, K. 2011. Reproductive Biology of a Natural Mangrove Hybrid *Rhizophora annamalayana* and its Parent Species (*R. apiculata* and *R. mucronata*) (Rhizophoraceae). *Bot. Marina.* **54**: 583–589.

Semeniuk, V. 1980. Mangrove Zonation along an Eroding Coastline in King Sound, Northwestern Australia. *J. Ecol.* **68**: 789–812.

Semeniuk, V. 1983. Mangrove Distribution in Northwestern Australia in Relationship to Regional and Local Freshwater Seepage. *Vegetatio* **53**: 11–31.

Semeniuk, V., Kenneally, K. F., and Wilson, P. G. 1978. *Mangroves of Western Australia*, Handbook No. 12. Perth: Western Australian Naturalists Club.

Semple, J. C. 1970. The Distribution of Pubescent Leaved Individuals of *Conocarpus erectus* (Combretaceae). *Rhodora* **72**: 544–546.

Seneski-de Lima, C., Torres-Boeger, M. R., Larcher-de Carvalho, L., et al. 2014. Sclerophylly in Mangrove Tree Species from South Brazil. *Rev. Mex. Biodivers.* **84**: 1159–1166.

Shaikh, S. S., Gokhale, M. V., and Charan, N. S. 2010. Seedling Morphology of Mangrove Species of Maharashtra and Goa States of India. *Phytomorphology* **60**: 77–90.

Sheridan, R. P. 1991. Epicaulous, Nitrogen-fixing Microepiphytes in a Tropical Mangal Community, Guadeloupe, French West Indies. *Biotropica* **23**: 530–541.

Sheridan, R. P. 1992. Nitrogen Fixation by Epicaulous Cyanobacteria in the Pointe de la Saline Mangrove Community, Guadeloupe, French West Indies. *Biotropica* **24**: 571–574.

Sheue, C.-R., Chesson, P., Chen, Y.-J., et al. 2003. Comparative Systemic Study of Colleters and Stipules of Rhizophoraceae with Implications for Adaptation to Challenging Environments. *Bot. J. Linn. Soc.* **172**: 449–464.

Sheue, C.-S., Liu H.-Y., and Yong, J. W. H. 2003. *Kandelia obovata* (Rhizophoraceae), a New Mangrove Species from Eastern Asia. *Taxon* **52**: 287–294.

Sheue, C.-S., Yang, Y.-P., Liu H.-Y., et al. 2009. Re-evaluating the Taxonomic Status of *Ceriops australis* (Rhizophoraceae) based on Morphological and Molecular Evidence. *Bot. Studies* **50**: 89–100.

Shi, S. H., Huang, Y. L., Zeng, K., et al. 2005. Molecular Phylogenetic Analysis of Mangroves: Independent Evolutionary Origin of Vivipary and Salt Secretion. *Molec. Phylog. Evol.* **34**: 159–166.

Shimony, C., Fahn, A., and Reinhold, L. 1973. Ultrastructure and Ion Gradients in the Salt Glands of *Avicennia marina* (Forssk.) Vierh. *New Phytol.* **72**: 27–36.

Sidhu, S. S. 1968. Further Studies on the Cytology of Mangrove Species of India. *Caryologia* **21**: 353–357.

Silva, C. A. R., Lacerda. L. D., and Rezende, C. E. 1990. Metals Reservoir in a Red Mangrove Forest. *Biotropica* **22**: 339–345.

Simberloff, D., Brown, B. J., and Lowrie, S. 1978. Isopod and Insect Root Borers may Benefit Florida Mangroves. *Science* **210**: 630–632.

Sinclair, J. 1968. Florae Malesianae Precursores, 17. The Genus *Myristica* in Malesia and outside Malesia. *Gard. Bull. Singapore* **23**: 1–540.

Sleumer, H. 1972. A Taxonomic Revision of the Genus *Scolopia* Schreb. (Flacourtiaceae). *Blumea* **20**(1): 25–64.

Smith, A. C. 1981. *Flora Vitiensis Nova*, Vol. 2. Lawai, Kauai, Hawaii: Pacific Tropical Botanical Garden.

Snedaker, S. C. 1982. Mangrove Species Zonation: Why? In *Contributions to the Ecology of Halophytes. Tasks for Vegetation Science*. Vol. 2. D. N. Sen and K. S. Rajpurohit, eds. The Hague: Junk; Chap. 1.

Snedaker, S. C., Jiménez, J. A., and Brown, M. S. 1981. Anomalous Aerial Roots in *Avicennia germinans* (L.) L. in Florida and Costa Rica. *Bull. Mar. Sci. Miami* **31**: 467–470.

Soltis, D. E., Smith, S. A., Cellinese, N., et al. 2011. Angiosperm Phylogeny: 17 Genes, 640 Taxa. *Amer. J. Bot.* **98**: 704–730.

Soto, R. and Jiménez, J. A. 1982. Analisis fisionomico estructural del manglar de Puerto Soley, La Cruz, Guanacaste, Costa Rica. *Rev. Biol. Trop.* **30**: 161–168.

Spalding, M., Kainuma, M., and Collins, L. 2010. *World Atlas of Mangroves*. Earthscan. LLC., Washington D.C. 20036.

Sperry, J. S. 1983. Observations on the Structure and Function of Hydathodes in *Blechnum lehmannii*. *Am. Fern J.* **73**: 65–72.

Sperry, J. S., Tyree, M. T., and Donnelly, J. R. 1988. Vulnerability of Xylem to Embolism in a Mangrove vs an Inland Species of Rhizophoraceae. *Physiol. Plant.* **74**: 276–283.

Stace, C. A. 1965a. Cuticular Studies as an Aid to Plant Taxonomy. *Bull. Br. Mus. Nat. Hist.* **4**: 3–78.

Stace, C. A. 1965b. The Significance of the Leaf Epidermis in the Taxonomy of the Combretaceae. 1. A General Review of Tribal Generic and Specific Characters. *J. Linn. Soc. Bot.* **59**: 229–252.

Start, A. N. and Marshall, A. G. 1976. Nectarivorous Bats as Pollinators of Trees in West Malaysia. In *Tropical Trees, Variation Breeding, and Conservation*. J. Burley and B. T. Styles, eds. New York, NY: Academic Press; pp. 141–150.

Stearn, W. T. 1958. A Key to West Indian Mangroves. *Kew Bull.* **1985**: 33–37.

Stehli, F. G. and Wells, J. W. 1971. Diversity and Age Patterns in Hermatypic Corals. *Syst. Zool.* **20**: 115–126.

Steinke, T. D. 1986. A Preliminary Study of Buoyancy Behaviour in *Avicennia marina* Propagules. *S. Afr. J. Bot.* **52**: 559–565.

Steinke, T. D. and Lambert, G. 1986. A Preliminary Study of the Phenology of *Scaevola plulmieri*. *S. Afr. J. Bot.* **52**: 43–46.

Stern, W. L. and Voigt, G. K. 1959. Effect of Salt Concentration on Growth of Red Mangrove in Culture. *Bot. Gaz.* **121**: 36–39.

Stevens, P. F. 1980. A Revision of the Old World Species of *Calophyllum* (Guttiferae). *J. Arnold Arbor.* **61**: 117–699.

Stokey, A. G. and Atkinson, L. R. 1952. The Gametophytes of *Acrostichum speciosum* Willd. *Phytomorphology* **2**: 105–113.

Studholme, W. P. and Philipson, W. R. 1966. A Comparison of the Cambium in Two Woods with Included Phloem: *Heimerliodendron brunonianum* and *Avicennia resinifera*. *N. Z. J. Bot.* **4**(4): 355–365.

Su, G.-H., Huang, Y.-L., Tan, F.-X., et al. 2006. Genetic Variation in *Lumnitzera racemosa*, a Mangrove Species from the Indo-West Pacific. *Aquat. Bot.* **84**: 341–346.

Sugaya, T., Yoshimaru, H., Takeuchi, T., et al. 2002. Development and Polymorphism of Simple Sequence Repeat DNA Markers for *Kandelia candel* (L.) Druce. *Mol. Ecol. Notes* **2**: 65–66.

Sugaya, T., Takeuchi, T., Yoshimaru, H., et al. 2003. Development and Polymorphism of Simple Sequence Repeat DNA Markers for *Bruguiera gymnorrhiza* (L.) Lamk. *Mol. Ecol. Notes* **3**: 88–90.

Sun, M. and Lo, E. Y. Y. 2011. Genomic Markers Reveal Introgressive Hybridisation in the Indo-West Pacific Mangroves; a Case Study. *PLoS ONE* **11; 6**(5): e19671.

Sun, M., Wang, K. C., and Lee, J. S. Y. 1998. Reproductive Biology and Population Genetic Structure of *Kandelia candel* (Rhizophoraceae), a Viviparous Mangrove Species. *Amer. J. Bot.* **85**: 1631–1637.

Sun, Q., Kobayashi, K., and Suzuki, M. 2004. Intercellular Space System in Xylem Rays of Pneumatophores in *Sonneratia alba* (Sonneratiaceae) and its Possible Functional Significance. *IAWA J.* **25**: 141–154.

Sussex, I. 1975. Growth and Metabolism of the Embryo and Attached Seedling of the Viviparous Mangrove, *Rhizophora mangle*. *Amer. J. Bot.* **62**: 948–953.

Swingle, W. T. 1943. The Botany of Citrus and the Wild Relatives of its Orange Subfamily. In *The Citrus industry*. Vol. 1,. H. J. Webber and L. D. Batchelor, eds. Berkeley, Los Angeles, CA: Division of Agricultural Sciences.

Szyszylowicz, I. 1893. Theaceae. In *Die natürlichen Planzenfamilien*. 1, t. **3**. A. Engler and K. Prantl, Leipzig: Wilhelm Engelmann; pp. 175–192.

Takayama, K., Tamura, M., and Tateishi, Y. 2008. Isolation and Characterisation of Microsatellite Loci in the Red Mangrove *Rhizophora mangle* L. (Rhizophoraceae) and its Related Species. *Conserv. Genet.* **9**: 1323–1325.

Takayama, K., Tamura, M., and Tateishi, Y., et al. 2009. Isolation and Characterisation of Microsatellite Loci in a Mangrove Species *Rhizophora stylosa* (Rhizophoraceae). *Conserv. Genet. Resour.* **1**: 175–178.

Takayama, K., Tamura, M., Tateishi, Y., et al. 2013. Strong Genetic Structure over the American Continents and Transoceanic Dispersal in the Mangrove Genus *Rhizophora* (Rhizophoraceae) Revealed by Broad-scale Nuclear and Chloroplast DNA Analysis. *Amer. J. Bot.* **100**: 1191–1201.

Tan, H. and Rao, A. N. 1981. Vivipary in *Opiorrhiza tomentosa* Jacq (Rubiaceae). *Biotropica* **13**: 232–233.

Tan, H. T. W. and Rao, A. N. 1988. Sporogenesis and Gametogaenesis in Scyphiphora Hydrophyllacea Gaertn. F. (Rubiaceae). *Flora* **180**: 413–416.

Tang, T., Zhong, Y., Jian, S., and Shi, S. 2003. Genetic Diversity of *Hibiscus tiliaceus* (Malvaceae) in China Assessed using AFLP Markers. *Ann. Bot.* **92**: 409–414.

Tattersfield, F., Martin, J. T., and Howes, F. N. 1940. Some Fish-poison Plants and their Insecticidal Properties. *Kew Bull.* **1940**: 169–180.

Teas, H. J. 1982. An Epidemic Dieback Gall Disease of *Rhizophora* Mangroves in the Gambia, West Africa. *Plant. Dis. Rep.* **66**: 522–523.

Thatui, H. W., Patra, J. K., and Das, S. K. 2014. Free Radical Scavenging and Antioxidant Potential of Mangrove Plants: A Review. *Acta Physiol. Plant.* **36**: 561–579.

Thibodeau, F. R. and Nickerson, N. H. 1986. Differential Oxidation of Mangrove Substrate by *Avicennia germinans* and *Rhizophora mangle*. *Amer. J. Bot.* **73**: 512–516.

Thom, B. G. 1967. Mangrove Ecology and Deltaic Geomorphology: Tobasco, Mexico. *J. Ecol.* **55**:301–343.

Thom, B. G. 1975. Mangrove Ecology from a Geomorphological Viewpoint. In *Proceedings of the International Symposium on Biology and Management of Mangroves.* G. E. Walsh, S. C. Snedaker, and H. J. Teas., eds. Gainesville, FL: Institute of Food and Agricultural Sciences, University of Florida; pp. 469–481.

Tobe, H. and Raven, P. H. 1988. Seed Morphology and Anatomy of Rhizophoraceae, Inter- and Infrafamilial Relationships. *Ann. Mo. Bot. Gard.* **75**: 1319–1342.

Tomlinson, P. B. 1971. The Shoot Apex and its Dichotomous Branching in the *Nypa* palm. *Ann. Bot. Lond.* **35**: 865–879.

Tomlinson, P. B. 1978. *Rhizophora* in Australasia – Some Clarification of Taxonomy and Distribution. *J. Arnold Arbor.* **59**: 156–169.

Tomlinson, P. B. 1982a. Field Collection and Study of Old World (Indo-Pacific) Mangroves. *Nat. Geog. Soc. Res. Rep.* **14**: 669–677.

Tomlinson, P. B. 1982b. Helobiae (Alismatidae). Vol. 7. Anatomy of the Monocotyledons. C. R. Metcalfe, ed. Oxford, UK: Clarendon Press.

Tomlinson, P. B. 2001. *The Biology of Trees Native to Tropical Florida.* Second edn. Petersham, MA: Published privately.

Tomlinson, P. B., Bunt, J. S., Primack, R. B., et al. 1978. *Lumnitzera rosea* (Combretaceae) – its Status and Floral Morphology. *J. Arnold Arbor.* **59**: 342–351.

Tomlinson P. B. and Cox P. A. 2000. Systematic and Functional Anatomy of Seedlings in Mangrove Rhizophoraceae: Vivipary Explained? *Bot. J. Linn. Soc.* **134**: 215–231.

Tomlinson, P. B., Horn, J. W., and Fisher, J. B. 2011. *The Anatomy of Palms.* Oxford, UK: Oxford University Press.

Tomlinson, P. B., Primack, R. B., and Bunt, J. S. 1979. Preliminary Observations in Floral Biology in Mangrove Rhizophoraceae. *Biotropica* **11**: 256–277.

Tomlinson, P. B. and Wheat, D. W. 1979. Bijugate Phyllotaxis in Rhizophoreae (Rhizophoraceae). *Bot. J. Linn. Soc.* **78**: 317–321.

Tomlinson, P. B. and Womersley, J. S. 1976. A Species of *Rhizophora* new to New Guinea and Queensland, with Notes Relevant to the Genus. *Contrib. Herb. Austral.* **19**: 1–10.

Tralau, H. 1964. The Genus *Nypa* van Wurmb. *K. Svensk Vetensk. Akad. Handl.*, Series 5, **10**(1): 5–29.

Triest, L. 2008. Molecular Ecology and Biogeography of Mangrove Trees, Towards Conceptual Insights on Gene Flow and Barriers: A Review. *Aquat. Bot.* **89**: 138–154.

Triono, T., Brown, A. H. D., West, J. G., and Crisp, M. D. 2007. A Phylogeny of *Pouteria* (Sapotaceae) from Malesia and Australia. *Austral. Syst. Bot.* **20**: 107–118.

Trochain, J. and Dulau, L. 1942. Quelques particularités anatomiques *d'Avicennia nitida* (Verbernaceae) de Ia mangrove ouest africaine. *Bull. Soc. Hist. Nat. Toulouse* **77**: 271–281.

Troll, W. 1930. Über die sogenannten Atemwurzel der Mangrove. *Ber. dtsch. Bot. Ges.* **48** (suppl.): 81–99.

Troll, W. 1933a. *Camptostemon schultzii* Mast. und *Camptostemon philippensis* (Vid.) Becc. als neue Vertreter der austral-asiatischen Mangrovevegetation. *Flora* n.f. **128**: 348–360.

Troll, W. 1933b. Über *Acrostichum aureum* L., *Acrostichum speciosum* Willd. und neotene Formen des lezteren. *Flora* n.f. **128**: 301–328.

Troll, W. 1943. *Vergleichende Morphologie den hoheren Pflanzen*, bd. 1 t. 3, Vegetationsorgane. Berlin: Gebruder Borntraeger.

Troll, W. and Dragendorff, O. 1931. Ueber die Luftwurzeln von *Sonneratia* L. und ihre biologische Bedeutung. *Planta* **13**: 311–473.

Tryon, R. M. and Tryon, A. F. 1982. *Ferns and Allied Plants, with Special Reference to Tropical America*. New York, NY: Springer-Verlag.

Tyagi, A. P. and Singh, V. V. 1998. Pollen Fertility and Intraspecific and Interspecific Compatibility in Mangroves of Fiji. *Sex Plant Reprod.* **11**: 60–63.

Uhl, N. W. 1972. Inflorescence and Flower Structure in *Nypa fruticans* (Palmae). *Amer. J. Bot.* **59**: 729–743.

Ulken, A. 1983. Distribution of Phycomycetes in Mangrove Swamps with Brackish Waters and Waters of High Salinity. In *Biology and Ecology of Mangroves. Tasks for Vegetation Science* 8. H. Teas, ed. The Hague: Junk; Chap. 12.

United Nations Educational, Scientific, and Cultural Organization. 1978. Secondary Successions. In *Tropical Forest Ecosystems*. A State-of-Knowledge Report Prepared by UNESCO/UNEP/FAO. Paris: UNESCO; Chap. 9.

Valeton, T. 1895. Les *Cerebera* du Jardin Botanique de Buitenzorg. *Ann. Jard. Bot. Buitenz.* **12**: 238–248.

Van Borassum Waalkes, J. 1966. Malesian Malvaceae Revised. *Blumea* **14**: 1–213.

Van Rheede tot Drakenstein, H. 1678–1703. *Hortus indicus Malabaricus*, 12 vols. Amsterdam: J. van Someren & J. van Dyck.

Van Slooten, P. F. 1924. Contributions a l'étude de Ia flore des Indes néerlandaises II. The Combretaceae of the Dutch East Indies. *Bull. Jard. Bot. Buitenz.* Series 3, **6**(1): 11–64.

Van Steenis, C. G. G. J. 1928. The Bignoniaceae of the Netherlands Indies. *Bull. Jard. Bot. Buitenz.* Series 3, **10**: 173–290.

Van Steenis, C. G. G. J. 1936. *Osbornia octodonta*, een weinig bekende mangrove-boom. *Trop. Nat.* **25**: 194–196.

Van Steenis, C. G. G. J. 1937. De soorten van het geschlacht *Acanthus* in Nederlandsch-Indie. *Trop. Nat.* **26**: 202–207.

Van Steenis, C. G. G. J. 1948. Plumbaginaceae. *Flora Malesiana* 1, **4**: 107–112.

Van Steenis, C. G. G. J. 1949. Vicarism in the Malaysian Flora. *Flora Malesiana* 1, **4**: 59.

Van Steenis, C. G. G. J. 1958. Ecology of Mangroves. *Flora Malesiana* 1, **5**: 431–441.

Van Steenis, C. G. G. J. 1962. The Distribution of Mangrove Plant Genera and its Significance for Palaeogeography. *Proc. Kon. Net. Amsterdam*, Series C, **65**: 164–169.

Van Steenis, C. G. G. J. 1977. Bignoniaceae. *Flora Malesiana* 1, **8**: 114–186.

Van Steenis-Kruseman, M. J. 1950. Malaysian Plant Collectors and Collections. *Flora Malesiana* **1**(1): 3–63a.

Van Tieghem, P. H. 1898. Avicenniacées et Symphoremacées. Place de ces deux nouvelles families dans Ia classification. *J. Bot. Paris.* **12**: 345–365.

Van Vliet, G. J. C. M. 1976. Wood Anatomy of Rhizophoraceae. *Leiden Bot. Ser.* **3**: 20–75.

Vezey E. L., Shah, V. P., Skvarla, J. J., and Raven, P. H. 1988. Morphology and Genetics of Rhizophoraceae Pollen. *Ann. Mo. Bot. Gard.* **75**: 1369–1386.

Verdcourt, B. 1979. A Manual of New Guinea Legumes. *Bot. Bull. Officer of Forests, Division of Botany*. Lae, Papua New Guinea.

Verheyden, A., Kauro J. G., Beeckman, H. et al. 2004. Growth Rings, Growth Ring Formation and Age Determination in the Mangrove *Rhizophora mucronata*. *Ann. Bot.* **94**: 59–66.

Von Faber, F. C. 1913. Ueber transpiration und osmotischen druck bei den Mangroven. *Ber. d. bot. Gesell.* **31**: 277–281.

Von Faber, F. C. 1923. Zur Physiologie der Mangroven. *Ber. d. bot. Gesell.* **41**: 227–234.

Wah Juliana, W. A., Shahril, M. H., and Abdul Rahman, N. A. N. 2012. *Gluta velutina*: An Important Component of Malaysian Mangroves. *Malay Appl. Biol. J.* **41**: 55–58.

Waisel, Y. 1972. *Biology of Halophytes*. New York, NY: Academic Press.

Walsh, G. E. 1974. Mangroves: A Review. In *Ecology of Halophytes*. New York, NY: Academic Press; pp. 51–174.

Walsh, G. E. 1977. Exploitation of Mangal. In *Ecosystems of the World. Vol. 1. Wet Coastal Ecosystems*. V. J. Chapman, ed. Amsterdam: Elsevier Scientific, Chap. 16.

Walsh, G. E., Ainsworth, K. A., and Rigby, R. 1979. Resistance of Red Mangrove *(Rhizophora mangle* L.) to Lead, Cadmium, and Mercury. *Biotropica* **11**: 22–27.

Walsh, G. E., Snedaker, S. C., and Teas, H. J. (eds.) 1975. *Proceedings of the International Symposium on Biology and Management of Mangroves*. 2 Vols. Gainesville, FL: Institute of Food and Agricultural Sciences, University of Florida.

Walter, H. and Steiner, M. 1936. Oekologie der ost-afrikanschen Mangroven. *Z. Bot.* **30**: 65–93.

Wang, B. S., Liang, S. C., Zhang, W. Y., et al. 2003. Mangrove Flora of the World. *Acta Bot. Sinica* **45**: 644–653.

Wang, R., Cheng, Z., Chen, E., and Zheng, X. 1999. Two Hybrids of the Genus *Sonneratia* (Sonneratiaceae) from China. *Guihaia* **19**: 199–206.

Wang, R. and Chen, Z. 2002. Systematic and Biogeographic Study on the Family Sonneratiaceae. *Guihaia* **22**: 214–219.

Wang, W., Yan, Z., Yiu, S., et al. 2011. Mangroves: Obligate or Facultative Halophytes? A Review. *Trees-Struct. Func.* **25**: 953–963.

Ward, C. J. and Steinke, T. D. 1982. A Note on the Distribution and Approximate Areas of Mangroves in South Africa. *S. Afr. J. Bot.* **1**: 51–53.

Warne, K. 2011. *Let Them Eat Shrimp*. Washington. DC: Island Press/Shearwater Books.

Watson, J. G. 1928. Mangrove Forests of the Malay Peninsula. *Malay For. Rec.* no. **6**.

Weber-El Ghobary, M. O. 1984. The Systematic Relationships of *Aegialitis* (Plumbaginaceae) as Revealed by Pollen Morphology. *Plant Syst. Evol.* **144**: 53–58.

Webster, G. L. 1967. The Genera of Euphorbiaceae in the Southeastern United States. *Arnold Arbor.* **48**: 303–430.

Webster, G. L. 1975. Conspectus of a New Classification of the Euphorbiaceae. *Taxon* **24**: 593–601.

Webster, G. L. 2014. Euphorbiaceae. In *1990–. Families and Genera of Vascular Plants*. K. Kubitzki, ed. Heidelberg: Springer; XI: 51–216.

Wells, A. G. 1983. Distribution of Mangrove Species in Australia. In *Biology and Ecology of Mangroves. Tasks for Vegetation Science* 8. H. J. Teas, ed. The Hague: Junk, Chap. 6.

West, R. C. 1977. Tidal Salt-Marsh and Mangal Formations of Middle and South America. In *Ecosystems of the World. Vol. 1. Wet Coastal Ecosystems*. Amsterdam: Elsevier Scientific, Chap. 9.

Westermaier, M. 1900. Zur Kenntnis der Pneumatophoren. *Botanische Untersuchungen im Amschluss an eine Tropenreise*, Heft I. Freiburg.

White. C. T. 1926. A Variety of *Ceriops tagal* C. B. Rob. (= *C. candollean* W. & A.). *J. Bot. Lond.* **64**: 220–221.

Whitmore, T. C. 1983. Secondary Succession from Seed in Tropical Rain Forests. *Common. For. Abstr.* **44**: 767–779.

Whitten, A. J. and Damanik, S. J. 1986. Mass Defoliation of Mangroves in Sumatra, Indonesia. *Biotropica* **18**: 176.

Wilson, P. G. 2011. Myrtaceae. In *1990–. Families and Genera of Vascular Plants*. K. Kubitzki, ed. Heidelberg: Springer; X: 212–271.

Wilkie, P., Clark, A., Pennington, R. T. et al. 2006. Phylogenetic Relations within the Subfamily Sterculiodeae (Malvaceae/ Sterculiaceae – Sterculiaeae) using the Chloroplast Gene *ndhF. Systematic Bot.* **31**: 160–170,

Wilkinson, H. P. 1981. The Anatomy of the Hypocotyls of *Ceriops* Amott (Rhizophoraceae), Recent and Fossil. *Bot. J. Linn. Soc.* **82**: 139–164.

Willis, J. C. 1922. *Age and Area*. Cambridge, UK: Cambridge University Press.

Winograd, M. 1983. Observaciones sobre el hallozgo de *Pelliciera rhizophorae* (Theaceae) en el Caribe Colombiano. *Biotropica* **15**: 297–298.

Wise, R. R. and Juncosa, A. M. 1989. Ultrastructure of the Transfer Tissues during Viviparous Seedling Development of *Rhizophora mangle* (Rhizophoraceae). *Amer. J. Bot.* **76**: 1286–1298.

Wium-Anderson, S. 1981. Seasonal Growth of Mangrove Trees in Southern Thailand, 3. Phenology of *Rhizophora mucronata* Lamk. and *Scyphiphora hydrophyllacea* Gaertn. *Aquat. Bot.* **10**: 371–376.

Wium-Anderson, S. and Christensen, B. 1978. Seasonal Growth of Mangrove Trees in Southern Thailand, 2. Phenology of *Bruguiera cylindrica, Ceriops tagal, Lumnitzera littorea*, and *Avicennia marina. Aquat. Bot.* **5**: 383–390.

Woodroffe, C. D. and Grindrod, J. 1991. Mangrove Biogeography: the Role of Quaternary Environmental and Sealevel Change. *J. Biogeog.* **18**: 479–492.

Wright, D. F. 1977. *A North Queensland Mangrove Pollen Flora*. B.Sc.Hons. Botany Thesis. Townsville, Queensland: James Cook University.

Wyatt-Smith, J. 1953a. The Malayan Species of *Bruguiera. Malay. For.* **16**: 156–161.

Wyatt-Smith, J. 1953b. The Malayan Species of *Sonneratia. Malay. For.* **16**: 213–216.

Wyatt-Smith, J. 1954. The Malayan Species of *Avicennia. Malay. For.* **17**: 21–25.

Wyatt-Smith, J. 1960. Field Key to the Trees of Mangrove Forests in Malaya. *Malay. For.* **23**: 126–132.

Yahya, A. F., Hyun, J. O., Lee, J. H., et al. 2014. Genetic Variation and Population Genetic Structure of *Rhizophora apiculata* (Rhizophoraceae) in the Greater Sunda Islands, Indonesia, using Satellite Markers. *J. Plant Res.* **127**: 287–297.

Yamashiro, M. 1961. Ecological Study on *Kandelia candel* (L.) Druce, with Special Reference to the Structure and Falling of the Seedlings. *Hikobia* **2**(3): 209–214.

Yanez-Espinosa, L., Terrazas, T., and Lopez-Mata, L. 2001. Effects of Flooding on Wood and Bark Anatomy of Four Species in a Mangrove Forest Community. *Trees* **15**: 91–97.

Yoshioka, H., Kondo, K., Segawa, M., et al. 1984. *La Kromosoma (Japan)* II. **35–36**: 1111–1116.

Zahran, M. A. 1977. Africa A, Wet Formations of the African Red Sea Coast. In Ecosystems of the World. Vol. 1. Wet Coastal Ecosystems. V. J. Chapman, ed. Amsterdam: Elsevier Scientific; Chap. 10.

Zamski, E. 1979. The Mode of Secondary Growth and the Three-dimensional Structure of the Phloem in *Avicennia. Bot. Gaz.* **140**: 67–76.

Zamski, E. 1981. Does Successive Cambia Differentiation in *Avicennia* Depend on Leaf and Branch Initiation? *lsr. J. Bot.* **30**: 57–64.

Zhang, X., Chua, V. P., and Cheong, H.-F. 2015. Geometrical and Material Properties of *Sonneratia alba* Roots. *Trees – Structure and Function* **29**: 285–297.

Zhou, R., Gong, X., Boufford, D., et al. 2008. Testing a Hypothesis of Unidirectional Hybridization in Plants: Observations on *Sonneratia alba*, *Bruguiera*, and *Ligularia*. *BMC Evol. Biol.* **8**: 149.

Zimmermann, M. H. 1983. *Xylem Structure and the Ascent of Sap*. Berlin: Springer Verlag.

Zimmermann, M. H., Wardrop, A. B., and Tomlinson, P. B. 1968. Tension Wood in Aerial Roots of *Ficus benjamina* L. *Wood Science and Technology* **2**: 95–104.

Zimmermann, U., Zhu, J. J., Meinzer, F. C., et al. 1994. High Molecular Weight Organic Compounds in the Xylem Sap of Mangroves: Implications for Long-distance Water Transport. *Bot. Acta.* **107**: 218–229.

Zolgharnein, H., Kamyab, M., Keyvanshokooh, S., et al. 2010. Genetic Diversity of *Avicennia marina* (Forsk.) Vierh. Populations in the Persian Gulf by Microsatellite Markers. *J. Fish. Aquat. Sci.* **5**: 223–229.

Index

Bold face numbers refer to main description; *italic* numbers refer to illustrations.

Synonyms are not indexed unless cited in text; plate figures (PL.) refer to plate numbers.